BLEOMYCIN: CHEMICAL, BIOCHEMICAL, AND BIOLOGICAL ASPECTS

Proceedings of a joint U.S.–Japan Symposium held at the East–West Center, Honolulu, July 18–22, 1978

Edited by
Sidney M. Hecht

With Contributions by
D. J. Burlett S. S. Cohen S. T. Crooke J. C. Dabrowiak
L. H. DeRiemer A. Fujii D. A. Goodwin F. T. Greenaway
A. P. Grollman C. W. Haidle S. M. Hecht M. Hori
S. B. Horwitz F. Hutchinson Y. Kuroda M. D. Levin
R. S. Lloyd J. W. Lown J. Lunec C. F. Meares
D. Mizuno H. Morishima Y. Muraoka Y. Mushika
H. Naganawa A. D. Nunn M. Ohno N. J. Oppenheimer
M. Otsuka J. Peisach L. F. Povirk I. H. Raisfeld
D. L. Robberson F. S. Santillo E. A. Sausville J. E. Strong
Y. Sugiura Y. Takabe M. Takeshita T. Takita T. Terasima
H. Umezawa Y. Umezawa M. Watanabe T. Yoshioka

Springer-Verlag   New York Heidelberg Berlin

Sidney M. Hecht
Department of Chemistry
Massachusetts Institute of Technology
Cambridge, Mass. 02139
USA

With 166 Figures.

Library of Congress Cataloging in Publication Data. Main entry under title: Bleomycin, chemical, biochemical, and biological aspects. Bibliography: p. Includes index. 1. Bleomycin—Congresses. 2. Chemistry, Pharmaceutical—Congresses. I. Hecht, Sidney M. RS431.B55B47 615.7   79-4295

Softcover reprint of the hardcover 1st edition 1979

9 8 7 6 5 4 3 2 1

ISBN-13: 978-1-4612-6193-3   e-ISBN-13: 978-1-4612-6191-9
DOI: 10.1007/978-1-4612-6191-9

This volume is dedicated to Dr. Hanno Umezawa, under whose direction bleomycin was first isolated.

Preface

This book records the proceedings of a joint U.S.-Japan symposium on the chemistry, biochemistry, and biology of bleomycin, an antitumor antibiotic shown to be effective therapeutically against, eg, squamous cell carcinomas, Hodgkin's lymphoma, and testicular tumors. Several important and previously unreported observations were presented and the status of experimental work in the United States and Japan was reviewed; a summary and interpretation of the scientific presentations at the meeting has been prepared by the editor and is included as the first contribution in this volume.

In addition to the symposium contributions, an experimental section has been included at the end of the book dealing with practical methods for the fractionation, modification, and assay of bleomycin. It is hoped that this section will facilitate progress in this area of scientific endeavor.

The symposium from which this book is derived was organized by Drs. Umezawa, Takita, and Hecht and supported financially by the National Science Foundation, the National Cancer Institute, and the Japan Society for the Promotion of Science.

S. M. Hecht

v

Contents

Status Reports

Summary of the Bleomycin Symposium. S. M. HECHT 1
Advances in Bleomycin Studies. H. UMEZAWA 24
Review of the Structural Studies on Bleomycin. T. TAKITA 37

Synthetic and Biosynthetic Studies

Studies on the Total Synthesis of Bleomycin. S. M. HECHT,
D. J. BURLETT, Y. MUSHIKA, Y. KURODA, and M. D. LEVIN 48
A Synthetic Approach to the Pyrimidine Moiety of Bleomycin.
Y. UMEZAWA, H. MORISHIMA, T. YOSHIOKA, M. OTSUKA, and M. OHNO . . 63
Biogenetic Aspects of Bleomycin–Phleomycin Group Antibiotics.
A. FUJII 75

Physical and Spectral Analysis of Bleomycin

Liquid Chromatography of Bleomycin. Y. MURAOKA 92
NMR Study of Bleomycin. H. NAGANAWA 106

Bleomycin as a Metal Chelator

^{1}H-NMR Study of the Metal Binding Sites of Bleomycin
N. J. OPPENHEIMER 124
The Metallobleomycins. J. C. DABROWIAK, F. T. GREENAWAY, and
F. S. SANTILLO 137
Metal Complex of Bleomycin and Its Implication for the Mechanism
of Bleomycin Action. T. TAKITA 156
Electron Spin Resonance Studies of 1:1:1 Bleomycin–Cobalt(II)–
Oxygen Adduct Complex. Y. SUGIURA 165

Interaction of Bleomycin with DNA

A Role for Iron in the Degradation of DNA by Bleomycin.
S. B. HORWITZ, E. A. SAUSVILLE, and J. PEISACH 170
Contribution of the Superoxide Anion-Hydroxyl Radical Pathway
to the Cleavage of DNA by Bleomycin. J. W. LOWN 184
Interaction of Bleomycin with DNA. M. HORI 195

A Molecular Basis for the Interaction of Bleomycin with DNA
M. TAKESHITA, and A. P. GROLLMAN 207
Molecular Aspects of Bleomycin-Promoted Damage of Covalently
Closed Circular DNA. C. W. HAIDLE, R. S. LLOYD,
and D. L. ROBBERSON 222
The Use of Covalently Closed Circular DNA to Investigate Properties
of Bleomycin and Its Analogs. J. E. STRONG and S. T. CROOKE 244
The Action of Bleomycin on DNA. F. HUTCHINSON
and L. F. POVIRK 255

Biological Effects of Bleomycin

Polyamines and the Toxicity of the Bleomycins. S. S. COHEN 267
Cytotoxicity of the Various Bleomycins to Cultured Mammalian Cells
A. D. NUNN and J. LUNEC 287
Upward-Concave Dose–Response Relationship in Bleomycin Lethality
of Mammalian Cells. T. TERASIMA, M. WATANABE, and Y. TAKABE 297
Conjugation of Bifunctional Chelating Agents to Bleomycin for Use in
Nuclear Medicine. C. F. MEARES, L. H. DeRIEMER, and D. A. GOODWIN . . 309
Bleomycin-Induced Pulmonary Toxicity: A Model for the Study of
Pulmonary Fibrosis. I. H. RAISFELD 324
Concluding Remarks at the Symposium on Bleomycin. D. MIZUNO 336

Practical Methods for the Manipulation and Assay of Bleomycin

A Practical Method for the Separation of Bleomycin Components.
A. FUJII 341
Preparation of Bleomycinic Acid from Bleomycin A_2. A. FUJII 343
A Practical Method for Preparing Copper-Free Bleomycin A_2. A. FUJII 345
A Simple Assay for Bleomycin Activity: Degradation of Radioactive
DNA to Acid-Soluble Products. S. B. HORWITZ, E. A. SAUSVILLE,
and M. TAKESHITA 346

Index 349

List of Contributors

BURLETT, D. J., Department of Chemistry, Massachusetts Institute of Technology, Cambridge, Massachusetts, USA.

COHEN, S. S., Department of Pharmacological Sciences, State University of New York, Stony Brook, New York, USA.

CROOKE, S. T., Department of Pharmacology, Baylor College of Medicine and Bristol Laboratories, Inc., Syracuse, New York, USA.

DABROWIAK, J. C., Department of Chemistry, Syracuse University, Syracuse, New York, USA.

DERIEMER, L. H., Department of Chemistry, University of California, Davis, California, USA.

FUJII, A., Research Laboratories, Pharmaceutical Division, Nippon Kayaku Company, Shimo 3-31-12, Kita-ku, Tokyo, Japan.

GOODWIN, D. A., Department of Nuclear Medicine, Veteran's Administration Hospital, 3801 Miranda Avenue, Palo Alto, California, USA.

GREENAWAY, F. T., Department of Chemistry, Syracuse University, Syracuse, New York, USA.

GROLLMAN, A. P., Department of Pharmacological Sciences, State University of New York, Stony Brook, New York, USA.

HAIDLE, C. W., Section of Molecular Biology, The University of Texas System Cancer Center, M. D. Anderson Hospital and Tumor Institute, Houston, Texas, USA.

HECHT, S. M., Department of Chemistry, Massachusetts Institute of Technology, Cambridge, Massachusetts, USA. Present address: Department of Chemistry, University of Virginia, Charlottesville, Virginia, USA.

HORI, M., Shōwa College of Pharmaceutical Sciences, 5-1-8, Tsurumaki, Setagaya-ku, Tokyo 154, Japan.

HORWITZ, S. B., Department of Molecular Pharmacology, Albert Einstein College of Medicine, Bronx, New York, USA.

HUTCHINSON, F., Department of Molecular Biophysics and Biochemistry, Yale University, New Haven, Connecticut, USA.

KURODA, Y., Department of Chemistry, Massachusetts Institute of Technology, Cambridge, Massachusetts, USA.

LEVIN, M. D., Department of Chemistry, Massachusetts Institute of Technology, Cambridge Massachusetts, USA.

LLOYD, R. S., Section of Molecular Biology, The University of Texas System Cancer Center, M.D. Anderson Hospital and Tumor Institute, Houston, Texas, USA.

LOWN, J. W., Department of Chemistry, University of Alberta, Edmonton, Alberta, Canada.

LUNEC, J., Cyclotron Unit, Hammersmith Hospital, Ducane Road, London W120HS, England.

MEARES, C. F., Department of Chemistry, University of California, Davis, California, USA.

MIZUNO, D., Faculty of Pharmaceutical Sciences, University of Tokyo, 7-3-1, Hongo, Bunkyo-ku, Tokyo 113, Japan.

MORISHIMA, H., Institute of Microbial Chemistry 3-14-23, Kamiosaki, Shingawa-ku, Tokyo 141, Japan.

MURAOKA, Y., Research Laboratories, Pharmaceutical Division, Nippon Kayaku Co., Shimo 3-31-12, Kita-ku, Tokyo, Japan.

MUSHIKA, Y., Department of Chemistry, Massachusetts Institute of Technology, Cambridge, Massachusetts, USA.

NAGANAWA, H., Institute of Microbial Chemistry, 3-14-23, Kamiosaki, Shinagawa-ku, Tokyo 141, Japan.

NUNN, A. D., Division of Nuclear Medicine, Department of Radiology, Upstate Medical Center, 750 E. Adams Street, Syracuse, New York, USA.

OHNO, M., Faculty of Pharmaceutical Sciences, University of Tokyo, 7-3-1, Hongo, Bunkyo-ku, Tokyo 113, Japan.

OPPENHEIMER, N. J., Department of Pharmaceutical Chemistry, University of California, San Francisco, California, USA.

OTSUKA, M., Faculty of Pharmaceutical Sciences, University of Tokyo, 7-3-1, Hongo, Bunkyo-ku, Tokyo 113, Japan.

PEISACH, J., Departments of Molecular Pharmacology and Molecular Biology, Albert Einstein College of Medicine, Bronx, New York, USA.

POVIRK, L. F., Department of Molecular Biophysics and Biochemistry, Yale University, New Haven, Connecticut, USA.

RAISFELD, I. H., Department of Pharmacological Sciences, State University of New York, Stony Brook, New York, USA.

ROBBERSON, D. L., Section of Molecular Biology, The University of Texas System Cancer Center, M.D. Anderson Hospital and Tumor Institute, Houston, Texas, USA.

SANTILLO, F. S., Department of Chemistry, Syracuse University, Syracuse, New York, USA.

SAUSVILLE, E. A., Department of Molecular Pharmacology, Albert Einstein College of Medicine, Bronx, New York, USA.

STRONG, J. E., Department of Pharmacology, Baylor College of Medicine, Houston, Texas, USA.

SUGIURA, Y., Faculty of Pharmaceutical Sciences, Kyoto University, Sakyo-ku, Kyoto 606, Japan.

TAKABE, Y., First Department of Medicine, Chiba University School of Medicine, Chiba 280, Japan.

TAKESHITA, M., Department of Pharmacological Sciences, State University of New York, Stony Brook, New York, USA.

TAKITA, T., Institute of Microbial Chemistry, 3-14-23, Kamiosaki, Shinagawa-ku, Tokyo 141, Japan.

TERASIMA, T., Department of Physiopathology, National Institute of Radiological Sciences, 4-9-1, Anakawa, Chiba-shi, Chiba 280, Japan.

UMEZAWA, H., Institute of Microbial Chemistry, 3-14-23, Kamiosaki, Shinagawa-ku, Tokyo 141, Japan.

UMEZAWA, Y., Institute of Microbial Chemistry, 3-14-23, Kamiosaki, Shinagawa-ku, Tokyo 141, Japan.

WATANABE, M., First Department of Medicine, Chiba University School of Medicine, Chiba 280, Japan.

YOSHIOKA, T., Institute of Microbial Chemistry, 3-14-23, Kamiosaki, Shinagawa-ku, Tokyo 141, Japan.

Summary of the Bleomycin Symposium

SIDNEY M. HECHT

The bleomycins are a family of glycopeptide-derived molecules isolated from cultures of *Streptomyces* as copper chelates and differing only in the specific substituents at the "C-terminus" of the common structural unit, bleomycinic acid. In 1972, on the basis of a detailed study of the (partial)hydrolysis products arising from bleomycin, Umezawa, Takita, and their co-workers were able to propose a structure for the antibiotic. Although the proposed structure included some unusual structural features, verification by total synthesis or X-ray crystallography has not been possible as yet due to the complexity of the molecule and the lack of any crystalline derivative (Fig. 1).

During the symposium, Drs. Takita and Fujii discussed the isolation and X-ray crystallographic analysis of P-3A, a biosynthetic intermediate structurally related to bleomycin. Although disorder in the crystal, and several solvent molecules, obscured the key region of interest, the analysis indicated the absence of a β-lactam and the probable presence of a substituted propionamide moiety. While simple extrapolation of the results obtained for P-3A to bleomycin might be regarded as inappropriate, as the two species differ considerably in extent of biosynthetic elaboration, the ^{13}C-chemical shift values for these compounds determined by

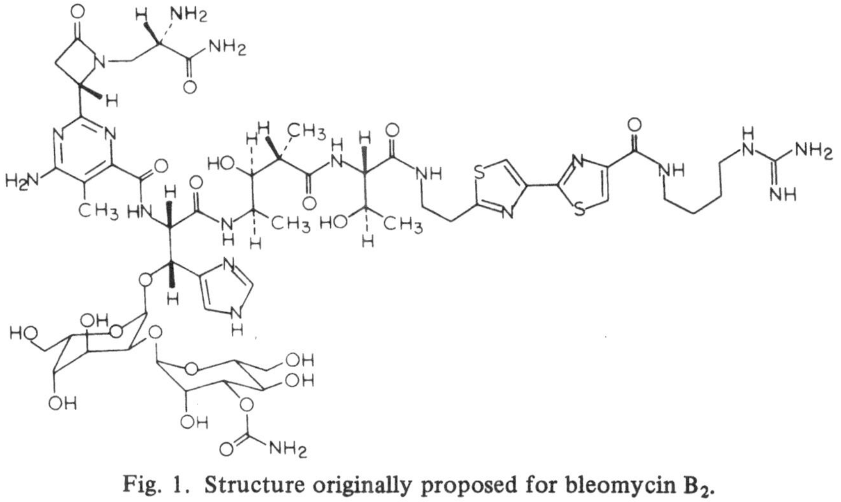

Fig. 1. Structure originally proposed for bleomycin B_2.

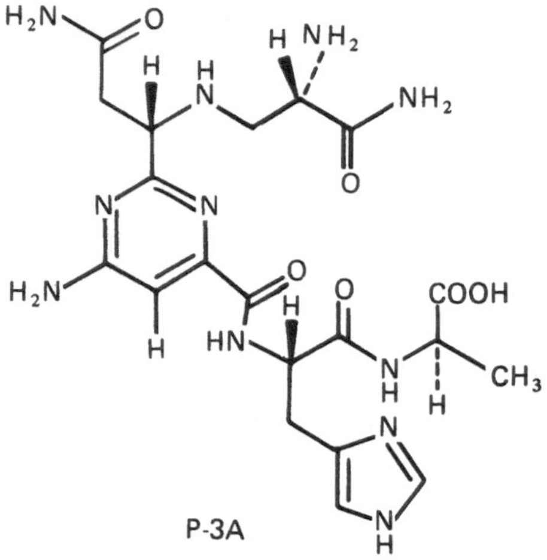

P-3A

Dr. Naganawa were remarkably similar, suggesting the need for revision of the structure of bleomycin (Fig. 2).

To facilitate the reevaluation of the structure of bleomycin, Dr. Ohno and his co-workers prepared a number of model compounds for comparison with bleomycin and P-3A. These included racemic samples of 1 and 2, the former of which was shown by X-ray crystallographic analysis to have a β-lactam moiety and the desired relative configurations at the two asymmetric centers. As shown in Table 1, the pK_a values of the primary (and secondary) amino groups in 1 and 2, when compared with those determined for P-3A and bleomycin, suggest strongly that bleomycin has a pyrimidine substituent analogous to that in 2. Although anomalously low for a secondary amine, the pK_a values of 2.7–3.4 observed for this

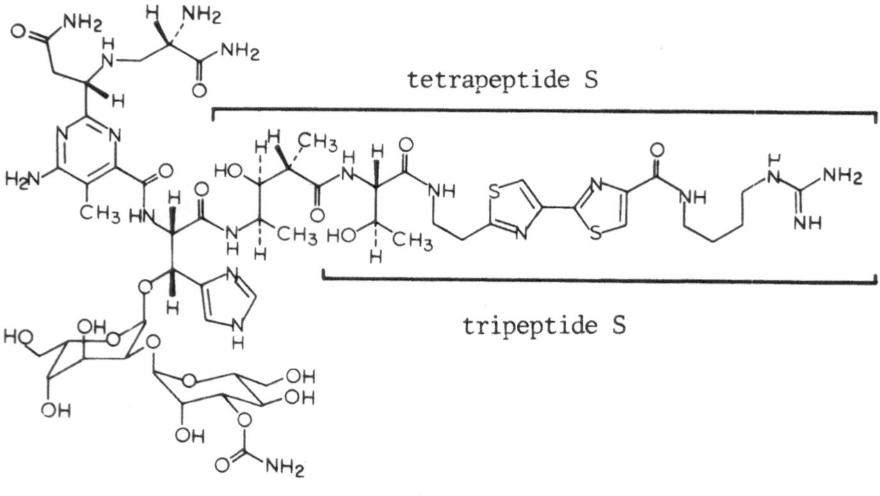

Fig. 2. Revised structure of bleomycin B_2.

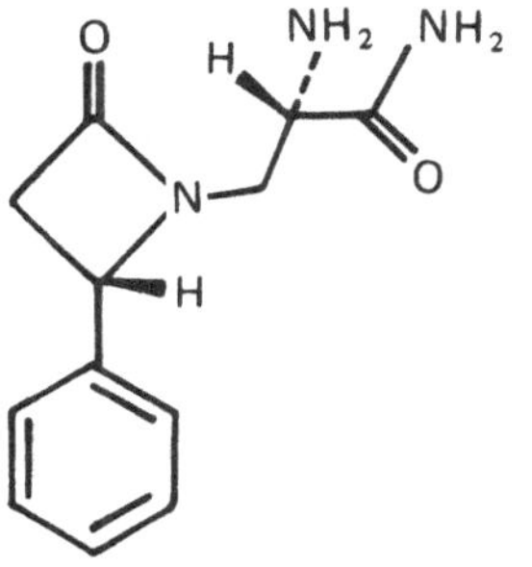

Table 1: pK_a Values of the Amino Groups in Bleomycin, P-3A, and Two Model Compounds

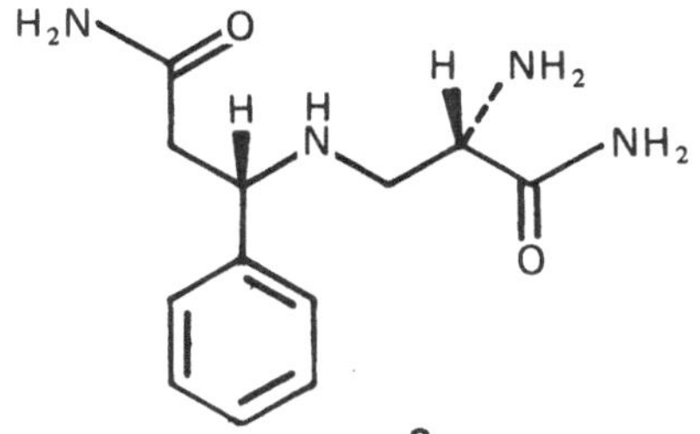

Amino group	Compound			
	1	2	P-3A	Bleomycin
Primary	5.7	7.4	7.5	7.5
Secondary	—	2.7	3.4	2.7

group correlate well for 2, P-3A, and bleomycin and have been attributed to H-bonding with the carboxamido moiety of the propionamide substituent. The assignment of the lower pK_a value to the secondary amine agrees well with the previously observed pH and Zn^{2+} dependence of the chemical shifts of both methylene groups in the pyrimidine substituent, which reflect proximity to a group having a pK_a of about 2.7. Reassignment of this pK_a from the pyrimidine moiety (actual pK_a 1.0) to the secondary amine also serves to explain the previously observed insensitivity of the chemical shift of the pyrimidine methyl group over this pH range, as described by Dr. Oppenheimer.

Also provided by Dr. Oppenheimer was an analysis of the ¹H-NMR coupling constants for 1, 3[1], and bleomycin. As shown in Table 2, these data were consistent with the revised, acyclic structure for bleomycin. Although not definitive, the ¹³C-NMR chemical shift data determined for 1, 2, P-3A, and bleomycin also suggested the need for revision of the structure originally proposed for bleomycin.

Table 2: Side Chain Coupling Constants[a]

Coupling	Compound		
	1	3	Bleomycin
³J	5.5	4.3	5.8
	2.8	2.5	7.9
²J	−15.0	−14.5	−14.8

[a]Expressed in Hz.

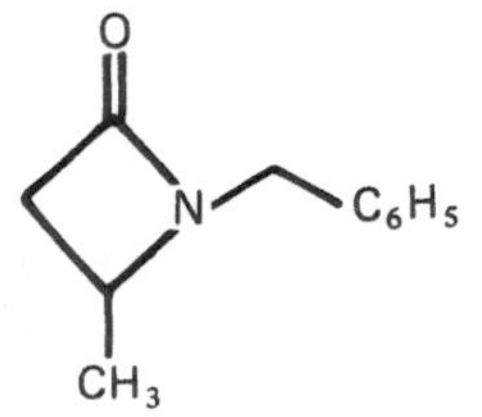

3

While the revised structure proposed for bleomycin was more nearly consistent with the available experimental data, the anomalously low pK_a value assigned to the secondary amine prompted investigation of alternative bleomycin structures immediately following the symposium[2]. One alternative structure considered involved attachment of the β-aminoalanineamide substituent through the α-, rather than the β-amino group, i.e.,

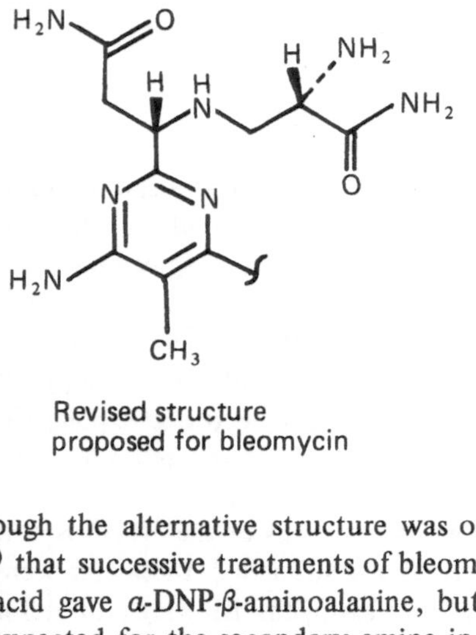

Revised structure
proposed for bleomycin

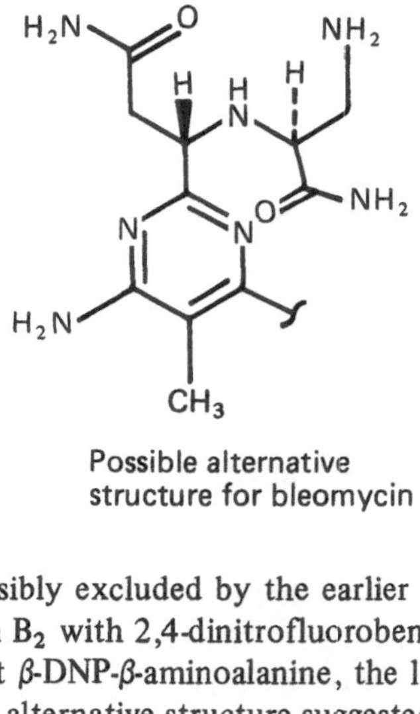

Possible alternative
structure for bleomycin

Although the alternative structure was ostensibly excluded by the earlier finding[3] that successive treatments of bleomycin B_2 with 2,4-dinitrofluorobenzene and acid gave α-DNP-β-aminoalanine, but not β-DNP-β-aminoalanine, the lower pK_a expected for the secondary amine in the alternative structure suggested the need for reinvestigation. Acetylation of bleomycin A_2 on the primary amino group [$(CH_3\overset{\text{O}}{\overset{\|}{C}})_2O$, H_2O, HCO_3^-] gave a monoacetyl derivative whose ^{1}H-NMR spectrum indicated a substantial (~ 0.5 ppm) downfield shift of a methine resonance. This observation served to exclude the possible alternative structure indicated above, but was consistent with the revised structure that has been proposed for bleomycin[2].

Although the reason(s) for the low apparent pK_a of the secondary amine is not altogether clear at present, it could conceivably be due to stabilization of the conjugate base via hydrogen bonding, e.g., as shown below. In addition to providing a satisfactory explanation for the observed pK_a (since this species would likely be difficult to protonate), a hydrogen-bonding scheme of this type would also account for the remarkable lack of reactivity of the secondary amine noted by Dr. Takita.

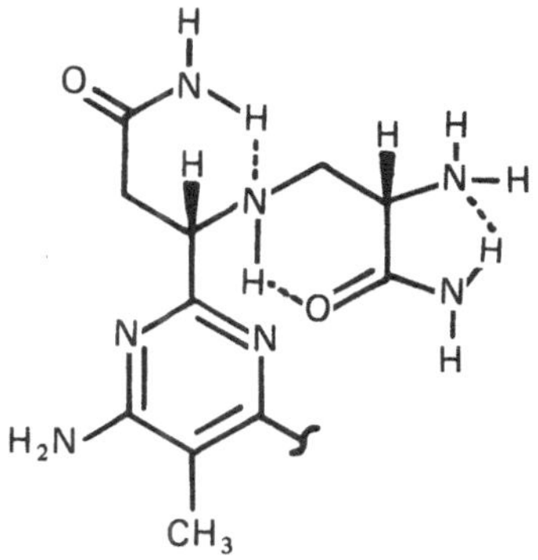

The syntheses of the model compounds utilized for NMR and pK_a studies was described by Dr. Ohno. The preparation of compound **1** involved elaboration of an intermediate (**4**) having a preformed β-lactam moiety. Separation of the diastereomers of **1** · HCl gave two racemic products with carbonyl stretching frequencies of 1710 and 1735 cm^{-1}, respectively. X-ray crystallography of the former isomer, having the $S,S(R,R)$-configuration, revealed an essentially orthogonal orientation of the β-lactam and pyrimidine rings.

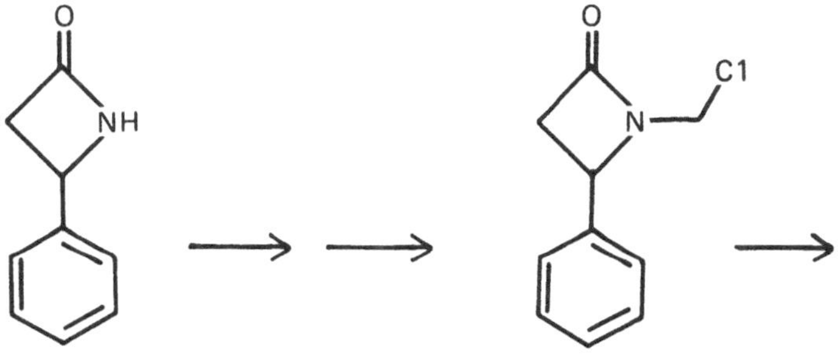

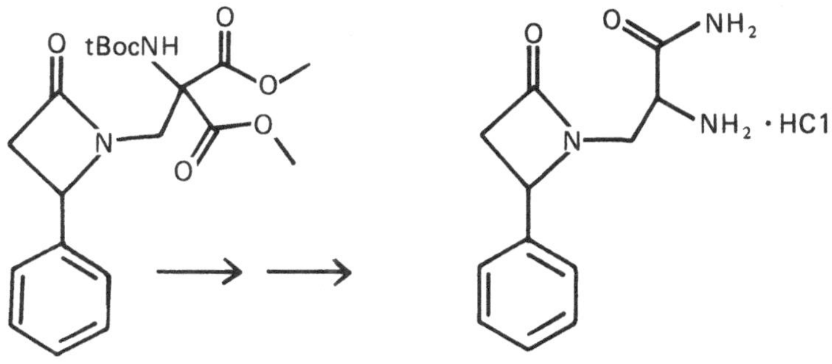

1 · HC1 (two diastereomers)

The preparation of model compound **2** was accomplished via conjugate addition of benzylamine derivative **5** to methyl acrylate **6**; subsequent ammonolysis and hydrogenolysis gave the desired product. Since a similar addition reaction involving pyrimidine **7** has previously been reported[4], and gave predominantly the desired $S,S(R,R)$ product(s), the pyrimidine moiety of bleomycin should now be readily accessible. Moreover, it should be possible to prepare related species useful in defining the reason(s) for the unusually low pK_a of the secondary amine and the strong influence of substituents on the pK_a value.

The total synthesis of bleomycin by an unambiguous route would permit verification of the proposed structure and provide access to a number of potentially useful analogs. In addition to the preparation of species having improved therapeutic properties, it might also be possible to prepare analogs having utility, e.g., as tumor radioimaging agents, or in defining the mechanism of action of bleomycin. Dr. Hecht described progress on the total synthesis of bleomycin and focused on the preparation of the bithiazole, L-gulose and $(2S,3S,4R)$-4-amino-3-hydroxy-2-methylvalerate moieties.

Synthetic elaboration of the bithiazole is a problem of more general interest since several antibiotics contain single or multiple thiazoles and thiazolines as structural components and there is a paucity of useful methods for the construction of all but the simplest thiazolines and thiazoles. Since, in the natural

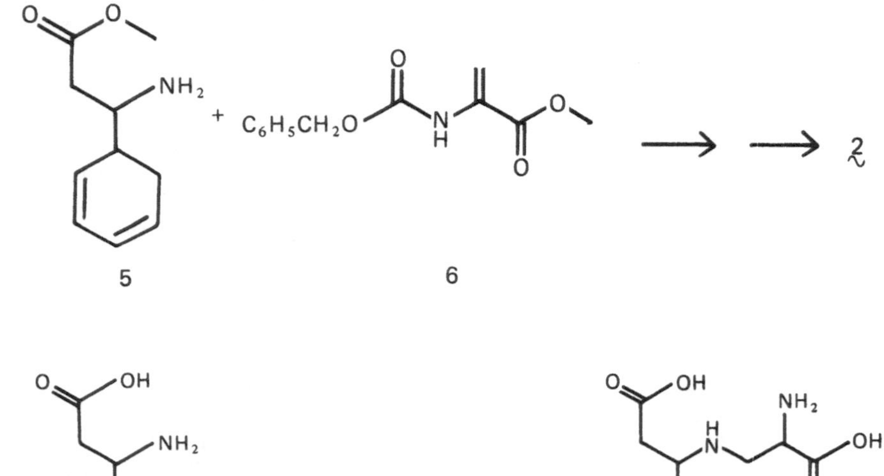

5   6

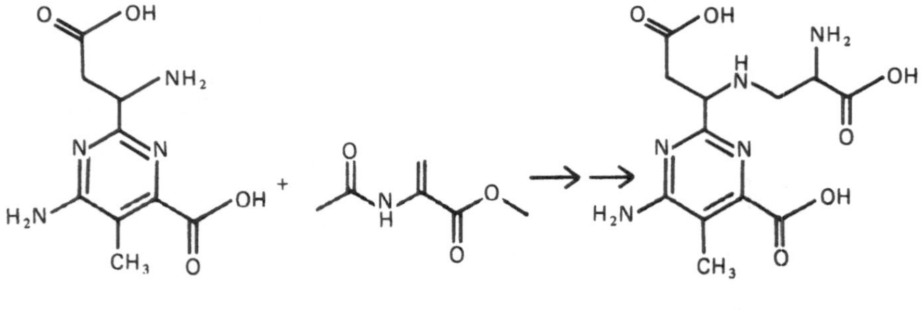

7   Pseudodipeptide

products containing such functionalities, biosynthetic derivation appears to involve dehydrative cyclization of cysteinyl peptides and subsequent dehydrogenation of the formed thiazolines, it seemed that a biomimetic synthesis of the bithiazole moiety of bleomycin might be more generally useful. *N*-acylated β-alanylcysteine derivatives 8 were converted to the corresponding thiazolines in good yield via the agency of HCl in $CHCl_3$; oxidation to the corresponding thiazoles was accomplished by treatment with NiO_2, a reagent not previously exploited for heterocyclic oxidations. After conversion of thiazole 9 ($R = C_6H_5$) to the requisite cysteinyl tripeptide derivative (10), successive treatments with $HCl/CHCl_3$ and NiO_2 gave the desired bithiazole.

The synthesis of the L-gulose moiety of bleomycin, blocked in a form suitable for construction of the carbohydrate moiety of the antibiotic, was accomplished by taking advantage of the well known relationship between this species and D-glucose. Thus 1,2-*O*-diacetyl-3,4-*O*-dibenzyl-β-D-glucopyranose (11), accessible in high yield from D-glucose, was converted to the respective 6-aldehydo sugar by treatment with *N*-chlorosuccinimide/$(CH_3)_2S/(C_2H_5)_3N$ at low temperature; the aldehyde was unstable but could be isolated as the crystalline dimethylhydrazone. Saponification of the acetate groups permitted facile conversion to the desired L-gulose derivative 12. Structural characterization of 12 included conversion to two crystalline 1,6-anhydro-L-gulose derivatives (13), the optical antipodes of which have been prepared previously. The preparation of L-gulose derivative 12 from D-glucose in high (42%) overall yield is also of more general interest, as it establishes a method for "epimerizations" of carbohydrates by functionalization of nonchiral carbon atoms.

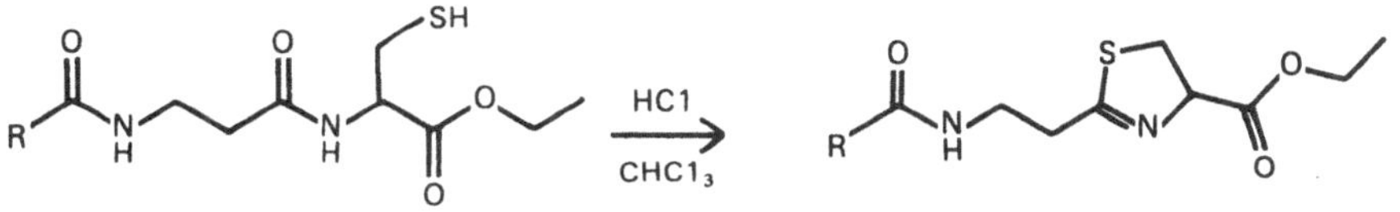

$R = CH_3, C_6H_5$

8

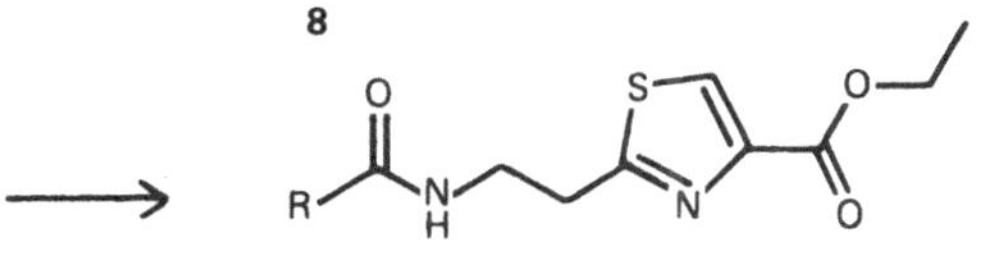

9

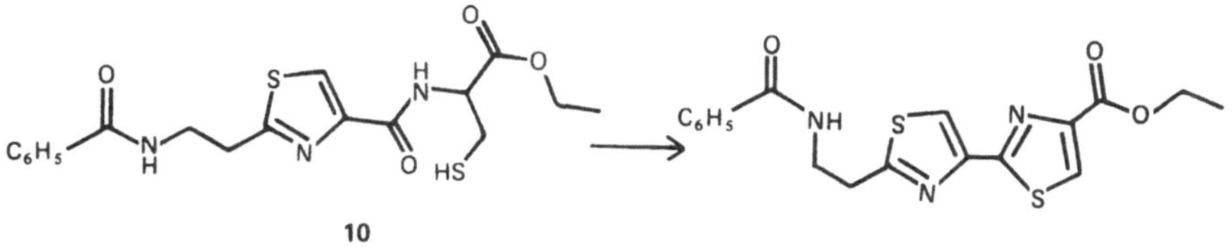

10

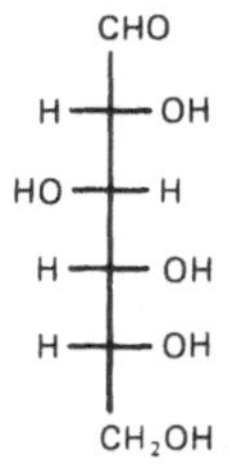

In addition to preparation via total synthesis, certain analogs of bleomycin differing only at the C-terminus have been prepared by addition of the desired "terminal amines" to the fermentation medium used to prepare bleomycin. The same species are also accessible chemically by condensation of bleomycinic acid with the appropriate amines in the presence of a water-soluble carbodiimide [see, e.g., Ref.[5]], followed by chromatographic purification as described by Dr. Muraoka. As virtually any amine can be coupled to bleomycinic acid in this fashion, and the latter is now accessible chemically from bleomycin demethyl-A_2 [by treatment with cyanogen bromide[6]] and enzymatically from bleomycin B_2 [by the use of an acylagmatine hydrolase activity from *Fusarium* sp.[7]], this method will undoubtedly find greater utility in the future. To date about 300 derivatives of bleomycin containing alterations at the C-terminus have been prepared and screened as potential anticancer agents. Several compounds with interesting biological properties were identified in this fashion, among them M5196, A5033, and PEP (pepleomycin). M5196 exhibited good organ distribution, was resistant to the enzyme activity that normally degrades bleomycins,

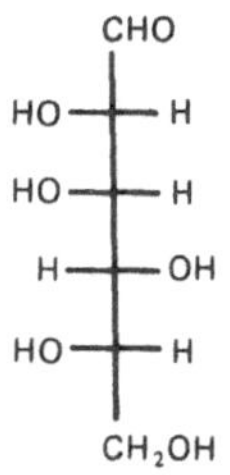

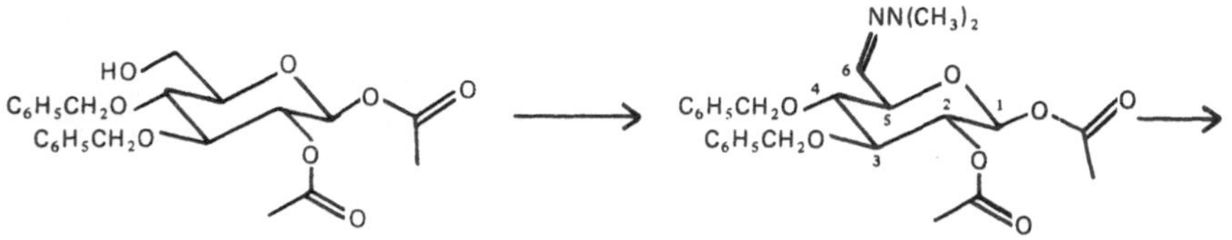

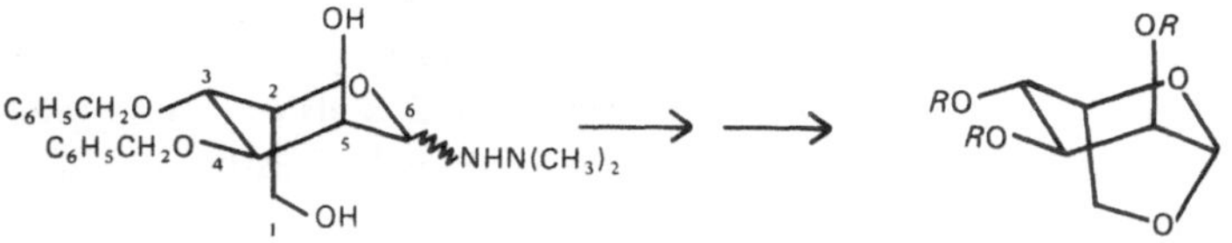

Bleomycin M5196:

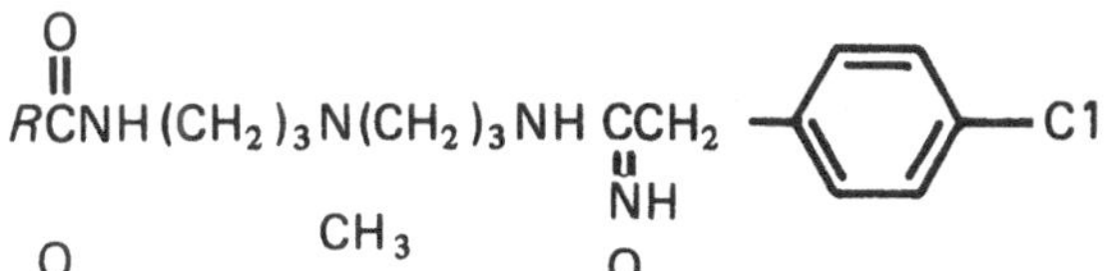

$$RCNH(CH_2)_3N(CH_2)_3NH\,CCH_2 \underset{}{\overset{}{\bigcirc}} C1$$

with $\overset{O}{\overset{\|}{}}$ above RCNH, CH_3 on the nitrogen, and $\overset{NH}{\overset{\|}{}}$ on CCH$_2$.

Bleomycin A5033:

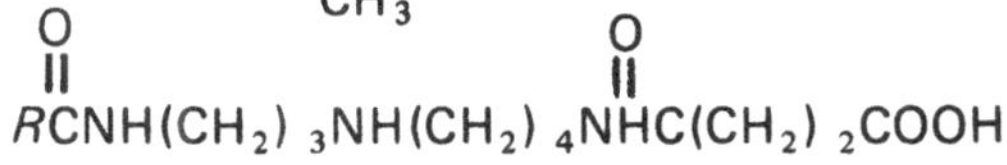

$$RCNH(CH_2)_3NH(CH_2)_4NHC(CH_2)_2COOH$$

Pepleomycin:

$$RCNH(CH_2)_3NH - \overset{H}{\underset{CH_3}{C}} \bigcirc$$

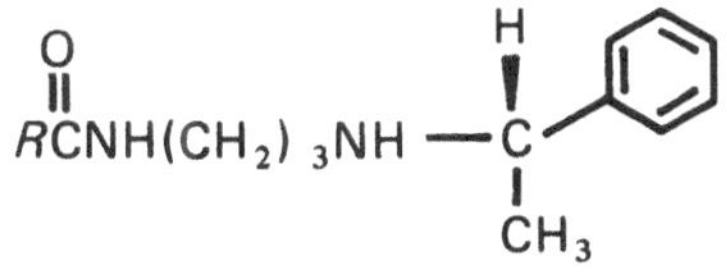

RCOOH = bleomycinic acid

and produced only 5% of the pulmonary fibrosis in mice normally caused by bleomycin. Although this compound was found to have strong renal toxicity in dogs, eliminating it as a candidate for clinical study, the greatly diminished pulmonary toxicity demonstrated that this activity need not parallel the anti-cancer activity of the bleomycins. A5033 had much lower pulmonary and renal toxicity than the original bleomycins, and was approximately equally as active in strand scission of SV40 DNA and inhibition of Ehrlich ascites carcinoma. While A5033 was also more active than bleomycin against Ehrlich solid carci-noma, it gave a weaker response in three other model systems and was found to be less effective clinically in the treatment of head, neck, and skin cancers as well as malignant lymphomas. On the other hand pepleomycin also exhibited good organ distribution and low pulmonary toxicity and was found to be strongly inhibitory to rat gastric carcinoma. Clinical trials involving several types of cancer are in progress.

A few additional analogs of bleomycin have been prepared by modification of preformed bleomycins. Dr. Takita described the preparation of epibleomycin and isobleomycin by treatment of Cu^{2+}-containing and Cu^{2+}-free bleomycin,

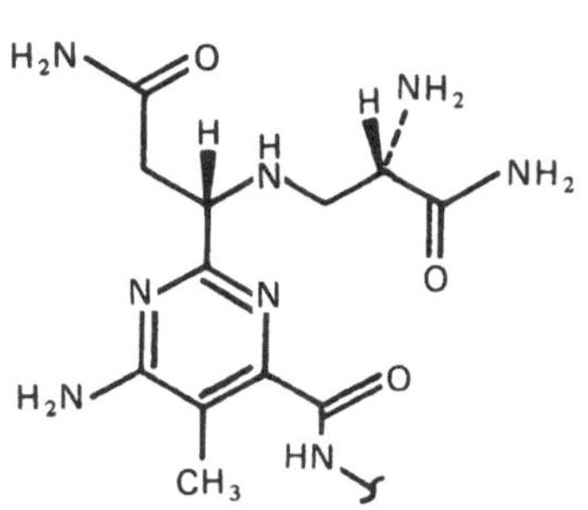

Epibleomycin 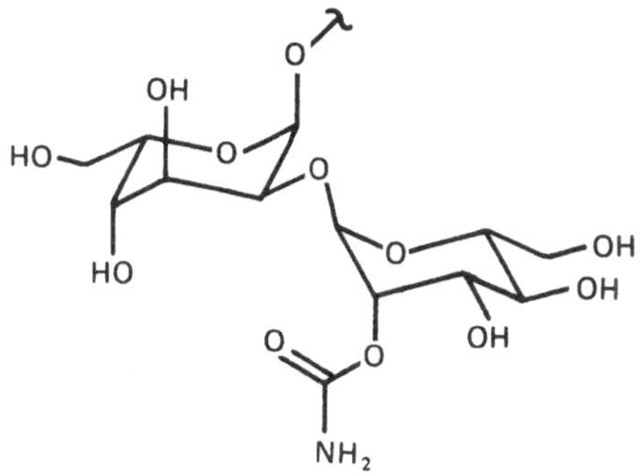   Isobleomycin

respectively, with triethylamine. Epibleomycin was found to be more active than bleomycin *in vitro* in causing DNA chain scission, but not active *in vivo*. Remarkably, isobleomycin did not nick DNA *in vitro*, but had considerable antibacterial activity, an observation that may be difficult to reconcile with the proposal that inhibition of bacterial growth by bleomycin involves DNA cleavage. A possible explanation, namely, that isobleomycin is converted to bleomycin *in vivo*, was suggested by Dr. Hori.

Dr. Meares described his work on the chemical modification of bleomycin A_2 to afford species conjugated to a bifunctional chelating agent. The modified bleomycins thus prepared were labeled with $^{111}In^{3+}$ and tested for organ distribution in tumor-bearing mice. The results indicated favorable tumor/organ radioactivity concentration ratios in all organs except the kidneys and the absence of indium release from the chelate. When the chelates were introduced into white rabbits bearing adenocarcinomas, scintillation scanning revealed the presence of tumors as small as 1 cm in diameter.

R = H, CH$_3$

In addition to the several bleomycins and phleomycins normally produced microbially, the structurally related species tallysomycin (Fig. 3) has recently been described[8]. Dr. Strong discussed the iron content and DNA strand scission capacity of this antibiotic in comparison with that of bleomycin. Although tallysomycin had a 20-fold greater affinity for salmon sperm DNA than bleomycin A_2, both were found to have similar effects in the disruption of supercoiled PM-2 DNA, as judged by the extent of fluorescence enhancement upon subsequent ethidium bromide binding to DNA. Tallysomycin was also reported to have 100-fold greater antibacterial activity than bleomycin and to be severalfold more active against certain experimental tumors, including Walker 256 carcinosarcoma and P-388 lymphocytic leukemia. However, its activity against B16 melanocarcinoma, the principal tumor used for comparison of bleomycin potencies, was slightly less than that of bleomycin.

The biological effects of bleomycin and its derivatives on bacterial and mammalian cells as well as certain experimental animals were described in some detail during the meeting. Dr. Terasima reported that both bacterial and mammalian cells gave biphasic dose–response curves when bleomycin toxicity was measured, an observation also made for certain mammalian cells by Dr. Nunn.

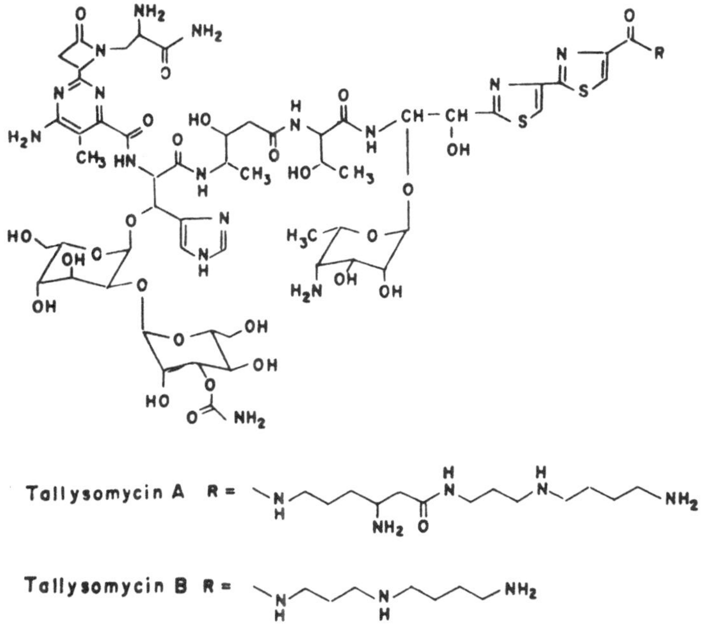

Fig. 3. Proposed structure of tallysomycin.

Dr. Terasima described experiments designed to elucidate the molecular mechanism(s) that gave rise to these biphasic dose–response curves. One of the experiments that he reported involved measurement of the repair of bleomycin-induced DNA damage *in vivo*, a parameter of considerable interest given the probability that bleomycin functions by DNA cleavage. Determination of the number of DNA breaks/cell and the average molecular weight of single-stranded DNA was carried out as a function of time after bleomycin treatment of mouse L cells; repair of DNA damage was observed, as shown in Fig. 4, but the molecular weight of the DNA did not recover fully during the course of the experiment. Dr. Hutchinson discussed a related series of experiments designed to determine whether bacterial cells with reduced DNA repair capacity were more sensitive to bleomycin than the corresponding wild type cells. Using *rec A* and *lex A* mutants of *Escherichia coli*, it was found that the repair-deficient cells were severalfold more sensitive to bleomycin and that the observed differences could be eliminated by employing a rich nutrient medium, consistent with the hypothesis that sensitivity of *E. coli* to bleomycin is mediated at the level of DNA damage.

Dr. Cohen has studied the toxicity of individual bleomycins to *E. coli* and to mouse L cells, and especially the contribution to these processes of the various organic amines that constitute the C-termini of the bleomycins. Bleomycin A_5 (containing spermidine at the C-terminus) was found to be at least

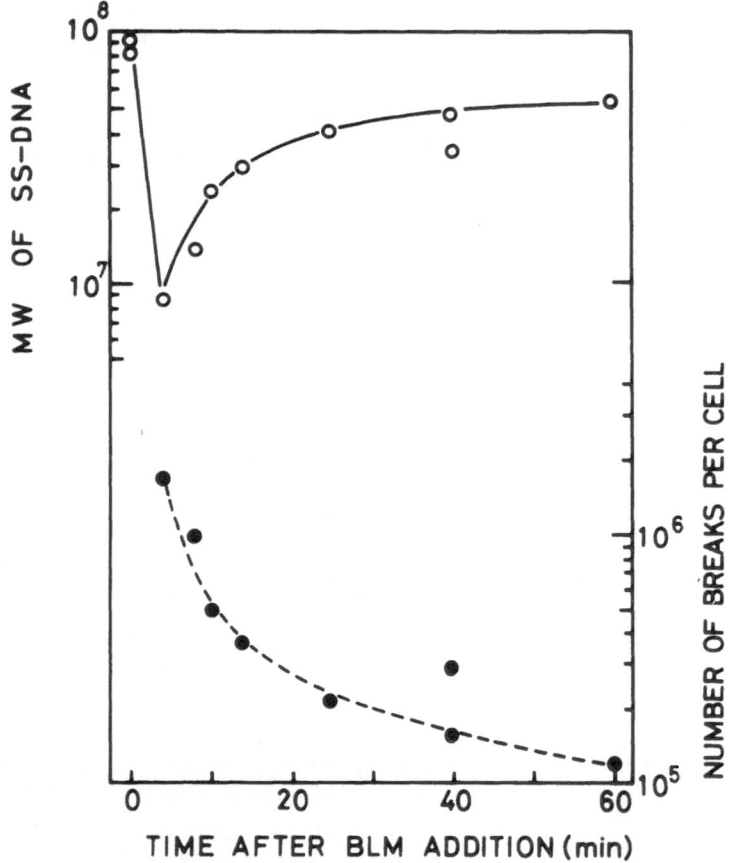

Fig. 4. Change in molecular weight of single-stranded DNA and L5 cells during incubation with bleomycin (100 μg/ml).

10-fold more toxic to all of the *E. coli* strains tested than were A_2, B_2, and A_2-b. Consistent with the hypothesis that the organic amines at the C-terminus of bleomycin facilitate the binding of the antibiotic to DNA, it was found that the toxicity of bleomycin A_5 to *E. coli* could be spared by high (10 mM) concentrations of spermidine. Repetition of these experiments in mouse L cells revealed somewhat less difference in lethality between the various bleomycins, but did establish the greater toxicity of bleomycin A_5 as compared with A_2 and B_2. While 10 mM spermidine had little effect on the killing of L cells by bleomycin A_5, at 10^{-4} M concentration the spermidine analog hirudonine did afford partial protection from killing. Interestingly, hirudonine did not protect against bleomycins A_2 or B_2.

Many bleomycins exhibit strong pulmonary toxicity that limits their clinical utility; the quantitative estimation of such toxicity is therefore of considerable

Hirudonine

importance. Dr. Raisfeld described the measurement of pulmonary toxicity in (28-week-old) mice following direct introduction of individual bleomycins to the lungs via syringe. At intermediate doses (1.6 μg/animal), bleomycins A_1, A_2, and B_2 caused interstitial fibrosis to about the same extent and no strain differences were noted in susceptibility to bleomycin. It was also found that bleomycin A_1 did not cause extensive disease when employed at low concentration and that blenoxane (the clinically used mixture of bleomycins consisting primarily of A_2 and B_2) was much more toxic than bleomycin A_2 or B_2 when used at a dose of 16 μg/animal. This test may prove to be the method of choice for determining the pulmonary toxicity of new bleomycin analogs since quantitative differences between bleomycins can be detected easily and single doses of toxic bleomycins produce lung lesions in virtually every animal treated.

One of the observations made by Dr. Cohen is that the bleomycins are substantially more toxic toward E. coli than L cells. For example, while the number of viable E. coli cells was reduced by a factor of 10^3 within two generation times by 1 μg/ml of bleomycin A_5, the same compound effected (99%) killing of a population of mouse L cells only when employed at a concentration of 20 μg/ml. This may be due to differences in the permeability of bleomycin to the specific cell lines employed, a suggestion also offered by Dr. Umezawa as a possible explanation for the remarkably different toxicities found for individual bleomycins in single strains of E. coli. Dr. Nunn described an aspect of bleomycin toxicity that is less easily explained, namely, that the doses required to inhibit mammalian cells grown in culture are very much greater than those needed to produce a response in the clinic. (While it is difficult to compare cells in culture with those in vivo, direct extrapolation suggests that the latter may be up to 500 times more sensitive to bleomycin.) Dr. Nunn noted further that the serum concentration of certain metals (including copper) can be altered in patients with, e.g., Hodgkin's disease, leukemias, and lung cancer, so that quite aside from its interaction with DNA, bleomycin (as an effective chelating agent) may produce clinical effects via transport of metals to or from a particular site.

The preparation of additional bleomycin analogs with improved antitumor activity and lower pulmonary and renal toxicity continues to be of interest and importance. Clearly the design of appropriate derivatives would be facilitated by an understanding of the molecular basis of the anticancer activity of bleomycin which, in turn, suggests the utility of an exact description of the three-dimensional structure of bleomycin and especially of its metal complexes. Drs. Oppenheimer and Dabrowiak described their physical studies of bleomycin and its Cu^{2+}, Zn^{2+}, and Fe^{2+} complexes. The binding of both Cu^{2+} and Zn^{2+} occurs with concomitant proton release; at pH 6.5, the addition of 1 equivalent of Zn^{2+} effected the release of 1.08 equivalents of H_3^+O, while at pH 5.0 about 1.3 equivalents of H_3^+O were released by 1 equivalent of Zn^{2+} or Cu^{2+}. In spite of the observed release of protons, the addition of Cu^{2+} or Fe^{2+} to bleomycin at low pH did not cause NMR line-broadening, suggesting the absence of formation of a stable complex under these conditions. Consistent with this interpretation was the observation that actual binding of Zn(II) to bleomycin A_2 caused changes in the chemical shifts of 42 of the 52 observable resonances in the ^{13}C-NMR spectrum and that binding of Cu(II) eliminated almost half of the resonances via line-broadening. Dr. Dabrowiak also studied the electrochemistry of bleomycin. Bleomycin A_2 exhibited a two-electron reduction wave at -1.22 V and a multielectron reduction wave at -1.48 V, corresponding to the pyrimidine and bithiazole moieties, respectively. Both reductions were irreversible electrochemically. Since the pyrimidine moiety is involved as a donor group in the formation of metal complexes of bleomycin, as discussed below, the electrochemical properties may be useful as a probe of the metal-binding properties of the antibiotic.

Drs. Dabrowiak, Takita, and Sugiura discussed the nature of the metal-binding sites in bleomycin; the six atoms involved are labeled in the structural formula to facilitate reference to each. Evidence for the participation of nitrogen-1 derives from potentiometric titration data. Specifically, the deprotonation of this nitrogen atom ($-NH_3^+ \rightarrow -NH_2$; $pK_a \sim 7.1$–7.4) is reflected in the titration curve of metal-free bleomycin A_2 but not in those of the Cu(II) and Zn(II) complexes. Also, it was found that acylation of nitrogen-1 (Schotten-Baumann conditions) in bleomycin A_2 was much less facile in the presence of Cu(II). In similar fashion, the deprotonation of the imidazole moiety (nitrogen-5; $pK_a \sim$ 4.5) was evident in the titration curve of bleomycin A_2, but not in the curves of the Cu(II) and Zn(II) complexes of bleomycin A_2. That nitrogen-5 is a donor atom in the formation of metal–bleomycin complexes may also be inferred from the shift of the two imidazole C–H proton resonances in bleomycin A_2 to lower field in the presence of Zn(II) and from the negative Pauly test obtained for Cu(II) bleomycin.

Evidence cited in support of the participation of nitrogen-3 as a donor atom in bleomycin–metal complexes included the shift of the pyrimidine CH_3 proton resonance to lower field upon addition of Zn(II), the presence of a peak at 253 nm in the difference ultraviolet spectrum of bleomycin and Cu(II)–bleomycin, and the formation of epibleomycin when Cu(II)–bleomycin (but not metal-free bleomycin) was treated with aqueous triethylamine. Similarly, ^{13}C-NMR indicated that the chemical shift of the carbon atom attached to oxygen-6 was altered upon addition of Zn(II) to metal-free bleomycin and that this resonance was also missing in the spectrum of Cu(II)–bleomycin A_2 (as expected), although the other carbohydrate resonances were essentially unchanged. Since it has also been shown that the carbamoyl moiety in bleomycin undergoes very rapid positional isomerization in the presence of triethylamine (to give isobleomycin), but that this transformation does not occur for Cu(II)–bleomycin, oxygen-6 is also believed to participate as a donor atom.

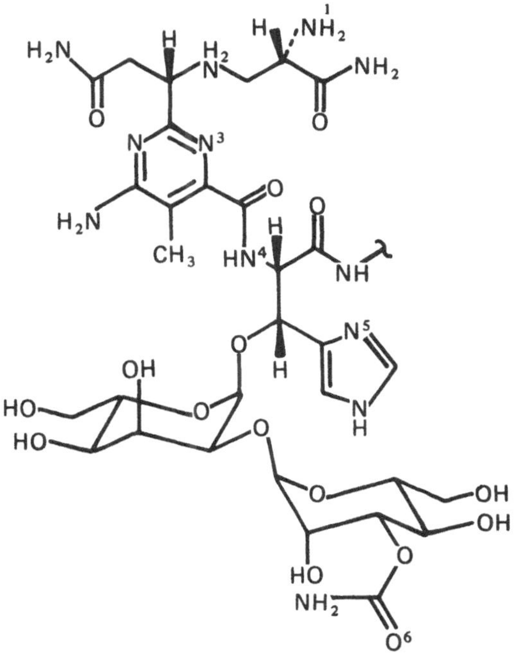

Aside from potentiometric titration data, the best evidence for the participation of nitrogens-2 and -4 in the formation of metal complexes with bleomycin is undoubtedly that obtained from X-ray crystallography of a Cu(II) complex of P-3A. Crystallographic analysis revealed that nitrogens-2 and -4 were equatorial ligands in a square pyramidal arrangement of the five nitrogen atoms about the copper, the latter of which was found to be displaced approximately 0.3 Å from the equatorial plane of the complex toward nitrogen-1 (Fig. 5). All of the available data are consistent with the interpretation that bleomycin forms metal complexes analogous to that determined for Cu(II)–P-3A and it is a structure of this type that is assumed to bind to and cleave DNA (Fig. 6).

The biological damage caused by bleomycin is similar to that obtained with ionizing radiation and other agents that function by producing lesions in DNA; effects observed have included inhibition of DNA synthesis, degradation of

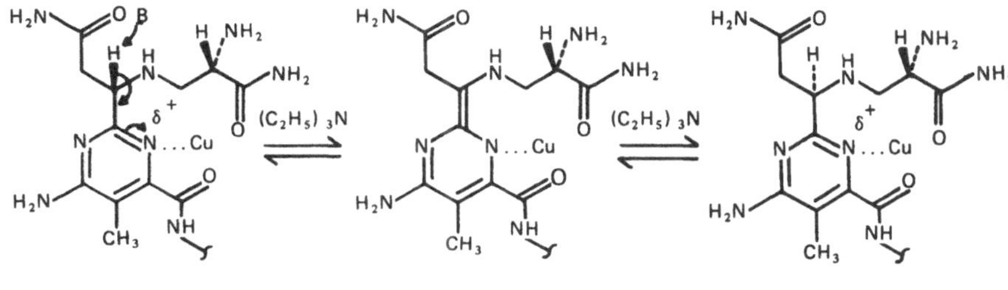

Bleomycin   Epibleomycin

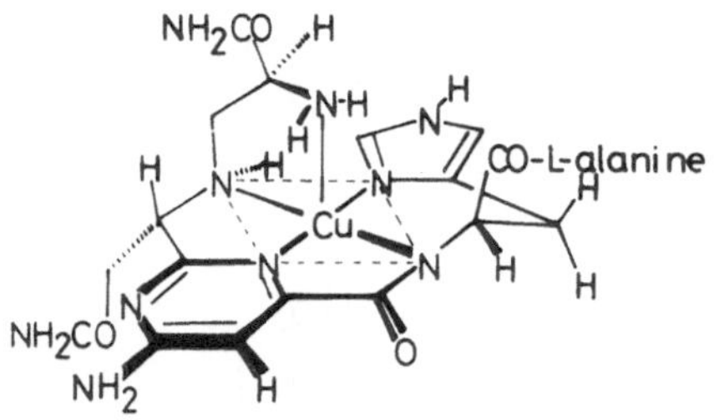

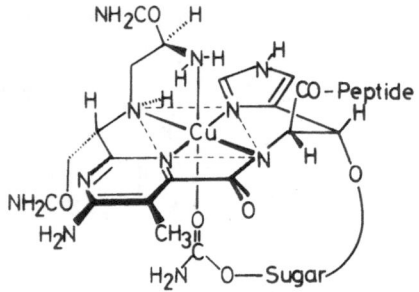

Fig. 5. Structure of Cu(II)–P-3A, as determined by X-ray crystallography.

DNA *in vivo* and *in vitro*, creation of chromosomal aberrations, loss of colony-forming ability, release of DNA from a membrane complex, and release of nucleosomes from chromatin. Thus, in spite of a few disquieting experimental observations (such as the insensitivity of cultured mammalian cells to bleomycin and the observed antibacterial activity of certain bleomycin derivatives — e.g., isobleomycin — that would not nick DNA *in vitro*) efforts at determining the mechanism by which bleomycin exerts its effect as an anticancer agent have focused on the binding and cleavage of DNA by the antibiotic.

Bleomycin has been shown to bind to and effect degradation of a variety of double-stranded DNAs, including chromatin, supercoiled covalently closed circular DNAs, linear duplex DNAs, and synthetic polydeoxynucleotides. Bleomycin creates single-strand breaks in DNA at neutral pH, as judged by velocity sedimentation analyses or gel electrophoresis of treated DNAs and by the change in fluorescence enhancement of ethidium bromide bound to superhelical covalently closed circular DNA after bleomycin treatment. Also created are DNA lesions that appear to be identical with apurinic (apyrimidinic) sites based on their susceptibility to hydrolysis by repair endoribonucleases and base[9] and the observations that free bases are released under conditions that result in little strand scission and that bleomycin treatment at neutral pH creates "alkali-labile" bonds more numerous than actual breaks observed at neutral pH. It should be noted, however, that Drs. Haseltine[10] and Grollman have found that the DNA fragments arising after treatment with bleomycin and alkali do not comigrate on polyacrylamide gels with authentic standards resulting from hydrolysis at apurinic

Fig. 6. Proposed structure of Cu(II)–bleomycin.

(apyrimidinic) sites according to the method of Maxam and Gilbert[11]. Double-strand breaks in DNA have also been detected and, as discussed by Drs. Haidle and Hutchinson, occur at a frequency well in excess of the predicted random coincidence of single-strand nicks on complementary DNA strands. Possibly relevant to the mechanism(s) by which double-strand breaks occur is the finding by Dr. Haidle that while double-strand scissions induced by bleomycin occur at many sites on the PM2 DNA genome, a disproportionate number occur in a single region of the genome, and the work carried out by Drs. Haseltine[10], Hori, and Grollman indicating that DNA cleavage by bleomycin is sequence dependent with the most facile cleavages occurring between ...GT... and ...GC...sequences. Dr. Hutchinson outlined three possible mechanisms for specific production of double-strand breaks, including a model originally proposed by Dr. Haidle that involves the binding of a bleomycin dimer to double-stranded DNA. Not explicitly considered was the obvious alternative that a single-strand break in DNA alters the local conformation of the molecule sufficiently to induce a disproportionate number of additional breaks in the same region, thus giving the appearance of specific double-strand cleavage.

The molecular basis of bleomycin binding to DNA has been studied by Drs. Chien, Grollman, and Horwitz[12] and by Dr. Hori; both laboratories have shown that addition of DNA to solutions containing various bleomycins results in quenching of the fluorescence emission associated with the bithiazole moiety. Chien et al.[12] have also investigated the interaction of bleomycin A_2 and DNA by proton magnetic resonance and demonstrated preferential broadening of the signals corresponding to the thiazole and S-methyl protons. That the C-terminal substituents of the various bleomycins may also contribute to DNA binding is supported by the observations made in Dr. Cohen's laboratory regarding the reversal of bleomycin toxicity (to varying extents) by certain polyamines structurally related to these C-terminal substituents, and also by Dr. Hori's finding that individual bleomycins (differing only at the C-terminus) vary substantially in their ability to effect strand scission of DNA. It is probably worth noting that the apparent equilibrium constants for the DNA complexes with bleomycin A_2 and the derived tripeptide S were quite similar (1.2 vs 1.4×10^5 M^{-1}), suggesting that additional structural components of bleomycin do not contribute significantly to DNA binding.

As first noted by Ishida and Takahashi[13], and demonstrated convincingly by Dr. Horwitz and her co-workers, Fe(II) is required for DNA chain scission by bleomycin. Although none of the early *in vitro* experiments actually involved the explicit addition of Fe to the incubation mixtures, Drs. Horwitz, Strong, and Lown have all shown that clinical grade bleomycin contains considerable "contaminating" Fe (estimated at 0.02% mole % by Dr. Lown) and that the degradation of DNA in the absence of added Fe is suppressed by chelating agents.

In addition to Fe(II) [or Fe(III) and a reducing agent], chain scission of DNA by bleomycin was also shown to be sensitive to the presence of O_2[14,15], suggesting that this species, or a reduction product thereof, might participate directly in the degradation of DNA. In fact radical scavengers have been shown to protect DNA from bleomycin-promoted damage. An ESR spectrum that seemed consistent with the presence of O_2^- was observed by Dr. Dabrowiak after exposure of one preparation of Fe(II)–bleomycin A_2 to oxygen and Drs. Horwitz and Lown have both shown that superoxide dismutase inhibits the degradation of DNA by moderate concentrations of bleomycin. Hydroxyl radicals are known to degrade DNA and the generation of ·OH in the presence of bleomycin has been inferred by Dr. Lown on the basis of spin trapping experiments with phenyl t-butylnitrone. It is likely that H_2O_2 is also present in significant concentrations, although the addition of catalase to incubation mixtures containing Fe(II)–bleomycin and DNA has been reported to both stimulate and inhibit DNA degradation.

Based on the requirements for Fe and O_2 in the degradation of DNA by bleomycin, and on the structure of the Cu(II) complex of P-3A, Dr. Takita proposed a structure for the "active" form of bleomycin. As shown in the structural formula (Fig. 7), the proposed structure involves the same octahedral arrangement of ligands about Fe(II) as was previously suggested for Cu(II), with the exception that O_2 is envisioned as an axial ligand in place of the carbamoyl group. The O_2 could thus be reduced by coordinated Fe(II) and utilized directly for DNA cleavage; if the Fe(II)–bleomycin complex were bound to DNA at the time, then specificity in the mode of DNA cleavage might derive from the generation of reduced O_2 in proximity to a particular reactive site. Though possibly not representing a serious conceptual difficulty in the context of this proposed "active" form, it should be noted that isobleomycin (which differs from bleomycin only in the position of a functionality – the carbamoyl moiety – that is ostensibly not involved as a ligand in the active complex) has been shown by Dr. Hori to be inactive in DNA chain scission.

At present, the actual form(s) of reduced O_2 involved in the degradation of DNA is unclear. Superoxide is obviously an attractive candidate since it could be generated by transfer of a single electron according to the scheme:

$$\text{Fe(II)–bleomycin–}O_2 \rightarrow \text{Fe(III)–bleomycin–}O_2^-$$

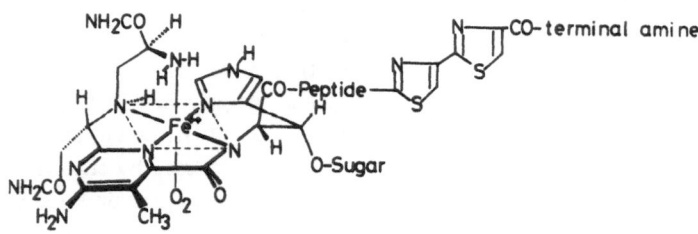

Fig. 7. Postulated structure of Fe(II)–bleomycin.

However, as pointed out by Drs. Horwitz and Lown, it is difficult to exclude the possibility that H_2O_2 and $\cdot OH$ function in the overall scheme leading to the degradation of DNA by bleomycin. The stoichiometry of O_2 utilization has been determined by Dr. Horwitz, who measured the effect of excess bleomycin on the oxidation of Fe(II) in the presence and absence of DNA. At very low concentrations of Fe(II) in the presence of DNA, about 1 mole of O_2 was consumed for each equivalent of Fe(II) added, but at all other metal ion concentrations the ratio of O_2 consumed/Fe(II) added was about 0.5.

The study of the molecular mechanism by which bleomycin degrades DNA has been a particularly challenging problem for several reasons. As noted by Drs. Hori and Hutchinson, e.g., it is difficult to obtain reproducible results from one experiment to another, even when individual runs are conducted under carefully controlled conditions. The observed differences may certainly be attributed in part to the dependence of the reaction on several variables, including O_2, reducing agents, Fe, pH, the choice of buffer, and trace metal contaminants that can inhibit DNA chain scission. It is less clear whether other factors, related more directly to the intrinsic chemical behavior of bleomycin in combination with DNA, also contribute to the apparent lack of reproducibility. Difficulties in the interpretation of experimental data can also be attributed to the fact that required components of the incubation mixtures [such as Fe(II) and thiols] degrade DNA to some extent even in the absence of bleomycin and, formerly, derived additionally from the lack of firm structures for bleomycin and its metal complexes.

Beyond this, it has also been observed by Dr. Hori that the amount of DNA strand scission obtained with bleomycin is not directly proportional to the concentration of the antibiotic, i.e., that in order to produce strand scission substantially in excess of that obtained with low concentrations of bleomycin requires much (approximately 10-fold) greater concentrations of the drug. Since it has also been shown that bleomycin creates "alkali-labile" lesions (possibly identical in structure with apurinic sites) as well as strand scission, it seems not unlikely that bleomycin degrades DNA by more than one mechanism.

Although the amount of data relevant to the molecular mechanism of DNA binding and cleavage is not great, certain facts have been established reasonably firmly and must be accommodated in any proposed mechanism(s). These include:

(a) Bleomycin binds to many types of DNA and has a special affinity for guanylic acid moieties; it does not bind effectively to RNA, however.

(b) All four heterocyclic components of DNA are released as the free bases. The loss of thymine occurs to the greatest extent, but it is unclear whether this is due to some special property of this base that facilitates its removal or is simply a consequence of the preferential binding of bleomycin in response to certain DNA base sequences. Additionally, Dr. Hutchinson has isolated small quantities of purines and pyrimidines that were modified during the process of base release from DNA by bleomycin and these may be pertinent to the mechanism of base release.

(c) DNA strand scission involves primarily ... GT ... and ... GC ... sequences, cleavage occurring with the release of the pyrimidine and within or to the 3'-side of the thymidine (cytidine) carbohydrate moiety. Products of the cleavage include polymers (oligomers) terminating primarily or exclusively with 5'-phosphates; no species containing 3'-phosphates were present, however, suggesting that part or all of the carbohydrate moiety originally associated with the released base may still be attached to the polymer.

(d) If the "alkali-labile" sites created by bleomycin treatment lack heterocyclic bases but contain the other structural components of the original nucleotides, then there is at least one mechanism for DNA degradation by bleomycin that does not involve concomitant base release and chain scission as an obligatory step.

(e) DNA degradation results in the release of a compound that reacts with 2-thiobarbituric acid to give an intensely colored species having λ_{max} 532 nm. Aldehydes are known to afford spectrally similar products when treated with thiobarbituric acid and the adduct containing the DNA fragment was shown to have the same ultraviolet spectrum as the product resulting from treatment of 2-thiobarituric acid with malondialdehyde. [On the basis of similar evidence, malondialdehyde has been postulated to form during the ostensibly analogous degradation of DNA by ionizing radiation[16,17].]

(f) When DNA is treated at limiting concentrations of bleomycin, the resulting DNA lesions are more numerous than bleomycin molecules.

Assuming, as seems reasonable on the basis of the accumulated data, that DNA chain scission is initiated by free radicals produced from O_2, then a number of mechanisms can be written to account for all or most of the experimental observations outlined above. Chemically reasonable processes would include one similar to that suggested by Dr. Grollman, in which an electron-deficient species (such as O_2^- or $HOO\cdot$) abstracted a hydrogen atom from C-4' to form a stabilized alkyl radical (i), the latter of which could react with O_2 and abstract a hydrogen atom to afford a C-4' hydroperoxide (ii). Base (e.g., H_2O, OH^-)-catalyzed collapse of species ii would afford ester iii, a precursor of a mixture of products including a purine or pyrimidine, malondialdehyde, a polymer terminating with a 5'-phosphate group and another polymer having a (blocked) 3'-phosphate group. This mechanism thus incorporates many of the requisite features noted above, although not the reported release of purines and pyrimidines without concomitant chain scission. Moreover, it is difficult to see why collapse of intermediate ii in this scheme (and [3] in the scheme outlined by Dr. Grollman) would not also lead to significant amounts of chain breakage without rupture of the C-3'–C-4' bond, as phosphate is a reasonable leaving group. (It is conceivable, though, that collapse might be initiated via Lewis acid coordination of the hydroperoxide moiety, which could facilitate cleavage of the C-3'–C-4' bond with the concomitant development of carbonium ion character at C-3').

Also outlined during the symposium by Dr. Lown was a mechanistic scheme for DNA degradation involving $\cdot OH$, analogous to that described by von Sonntag

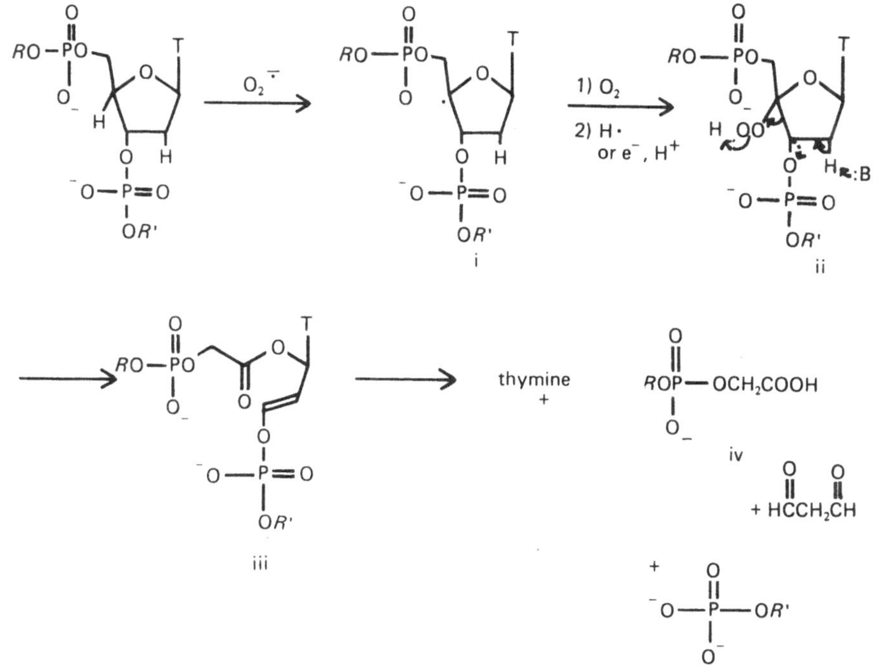

$O_2^{\overline{\cdot}}$
1) O_2
2) H·
or e^-, H^+
i
ii
iii
thymine
+
$ROP—OCH_2COOH$
iv
+ $HCCH_2CH$

and his co-workers[18] as resulting from γ-irradiation of herring sperm and calf thymus DNAs. It may be noted that this scheme also fails to provide for the release of bases in the absence of chain scission.

Release of bases with formation of "alkali-labile" sites, but without explicit chain scission, could probably be accomplished in several ways, including some

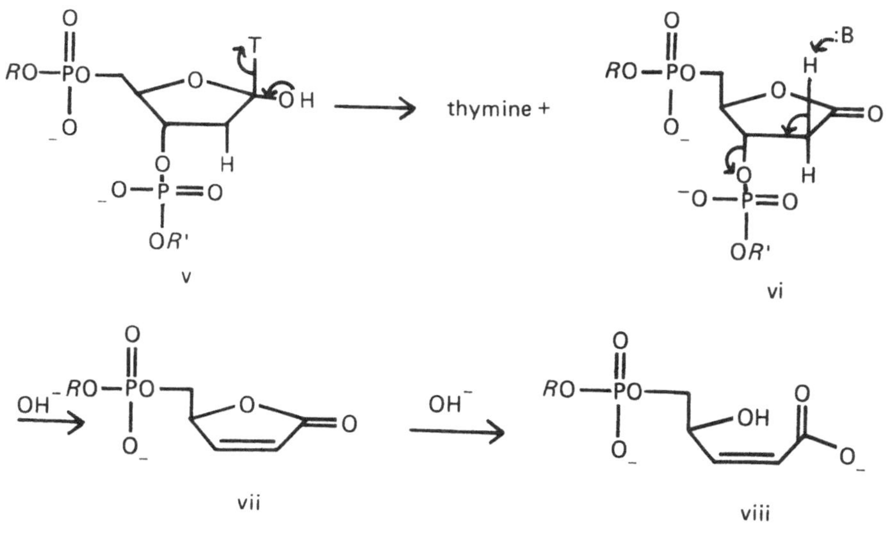

thymine +
v
vi
vii
viii
OH^-

that involve (transient) modification of the purine and pyrimidine moieties. Although it is entirely in the realm of speculation, one obvious route would entail oxidation at C-1', presumably via initial abstraction of a hydrogen atom from that position. If the C-1'–OH species v were to form by recombination of a C-1' radical with ·OH, base release would occur with simultaneous formation of vi[19–21]. The latter has an acidic proton at C-2 which would render this site "alkali labile." Moreover vi might be sufficiently similar to an apurinic site in structure to act as a substrate for repair endoribonucleases, as has been noted experimentally. Treatment of vii with strong base might be expected to aford viii, which would have properties similar to those of iv on polyacrylamide gels. One may note, however, that this scheme would not account for the formation of malondialdehyde and that vii has an acidic proton at C-4', which could result in β-elimination of a polymer terminating in a 3'-phosphate group (which is not observed experimentally).

At present there is insufficient evidence to establish any mechanistic scheme firmly. Experiments that would help to define the relevant pathway(s) include:

(a) Characterization of the structures of the reaction products, including the nature of the modified pyrimidines and carbohydrate moieties.

(b) Identification of the structure of the adduct(s) resulting from treatment with thiobarbituric acid and careful assay of the stoichiometry of its formation.

(c) Estimation of the rate and extent of generation of O_2^-, H_2O_2, and ·OH by Fe(II)–bleomycin and O_2, and of the propensity of each of the species to effect the postulated transformations at the observed rate.

(d) Repetition of many of the mechanistic studies in the absence of added nucleophiles such as thiols or Tris, since certain thiols are known to react reversibly with pyrimidines in a fashion that would increase the probability of depurination[22] and can also autoxidize with the formation of H_2O_2[23], while Tris has been employed to effect chain scission of nucleic acids at apurinic sites[24,25].

Acknowledgments

This summary is based on the presentations of the participants at the Bleomycin Symposium held at the East–West Center and those participants in an earlier meeting held at Yale University on April 8, 1978. Additional discussions with several of the meeting participants, including Profs. Norman Oppenheimer and John Kozarich, were especially helpful.

References

1. P. K. Wang, M. Madhavarao, D. F. Martent, and M. Rosenblum, *J. Am. Chem. Soc.*, **99**, 2823 (1977).
2. N. J. Oppenheimer, L. O. Rodriguez, and S. M. Hecht, *Biochemistry*, in press.
3. T. Takita, Y. Muraoka, T. Yoshioka, A. Fujii, K. Maeda, and H. Umezawa, *J. Antibiot. (Tokyo)*, **25**, 755 (1972).

4. T. Yoshioka, Y. Muraoka, T. Takita, K. Maeda, and H. Umezawa, *J. Antibiot. (Tokyo)*, **25**, 625 (1972).
5. W. Tanaka, *J. Antibiot. (Tokyo)*, **30** (Suppl.), S-41 (1978).
6. A. Fujii, this volume, p. 000.
7. H. Umezawa, Y. Takahashi, A. Fujii, T. Saino, T. Shirai, and T. Takita, *J. Antibiot. (Tokyo)*, **26**, 117 (1973).
8. M. Konishi, K. Saito, K. Numata, T. Tsuno, K. Asama, H. Tsukiura, T. Naito, and H. Kawaguchi, *J. Antibiot. (Tokyo)*, **30**, 789 (1977).
9. R. Schyns, M. Mulquet, and W. G. Verly, *FEBS Lett.*, **93**, 47 (1978).
10. A. D. D'Andrea and W. A. Haseltine, *Proc. Nat. Acad. Sci. USA*, **75**, 3608 (1978).
11. A. Maxam and W. Gilbert, *Proc. Nat. Acad. Sci. USA*, **74**, 560 (1977).
12. M. Chien, A. P. Grollman, and S. B. Horwitz, *Biochemistry*, **16**, 3641 (1977).
13. R. Ishida and T. Takahashi, *Biochem. Biophys. Res. Commun.*, **66**, 1432 (1975).
14. J. Onishi, H. Iwata, and Y. Takagi, *J. Biochem. (Tokyo)*, **77**, 745 (1975).
15. E. A. Sausville, J. Peisach, and S. B. Horwitz, *Biochem. Biophys. Res. Commun.*, **73**, 814 (1976).
16. N. P. Krushinskaya and M. I. Shal'nov, *Radiobiology*, **7**, 36 (1967).
17. D. S. Kapp and K. C. Smith, *Radiat. Res.*, **42**, 34 (1970).
18. M. Dizdaroglu, C. VonSonntag, and D. Schulte-Frohlinde, *J. Am. Chem. Soc.*, **97**, 2277 (1975).
19. B. Singer and H. Fraenkel-Conrat, *Biochemistry*, **4**, 227 (1965).
20. H-J. Raese and E. Freese, *Biochim. Biophys. Acta*, **155**, 476 (1968).
21. H-J. Raese, E. Freese, and M. S. Melzer, *Biochim. Biophys. Acta*, **155**, 491 (1968).
22. Y. Wataya, H. Hayatsu, and Y. Kawazoe, *J. Am. Chem. Soc.*, **94**, 8927 (1972).
23. P. P. Trotta, L. M. Pinkus, and A. Meister, *J. Biol. Chem.*, **249**, 1915 (1974).
24. R. Thiebe and H. G. Zachau, In *Methods in Enzymology* (K. Moldave and L. Grossman, eds.), Vol. 20, p. 179. Academic Press, New York (1971).
25. M. Simsek, G. Petrissant, and U. L. RajBhandary, *Proc. Nat. Acad. Sci. USA*, **70**, 2600 (1973).

Advances in Bleomycin Studies

Hamao Umezama

The clinical effect of bleomycin was first observed in the treatment of squamous cell carcinomas in 1966. The size of this kind of tumor decreases with bleomycin treatment and even cases of complete cure have been reported. Among the tumors belonging to this group, the well-differentiated type responds to bleomycin treatment with a very high frequency. There are patients who have lived more than 10 years after several courses of treatment with bleomycin alone. The rate of cure from squamous cell carcinoma has been increased by the simultaneous use of bleomycin and radiation or by applying radiation after bleomycin treatment. Most Hodgkin's lymphomas are very sensitive to bleomycin treatment. This tumor decreases in size rapidly with such treatment, and complete cures have been obtained by a combination of bleomycin with other antitumor agents. Even a daily dose of bleomycin as small as 1 mg exhibits a therapeutic effect on Hodgkin's lymphoma. Recently, treatment by a combination of bleomycin and a *Vinca* alkaloid or a platinum compound was found to have a therapeutic effect on testis tumors. Bleomycin treatment does not suppress immune function and does not damage bone marrow; however, the amount of bleomycin used in one course of treatment is presently limited to 200–300 mg per person because of its pulmonary toxicity. Therefore, a bleomycin with lower pulmonary toxicity could exhibit stronger therapeutic action against the tumors described above than the bleomycin presently employed. One hopes that the current studies of the chemistry, biochemistry, pharmacology, and biology of bleomycin will contribute to the development of more effective bleomycins and permit the cure of squamous cell carcinomas, Hodgkin's lymphomas and testis tumors [1].

This chapter reviews the studies on bleomycin and outlines those problems requiring future resolution.

Discovery and Chemical Studies on Bleomycin

In 1956, the author and his colleagues discovered kanamycin and phleomycin[2] during the course of a study of water-soluble basic antibiotics; Nihon Kayaku Co., Tokyo, and Bristol Laboratories, Syracuse, New York, collaborated with us in the development of this compound. In 1959, after sufficient quantities of phleomycin had been prepared, this antibiotic was found to inhibit Ehrlich carcinoma with an unusually high therapeutic index[3,4]. Soon thereafter it was found that phleomycin inhibited DNA synthesis in HeLa cells and *E. coli*[5]. On the basis of its high therapeutic index in the treatment of Ehrlich carcinoma, we felt that phleomycin might be worthy of clinical study and tested its toxicity toward dogs. We found that phleomycin had strong renal toxicity[6], precluding its clinical study; therefore, in parallel with the chemical study of phleomycin[7], we searched for phleomycin-like antibiotics and found a new one in 1963.

The antibiotics produced by the strain elaborating this phleomycin-like antibiotic were separated into A and B groups and the A group antibiotics, which gave a negative Sakaguchi reaction, were shown to be different than the known phleomycins by paper chromatography and by their acid stability[8]. Moreover, the bleomycin A group derivatives did not cause renal damage in dogs. The lethal dose caused hepatotoxicity[9]. The mixture of bleomycins extracted from culture filtrates was separated into the individual bleomycin copper complexes by CM–Sephadex chromatography[10]. Although the reason is unclear, the therapeutic effect of each of the individual bleomycin copper complexes on Ehrlich carcinomas was inferior to that obtained with a mixture of bleomycins. Among individual bleomycin copper complexes, A_5 showed the best therapeutic activity, but was still inferior to the mixture[11]. Therefore, in collaboration with Nihon Kayaku Co. Laboratory, we decided to use the mixture for clinical study. This mixture contained 55–70% A_2, 25–32% B_2, and very small quantities of other bleomycins, such that B_4 and B_6 which had strong renal toxicity constituted less than 1.0% in the mixture. Soon after the clinical study started in September 1965, the copper complex was found to cause damage to the vein into which bleomycin was injected. Therefore, copper-free bleomycin was supplied for clinical study. As is well known at present, copper-free bleomycin hydrochloride or sulfate can be applied not only intravenously but also subcutaneously or intramuscularly.

The structural study was started in 1965 and continued in the author's institute and in the laboratories of Nihon Kayaku Co. As reported in a symposium in 1971[12], before the end of 1970 the structures of all of the partial hydrolysis products of bleomycin A_2 were established:

Tripeptide S:

Tripeptide A:

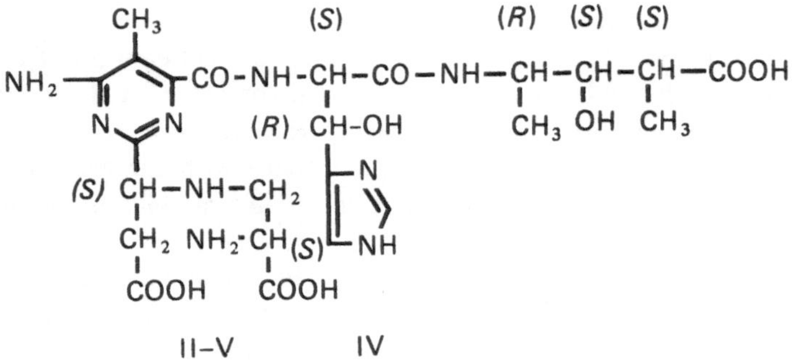

$$\text{II-V} \qquad \text{IV}$$

Sugar moiety, 2-O-(3-O-carbamoyl-α-D-mannopyranosyl)-L-gulopyranose:

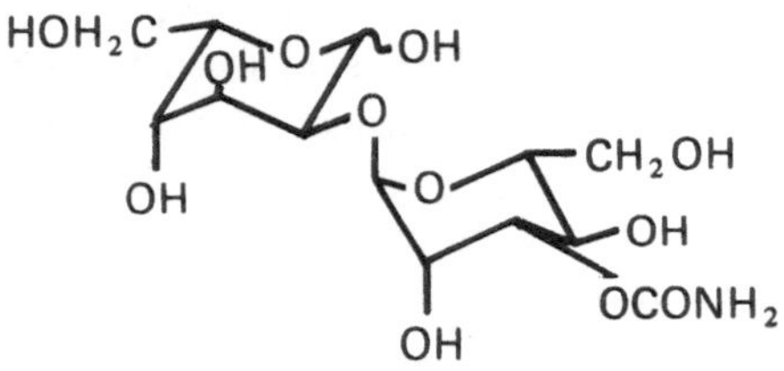

It was found that in the bleomycin molecule the carboxyl group of the 4-amino-3-hydroxy-2-methylpentanoic acid moiety of tripeptide A (which was also called pseudotetrapeptide because its total hydrolysis gave four amino acids, II, V, IV, III as shown above) was bound to the amino group of the threonine moiety of tripeptide S and that the various bleomycins differed from one another in the terminal amine as follows:

A_1:

$$R\text{CO-NH-CH}_2\text{-CH}_2\text{-CH}_2\overset{\overset{\text{O}}{\|}}{\text{-S}}\text{-CH}_3;$$

Dimethyl A_2:

$$R\text{CO-NH-CH}_2\text{-CH}_2\text{-CH}_2\text{-S-CH}_3;$$

A_2:

$$R\text{CO-NH-CH}_2\text{-CH}_2\text{-CH}_2\overset{\overset{\text{CH}_3}{|}}{\underset{+}{\text{-S}}}\text{-CH}_3;$$

A_2'–a:

$RCO-NH-CH_2-CH_2-CH_2-CH_2-NH_2$;

A_2'–b:

$RCO-NH-CH_2-CH_2-CH_2-NH_2$;

A_5:

$RCO-NH-CH_2-CH_2-CH_2-NH-CH_2-CH_2-CH_2-CH_2-NH_2$;

A_6:

$RCO-NH-CH_2-CH_2-CH_2-NH-CH_2-CH_2-CH_2-CH_2-NH-CH_2-CH_2-CH_2-NH_2$;

B_2:

$$RCO-NH-CH_2-CH_2-CH_2-CH_2-NH-\underset{\underset{NH}{\|}}{C}-NH_2;$$

B_4:

$$RCO-NH-CH_2-CH_2-CH_2-CH_2-NH-\underset{\underset{NH}{\|}}{C}-NH-CH_2-CH_2-CH_2-CH_2-NH-\underset{\underset{NH}{\|}}{C}-NH_2;$$

$RCOOH$ = Bleomycinic acid

These differences in the terminal amine moieties of the bleomycins suggested that an amine added to the fermentation medium might be incorporated into bleomycin. As reported in the same symposium[12], in fact, a [^{14}C]-labeled 3-aminopropyldimethylsulfonium salt was incorporated into bleomycin A_2, and various artifical bleomycins were produced by addition of amines to the fermentation media[13].

Structures were proposed for the bleomycins and phleomycins in 1972[13-15]. In bleomycin the carboxyl group of the β-aminoalaine moiety (V in the formula shown above) was found to exist as the carboxamide (–$CONH_2$), the disaccharide was shown to form an α-glycosidic linkage with the hydroxyl group of the β-hydroxyhistidine moiety, and the carboxyl group of the β-amino-β-(4-amino-6-carboxy-5-methylpyrimidin-2-yl)propionic acid (II) was thought to form a β-lactam with the β-amino group[16]. The structural difference between the phleomycins and bleomycins was the oxidation state of the sulfur heterocycles:

bleomycin:

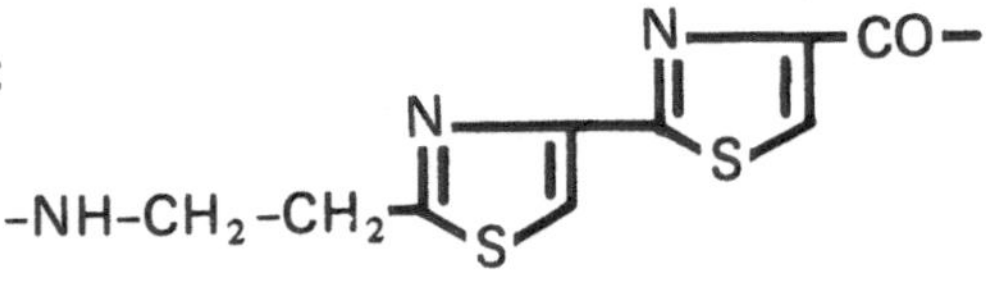

$-NH-CH_2-CH_2$

phleomycin:

$-NH-CH_2-CH_2$

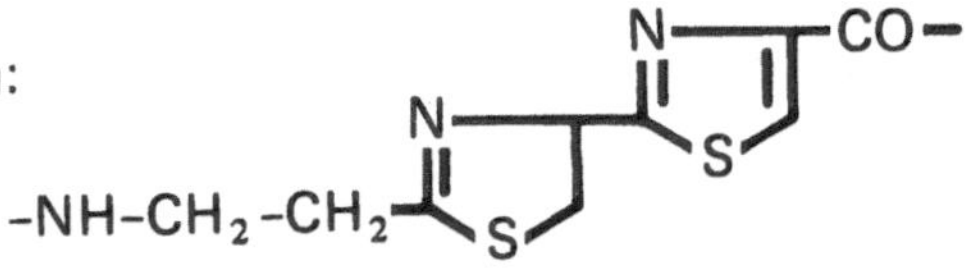

The main components of the phleomycins produced microbially contain the same terminal amines as bleomycins B_2, B_4, and B_6[17] and phleomycins containing two or more guanidine groups were shown to produce strong renal toxicity. Thus, phleomycins which have no renal toxicity can be produced by addition of the proper amines to the fermentation medium.

Furthermore, the absolute configurations of all asymmetric carbon atoms in the bleomycin molecule have been determined[13,18] as follows: V=L-β-amino-alanine; II=S-β-amino-β-(4-amino-6-carboxy-5-methylpyrimidin-2-yl)propionic acid; IV=erythro-β-hydroxy-L-histidine; III=(2S, 3S, 4R)-4-amino-3-hydroxy-2-methylvaleric acid; I=L-threonine; the sugar moiety=2-O-(3-O-carbamoyl-α-D-mannopyranosyl)-L-gulopyranose.

As reported by Dr. Fujii, a series of biosynthetic intermediates have recently been isolated from culture filtrates of a bleomycin-producing strain. If we name the first amino acid shown above (II–V) pyrimidoblamic acid, the biosynthesis of the peptide moiety is suggested to start with the synthesis of demethyl-pyrimidoblamylhistidine. The methylation of the demethylpyrimidoblamic acid moiety and the β-hydroxylation of the histidine moiety seem to occur after these species are separated from a multienzyme system far enough to undergo enzyme reactions for methylation or β-hydroxylation. An attempt to extract a multienzyme system for synthesis of the peptide moiety might be an interesting subject for study. Although the chromosomal control of the biosynthesis of zorbamycin and zorbonomycins, which are structurally related to the bleo-mycins and phleomycins, has been studied by Coats and Roesner[19], it also seems of interest to study the genetic control of the biosynthesis of the multi-enzyme system probably involved in the biosynthesis of bleomycin peptide.

As reported by Dr. Fujii, a copper complex of an intermediate in bleomycin biosynthesis was crystallized; by X-ray crystallographic analysis Dr. Iitaka and his co-workers have determined the molecular structure of this intermediate, except for the atoms attached to the carboxylate moiety of β-amino-β-(4-amino-6-carboxy-5-methylpyrimidin-2-yl)propionic acid. The structure of the pyrimi-doblamic acid moiety can now be proposed as follows:

H₂NCO ... NH₂ ... NH ... CONH₂ ... N ... N ... H₂N ... CO-*R* ... CH₃

We had established previously that the α-amino group and the N-1 nitrogen atom of the pyrimidoblamic acid moiety, as well as the N^{π} nitrogen atom of the imidazole group of the β-hydroxyhistidine moiety, were involved in copper complex formation [20]. The possible involvement of the carbamoyl group in the mannose moiety was also suggested. The structure of the copper complex of pyrimidoblamic acid suggested a possible structure for the bleomycin copper complex [which is different than that recently proposed by Dabrowiak *et al.* [21]]. Contributions at this meeting concerned with the bleomycin metal complex should provide information useful in elucidation of the mechanism of reaction of bleomycin with DNA.

Mechanism of Action of Bleomycin in Inhibiting the Growth of Cells

In 1964, we found that phleomycin inhibited DNA synthesis in *E. coli* and HeLa cells [5], and Kornberg reported in 1964 that phleomycin copper complex increased the T_m of DNA [22]. This study by Kornberg stimulated us to study further the antibiotics of the bleomycin–phleomycin group. The phleomycin used by Kornberg for his experiment was a mixture of more than three phleomycin copper complexes containing one, two, or more guanidine groups in their terminal amine moieties. As already described, the bleomycins and phleomycins are very similar and it was therefore anticipated that bleomycin copper complex might also increase the T_m of DNA. Contrary to this expectation, copper complexes of all of the bleomycins did not increase the T_m of DNA. Moreover, the copper complex of a phleomycin containing a (3-aminopropyl) dimethylsulfonium substituent (the amine moiety of bleomycin A_2) did not increase the T_m of DNA. Bleomycin B_4, containing two guanidine groups in the terminal amine, also failed to increase the T_m of DNA. The reason why the phleomycin copper complex prepared for the early study increased the T_m of DNA remains an interesting question that will be studied in more detail in the future.

As first reported in 1969, we found that copper-free bleomycin caused DNA strand scission in the presence of sulfhydryl compounds, peroxide, or ascorbic acid, and lowered the T_m of DNA in the same reaction mixture[13,23-25]. Thereafter, we continued our study of the reaction mechanism using SV40 DNA[26, 27]. The reaction mechanism has also been studied by Haidle[28-32], Müller[33-36], and others. The results may be summarized as follows: bleomycin first binds to DNA; there are base sequences to which bleomycin can bind selectively; after the binding a reaction which releases thymine occurs, and if the concentration of bleomycin and the concentration of a sulfhydryl compound are high, other bases are also released; under alkaline conditions strand scission occurs at the same time as the release of a base or else after the release of a base[37].

As reported by Dr. Hori, SV40 strand scission by bleomycin can be inhibited by double-stranded DNA or double-stranded oligodeoxynucleotides, indicating a requirement for a double-stranded structure. The equilibrium constant for bleomycin–DNA interaction has been determined by Horwitz and Grollman[38] to be about $10^5 M^{-1}$. As shown by Dr. Hori, if the binding is measured in terms of fluorescence quenching of the bithiazole moiety by DNA, both copper-free bleomycin and bleomycin copper complex bind not only to DNA but also to RNA. The discrepancy between the inhibition of SV40 DNA strand scission by double-stranded polydeoxynucleotides and the binding binding shown by quenching of bithiazole fluorescence is discussed by Dr. Hori (this volume, p. 203).

We have found that bleomycin is inactivated in aqueous solution by a sulfhydryl compound. We were able to show that this inactivation was caused by ferrous ion and oxygen and promoted by a sulfhydryl compound that reduced the ferric ion produced to ferrous ion. The reaction conditions that result in inactivation of bleomycin are the same as those that cause DNA strand scission. Copper-free bleomycin is inactivated under these conditions but bleomycin copper complex is not. (Copper-free bleomycin causes DNA strand scission but bleomycin copper complex does not.) We are continuing our study to elucidate the structure of the inactivated bleomycin; the pyrimidoblamic acid moiety of bleomycin is the most sensitive to this inactivation. In particular, the side chain [1-N-(2-amino-2-carbamoylethyl)amino-2-carbamoylethyl group] is split from the pyrimidine ring by this inactivation process.

Horwitz and her co-workers[39] have confirmed that the reaction of bleomycin with DNA, which leads to strand scission, requires ferrous ion and oxygen. As is well known, this reaction is promoted by a sulfhydryl compound. The role of ferrous ion in DNA strand scission by bleomycin is also discussed by Dr. Hori. The results hitherto reported suggest the possible involvement of bleomycin ferrous ion complex in DNA strand scission and in the inactivation of bleomycin. If an oxygen associated with ferrous ion in the bleomycin ferrous ion complex is brought close to a sensitive group in DNA, the reaction which results in strand scission may occur. In the absence of DNA, the oxygen associated with the ferrous ion may react with the pyrimidoblamic acid moiety of bleomycin. As already discussed, the studies of the metal complexes of bleomycin presented in the symposium should contribute to the elucidation of the chemical mechanism of bleomycin reaction with DNA. Also useful in this context would be an investigation of the state of the bleomycin molecule after reaction with DNA, i.e., whether it remains intact. The information available up to 1975 on the mechanism of action of bleomycin has been reviewed[13].

Actinomycin[40], steffimycin[41], etc. intercalate into DNA, and the local double-helical structure is influenced by this intercalation. It is possible that this change in DNA structure may cause a change in the affinity of bleomycin for DNA. The effect of actinomycin[27,40] and steffimycin[41] on the binding of bleomycin to DNA has been reported.

Mechanism of Therapeutic Action of Bleomycin against Squamous Cell Carcinomas

In the study of cancer chemotherapy, the mechanism of therapeutic action should be studied in parallel with the mechanism of cytotoxic action. If there are cells which lack the ability to repair the damage caused by bleomycin, these cells should be most sensitive to bleomycin treatment. As described above, Yagoda and Krakoff[42] have emphasized that Hodgkin's lymphoma responds to a very low daily dose of bleomycin. Although we have not yet studied this phenomenon, it should be interesting to learn the mechanism of therapeutic action against this tumor.

We have studied the mechanism of therapeutic action against squamous cell carcinomas. Immediately after the effect of bleomycin on penile cancer was first observed by Ichikawa et al. in 1966, we confirmed that bleomycin distributed in the skin of mice at high concentration compared with other organs[43]. Bleomycin was also shown to be effective against methylcholanthrene-induced squamous cell carcinomas in mouse skin but not against sarcomas in mouse skin induced by the same agent. In one series of experiments, an hour after the subcutaneous injection of copper-free [^{3}H]bleomycin, we measured radioactivity (indicating the total amount of bleomycin per gram of skin) and antibacterial activity (indicating the amount of bleomycin present in active form). In triplicate experiments, the radioactivity measurements indicated a concentration of 15.9–22.7 μg/g of bleomycin in carcinoma and a lower concentration (4.5–14.4 μg/g) in sarcoma. Moreover, a marked difference was recognized in the concentrations of bleomycin not inactivated: 11.0–17 μg/g in the carcinoma and 0–2.8 μg/g in the sarcoma. Thus, the therapeutic effect of bleomycin on squamous cell carcinomas was shown to be due to a high distribution of bleomycin and a low inactivation of this antibiotic in this tumor[13,44].

On the other hand, we found that all tissues contained an enzyme that inactivated bleomycin. As reported in 1971[12], we extracted this enzyme from mouse liver and found that 1 mole of ammonia was released during inactivation of 1 mole of copper-free bleomycin; the inactivated bleomycin had a slightly lower basicity than intact bleomycin, suggesting the hydrolysis of a carboxamide bond during inactivation. Later, when we proposed a structure for bleomycin in 1972, this enzyme (which we named bleomycin hydrolase) was shown to hydrolyze the α-aminocarboxamide group of the pyrimidoblamic acid moiety. The pK 7.3 of bleomycin shifts to pK 9.4 after inactivation. This enzyme was purified and confirmed to be an aminopeptidase-like enzyme. It hydrolyzed arginine β-naphthylamide and lysine β-naphthylamide but not leucine β-naphthylamide. The hydrolysis of arginine or lysine β-naphthylamide was competitively inhibited by bleomycin. The content of bleomycin hydrolase in squamous cell carcinomas in mouse skin, which is sensitive to bleomycin hydrolase, has been shown to be significantly less than that in sarcomas. It was also shown that the content of this enzyme was significantly lower in a rat hepatoma sensitive to bleomycin treatment than in a bleomycin-resistant rat hepatoma. Copper-free bleomycin is hydrolyzed by bleomycin hydrolase but bleomycin copper complex is resistant to this enzyme.

Information on the behavior of bleomycin in animals can be summarized as shown in Fig. 1[45]. A portion or all of the injected bleomycin binds to cupric ion in blood[46]. After penetration into cells, the cupric ion in bleomycin copper complex is reduced by sulfhydryl compounds which can be found in the low-molecular-weight fraction of the 105,000 × g supernatant of cell homogenates: the treatment of bleomycin copper complex with this fraction or with a sulfhydryl compound like cysteine in the presence of neocuproine (a cuprous ion-binding reagent) gives copper-free bleomycin and cuprous ion in equimolar

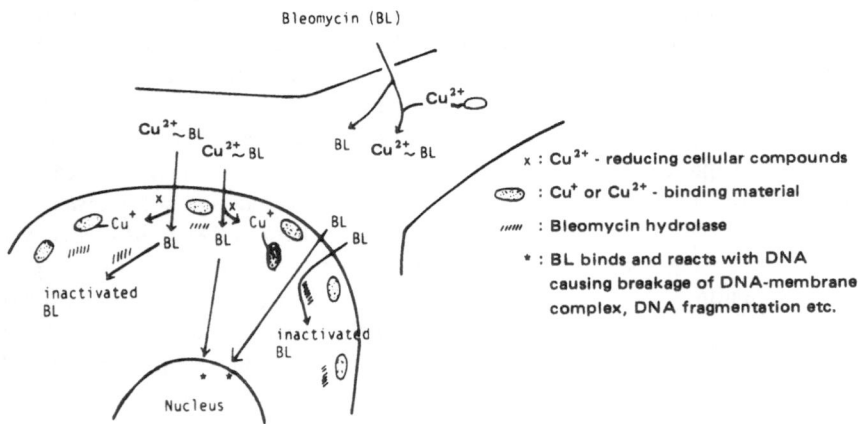

Fig. 1. Behavior of bleomycin *in vivo*.

ratio, and this reaction is inhibited by *N*-ethylmaleimide. Cuprous ion is easily converted by oxygen to cupric ion which readily binds to copper-free bleomycin. Therefore, in removing cupric ion from bleomycin copper complex, cuprous ion must be captured by some cellular component. As a result of this study a cuprous ion-binding protein was found in cells for the first time. AH66 rat hepatoma cells were used in the experiment[47]. The ability of this protein to capture cuprous ion was eliminated by preincubation with *N*-ethylmaleimide and the active group of the protein was suggested to be a sulfhydryl group.

As described above, the cupric ion of bleomycin copper complex is reduced in cells and the cuprous ion thus produced is captured by a cellular protein. Copper-free bleomycin thus formed may be inactivated by bleomycin hydrolase. Those copper-free bleomycin molecules that are not inactivated reach the nucleus where they bind to and react with DNA. The incubation of rat hepatoma cells with copper-free bleomycin at $0°C$ caused DNA fragmentation but incubation with bleomycin copper complex at $0°C$ had no effect[47]. In some cells, it is possible that bleomycin can also be inactivated by low-molecular-weight compounds in the presence of a trace amount of ferrous ion and oxygen. Bleomycin ^{57}Co complex has been shown to bind to DNA[48] and it is also possible that some bleomycin copper complex from which the cupric ion was not removed reaches the nucleus and binds to DNA. In this case, the cupric ion could be reduced and removed after the binding of the bleomycin to DNA, allowing strand scission to occur.

As already described, the therapeutic effect of bleomycin on squamous cell carcinomas is due to a low content of bleomycin hydrolase and a high distribution of this antibiotic in this tumor. The effect in skin occurs for the same reason. Bleomycin does not cause bone marrow depression. This has been suggested to be due to a high content of some high-molecular-weight material, probably bleomycin hydrolase.

Studies on Artificial Bleomycins

As already described, the various bleomycins differ from each other in their terminal amine moieties. If an amine is added to the fermentation medium and a bleomycin-producing strain is cultured, a bleomycin containing the added amine is produced and the production of other bleomycins is suppressed. By this fermentation method about 250 bleomycins have been prepared. The hydrolysis of bleomycin B_2 by a Fusarium enzyme[49] named acylagmatine amidohydrolase or the application of the cyanogen bromide method to bleomycin demethyl-A_2[50] affords bleomycinic acid. Bleomycinic acid is the common molecular component of all bleomycins; therefore, a variety of bleomycins can be synthesized from bleomycinic acid. Aminoalkyl esters of bleomycinic acid also cause single-strand scission of DNA. We have confirmed that various bleomycins are different in their degree of renal and pulmonary toxicity[13].

Bleomycin does not show pulmonary toxicity in young mice, because this antibiotic is rapidly inactivated in the lungs of these animals. This inactivation is much slower in older mice. A method for testing pulmonary toxicity using mice older than 15 weeks has been established by Matsuda and his co-workers at Nihon Kayaku Co. Among those bleomycins which have lower pulmonary toxicity than the present generation of bleomycins, bleomycin PEP [which is named pepleomycin and contains N-(3-aminopropyl)-α-phenethylamine as the terminal amine] has been sutdied clinically in the greatest detail. It inhibits Ehrlich carcinomas and squamous cell carcinomas induced by methylcholanthrene in mouse skin at least as well as the present bleomycins. In high dose, it also inhibits adenocarcinomas in rat stomach induced by N-methyl-N'-nitro-N-nitrosoguanidine, a type of tumor refractory to bleomycins presently in use. In pulmonary toxicity tests using aged mice and dogs, pepleomycin has been shown to have significantly lower toxicity than the bleomycins presently employed. Clinical study has also suggested that this bleomycin has lower pulmonary toxicity and may be more useful in the treatment of bleomycin-sensitive tumors.

Zorbamycin, zorbonomycin, YA56, YA56X[13], and tallysomycin[51] are antibiotics of the bleomycin–phleomycin group. The study of these species should also contribute to the development of useful antibiotics within the group.

References

1. S. K. Carter, T. Ichikawa, G. Mathé, and H. Umezawa, Fundamental and clinical studies of bleomycin. GANN Monograph on Cancer Research, No. 19 (Japanese Cancer Association), University of Tokyo Press (1976).
2. K. Maeda, H. Kosaka, K. Yagishita, and H. Umezawa, *J. Antibiot. (Tokyo)*, Ser. A, 9, 82 (1956).
3. W. T. Bradner and M. H. Pindell, *Nature (London)*, 196, 682 (1962).
4. H. Umezawa, M. Hori, M. Ishizuka and T. Takeuchi, *J. Antibiot. (Tokyo)*, Ser. A, 15, 274 (1962).
5. N. Tanaka, H. Yamaguchi, and H. Umezawa, *J. Antibiot. (Tokyo)*, Ser. A, 16, 86 (1963).
6. M. Ishizuka, H. Takayama, T. Takeuchi, and H. Umezawa, *J. Antibiot. (Tokyo)*, Ser. A, 19, 260 (1966).
7. M. Ikekawa, F. Iwami, H. Hiranaka, and H. Umezawa, *J. Antibiot. (Tokyo)*, Ser. A, 17, 194 (1964).
8. H. Umezawa, K. Maeda, T. Takeuchi, and Y. Okami, *J. Antibiot. (Tokyo)*, Ser. A, 19, 200 (1966).
9. M. Ishizuka, H. Takayama, T. Takeuchi, and H. Umezawa, *J. Antibiot. (Tokyo)*, Ser. A, 20, 15 (1967).
10. H. Umezawa, Y. Suhara, T. Takita, K. Maeda, *J. Antibiot (Tokyo)*, Ser. A, 19, 210 (1966).

11. H. Umezawa, M. Ishizuka, K. Kimura, J. Iwanaga, and T. Takeuchi, *J. Antibiot. (Tokyo)*, **21**, 592 (1968).
12. H. Umezawa, *Pure Appl. Chem.*, **28**, 665 (1971).
13. H. Umezawa, Fundamental and clinical studies of bleomycin. GANN Monograph on Cancer Research, No. 19 (Japanese Cancer Association), University of Tokyo Press, pp. 3–36 (1976).
14. T. Takita, Y. Muraoka, T. Yoshioka, A. Fujii, K. Maeda, and H. Umezawa, *J. Antibiot. (Tokyo)*, **25**, 755 (1972).
15. H. Naganawa, Y. Muraoka, T. Takita, H. Umezawa, *J. Antibiot. (Tokyo)*, **30**, 388 (1977).
16. Y. Muraoka, A. Fujii, T. Yoshioka, T. Takita, and H. Umezawa, *J. Antibiot. (Tokyo)*, **30**, 178 (1977).
17. T. Takita, Y. Muraoka, A. Fujii, H. Itoh, K. Maeda, and H. Umezawa, *J. Antibiot. (Tokyo)*, **25**, 197 (1972).
18. H. Nakamura, T. Yoshioka, T. Takita, H. Umezawa, Y. Muraoka, and Y. Iitaka, *J. Antibiot. (Tokyo)*, **29**, 762 (1976).
19. J. H. Coats and J. Roeser, *J. Bacteriol.*, **105**, 880 (1971).
20. Y. Muraoka, H. Kobayashi, A. Fujii, M. Kunishima, T. Fujii, Y. Nakayama, T. Takita, and H. Umezawa, *J. Antibiot. (Tokyo)*, **29**, 853 (1976).
21. J. C. Dabrowiak, F. T. Greenaway, W. E. Longo, M. VanHusen, and S. T. Crooke, *Biochim. Biophys. Acta*, **517**, 517 (1978).
22. A. Falaschi and A. Kornberg, *Fed. Proc.*, **23**, 940 (1964).
23. K. Nagai, H. Suzuki, N. Tanaka, and H. Umezawa, *J. Antibiot. (Tokyo)*, **22**, 569 (1969).
24. K. Nagai, H. Suzuki, N. Tanaka, and H. Umezawa, *J. Antibiot. (Tokyo)*, **22**, 624 (1969).
25. K. Nagai, H. Suzuki, N. Tanaka, and H. Umezawa, *J. Antibiot. (Tokyo)*, *Biophys. Acta*, **179**, 165 (1969).
26. H. Umezawa, H. Asakura, and M. Hori, *J. Antibiot. (Tokyo)*, **26**, 521 (1973).
27. H. Asakura, M. Hori, and H. Umezawa, *J. Antibiot. (Tokyo)*, **28**, 587 (1975).
28. C. W. Haidle, *Mol. Pharmacol.*, **7**, 645 (1971).
29. C. W. Haidle, M. T. Kuo, and K. K. Weiss, *Biochem. Pharmacol.*, **21**, 3308 (1972).
30. C. W. Haidle, K. K. Weiss, and M. T. Kuo, *Mol. Pharmacol.*, **8**, 531 (1972).
31. C. W. Haidle, K. K. Weiss, and M. L. Mace Jr., *Biochem. Biophys. Res. Commun.*, **48**, 1179 (1972).
32. M. T. Kuo and C. W. Haidle, *Biochim. Biophys. Acta*, **335**, 109 (1973); *Biophys. J.*, **13**, 1296 (1973).
33. W. E. G. Müller, Z. Yamazaki, H. Breter, and R. K. Zahn, *Eur. J. Biochem.*, **31**, 518 (1972).
34. W. E. G. Müller, Z. Yamazaki, and R. K. Zahn, *Biochem. Biophys. Res. Commun.*, **46**, 1667 (1972).
35. W. E. G. Müller, Z. Yamazaki, J. E. Zoellner, and R. K. Zahn, *FEBS Lett.*, **31**, 217 (1973).
36. Z. Yamazaki, W. E. G. Müller, and R. K. Zahn, *Biochim. Biophys. Acta*, **308**, 412 (1973).
37. S. L. Ross and R. E. Moses, *Biochemistry*, **17**, 581 (1978).
38. M. Chien, A. P. Grollman, and S. B. Horowitz, *Biochemistry*, **16**, 3641 (1977).
39. E. A. Sausville, J. Peisach, and S. B. Horowitz, *Biochem. Biophys. Res. Commun.*, **73**, 814 (1976).
40. J. Bearden and C. W. Haidle, *Biochem. Biophys. Res. Commun.*, **65**, 371 (1975).

41. W. E. G. Müller, H. J. Rohde, R. Steffen, A. Maidhof, and R. K. Zahn, *Cancer Lett.*, **1**, 127 (1976).
42. Y. Yagoda and H. Krakoff, Fundamental and clinical studies on bleomycin. GANN Monograph on Cancer Research, No. 19 (Japanese Cancer Association), University of Tokyo Press, pp. 255–268 (1976).
43. T. Ichikawa, A. Matsuda, K. Yamamoto, M. Tsubosaki, T. Kaihara, K. Sakamoto, and H. Umezawa, *J. Antibiot. (Tokyo)*, Ser. A, **20**, 149 (1967).
44. H. Umezawa, T. Takeuchi, S. Hori, T. Sawa, M. Ishizuka, T. Ichikawa, and T. Komai, *J. Antibiot. (Tokyo)*, **25**, 409 (1972).
45. H. Umezawa, *Lloydia*, **40**, 67 (1977).
46. M. Kanao, S. Tomita, S. Ishihara, A. Murakami, and H. Okada, *Chemotherapy*, **21**, 1305 (1973).
47. K. Takahashi, O. Yoshioka, A. Matsuda, and H. Umezawa, *J. Antibiot. (Tokyo)*, **30**, 861 (1977).
48. A. Kono, *Chem. Pharm. Bull. (Tokyo)*, **25**, 2882 (1977).
49. H. Umezawa, Y. Takahashi, A. Fujii, T. Saino, and T. Takita, *J. Antibiot. (Tokyo)*, **26**, 117 (1973).
50. T. Takita, A. Fujii, T. Fukuoka, and H. Umezawa, *J. Antibiot. (Tokyo)*, **26**, 252 (1973).
51. M. Konishi, K. Saito, K. Numata, T. Tsuno, K. Asama, H. Tsukiura, T. Naito, and H. Kawaguchi, *J. Antibiot. (Tokyo)*, **30**, 789 (1977).

Review of the Structural Studies on Bleomycin

Tomohisa Takita

More than 12 years have passed since we started our structural study of bleomycin (BLM). These studies are summarized in this chapter. BLM was isolated as a mixture of more than 10 components from the culture filtrate of *Streptomyces verticillus*. These were separated by CM–Sephadex column chromatography[1] and found to be different from one another in their terminal amine moieties[2]. BLMs thus isolated were blue-colored equimolar copper complexes. BLM has not yet been crystallized, so that we could not utilize X-ray crystallographic analysis. In the initial stages of our structural studies, we used exclusively metal-free BLM A_2, the main component of the natural BLMs. Metal-free BLM was prepared by treatment of the copper complex with hydrogen sulfide in methanolic solution.

The strong absorptions at 1650 and 1550 cm^{-1} and a broad absorption centered at 1050 cm^{-1} in the IR spectrum of BLM suggested the presence of peptide and sugar moieties, respectively. The molecular weight of BLM A_2 was in the range 1450–1600 as shown by ultracentrifuge and potentiometric titration. Mass spectrometric analysis by electron impact ionization and chemical ionization, and even in the field desorption mode, did not give any information about the molecular weight. The elemental analysis did not give a definite value due to the hygroscopic, amorphous, and polybasic nature of the molecule; therefore a definite molecular formula could not be established. This turned out to be a serious problem in the final stage of the structural elucidation.

Amine Components of Bleomycin

The total acid hydrolysis of BLM A_2 with $6 M$ HC1 gave seven ninhydrin-positive products[3]. They were separated by ion-exchange chromatography and isolated as crystalline materials. The structures and configurations were determined chemically or by X-ray analysis as shown in Fig. 1[4-9]. The Roman numbers in the left column are used to designate each amino acid in our published work. They were named in order of elution from cation-exchange columns. Compound I is L-threonine, the only common amino acid in the hydrolyzate. Compounds II, III, IV, VI, and VII are new amino acids and an amine not previously found in nature. We chemically synthesized compounds II[10], III[11], IV[6], and VII[12]; compound VI was first prepared by Cheng and Cheng[13] and recently by McGowan *et al.*[14] through a biomimetic route.

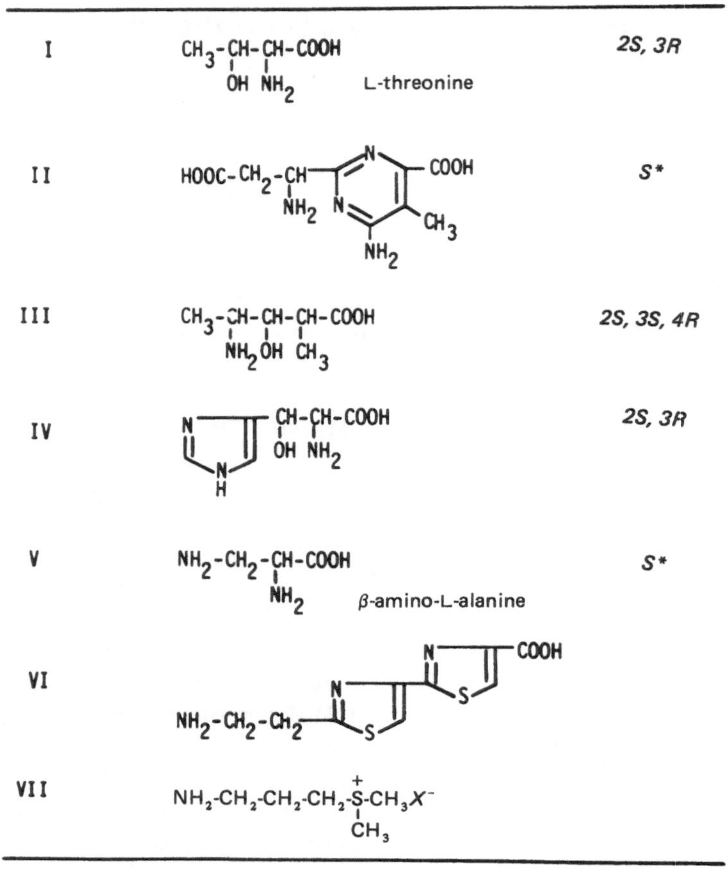

I $CH_3-CH-CH-COOH$
         | |
         OH NH_2 L-threonine   2S, 3R

II $HOOC-CH_2-CH$ [pyrimidine ring with N, COOH, CH_3, NH_2]
              |
             NH_2 S*

III $CH_3-CH-CH-CH-COOH$
         | | |
        NH_2 OH CH_3 2S, 3S, 4R

IV [imidazole ring]$-CH-CH-COOH$
                      | |
                      OH NH_2 2S, 3R

V $NH_2-CH_2-CH-COOH$
              |
             NH_2 β-amino-L-alanine   S*

VI $NH_2-CH_2-CH_2$-[bithiazole ring]-COOH

VII $NH_2-CH_2-CH_2-CH_2-\overset{+}{S}-CH_3 X^-$
                          |
                         CH_3

Fig. 1. Amine components of bleomycin A_2. *, Partially racemized during acid hydrolysis.

Sequence of Amine Components of Bleomycin

As BLM consisted of many polyfunctional amino acids, it was obvious that for sequence determination we would need several strategies for selective peptide bond cleavage. We noted first the presence of three amino acids, I, III, and IV, with a hydroxy group vicinal to an amino group. Therefore, we first attempted selective acid hydrolysis involving N to O acyl migration. BLM was dissolved in 6 M HCl and maintained at 37°C for 4 days. Three peptides denoted tripeptide S and pseudotetrapeptides A and B were formed in excellent yield[15]. Tripeptide S was obtained as crystals and shown to have the structure depicted in Fig. 2.

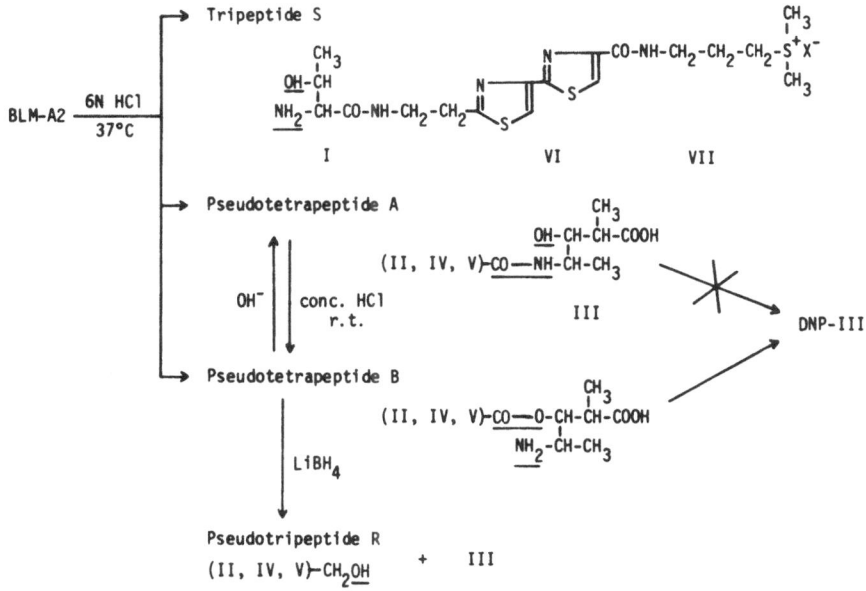

Fig. 2. Selective cleavage of bleomycin (1).

Pseudotetrapeptides A and B had the same components – II, III, IV, and V. They were interconvertible. Pseudotetrapeptide B gave DNP-III upon treatment with dinitrofluorobenzene followed by acid hydrolysis, but A did not. Therefore, the transformation of A to B was shown to be due to acyl migration from the N to O of compound III. To determine whether the carboxyl group of compound II, IV, or V had participated in this migration, pseudotetrapeptide B was hydrogenolyzed with LiBH₄. This hydrogenolysis not only cleaved compound III selectively, but also reductively labeled the carboxyl group participating in this acyl migration. Furthermore, the newly formed hydroxyl group of pseudotripeptide R played an important role in the following selective cleavage.

When pseudotripeptide R was maintained at 37°C in conc. HCl, a pseudodipeptide was obtained in almost quantitative yield (Fig. 3) by participation of the newly formed hydroxyl group. This pseudodipeptide was only obtained in good yield by this series of selective cleavages, because it was easily decomposed by competitive β-elimination to give ninhydrin-positive II or V when the temperature was raised. The structure of the pseudodipeptide was confirmed by synthesis, which involved condensation of compound II and N-acetyldehydro-alanine methyl ester followed by mild acid hydrolysis to remove the protecting groups. The linkage between pseudodipeptide and amine component IV was confirmed as follows. Pseudotripeptide R was esterified with methanol–conc.

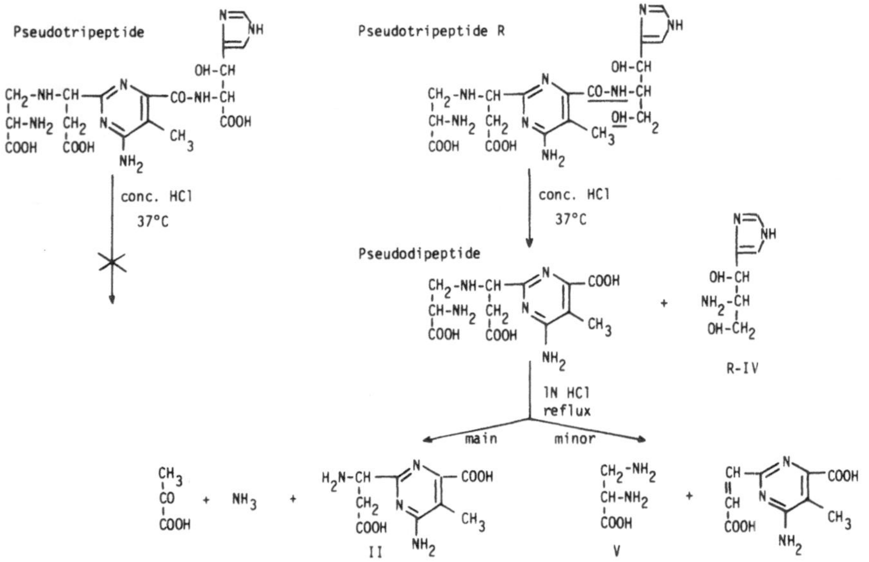

Fig. 3. Selective cleavage of bleomycin (2).

HCl, and then treated with $LiBH_4$ to reduce the methyl esters. Finally it was hydrolyzed to give pseudodipeptide diol having the pyrimidine-ring carboxyl. From this series of reactions, the carboxyl group on the pyrimidine ring of pseudodipeptide was shown to be connected to the amino group of IV.

In this fashion the structure of pseudotetrapeptide A was determined. Pseudotetrapeptide A and tripeptide S contained all of the amine components of BLM without overlapping. The next problem was to determine the linkage of tripeptide S with pseudotetrapeptide A. We tried oxidative cleavage with N-bromosuccinimide to cleave the β-hydroxyhistidyl carboxypeptide bond selectively[16]. Tetrapeptide S, composed of I, III, VI, and VII, was obtained in 45% yield. The N-terminal was compound III, as expected. Thus, the sequence of all of the amine components was established.

The Structure of the Sugar Moiety of Bleomycin[17, 18]

The sugar component of BLM was effectively isolated by methanolysis catalyzed with Amberlyst 15, which is a sulfonic acid ion-exchange resin for nonaqueous reactions. The sugar components thus isolated were found to be 3-O-carbamoyl-D-mannose and L-gulose. As regards the connection of the sugar to the peptide moiety, when BLM was maintained in 0.1 M NaOH at room temperature, the UV absorption at 290 nm increased, accompanied by liberation of disaccharide. This phenomenon could be explained by β-elimination of an O-glycosidic sugar from the β-hydroxyhistidine moiety (Fig. 4). This was confirmed by the appearance of a one proton signal at $\delta 8.52$ and by formation of DL-histidine by hydrogenation with palladium-on-carbon, followed by acid hydrolysis. The configuration of the glycosidic linkage of mannose was determined to be α by application of Hudson's rule and that of gulose was also shown to be α by ^{1}H-NMR analysis.

Completion of the Elucidation of the Structure of Bleomycin

Figure 5 shows an assembly of all of the fragments already described. At this stage of the structure determination, ^{13}C-NMR spectroscopy of BLM gave us the most important information[19,20], namely that this partial structure contained all of the carbon atoms present in BLM. The results of elemental analysis together with this carbon number suggested that one more nitrogen atom should be added to this structure of BLM.

In the terminal group analysis of BLM, only α-DNP-V, but neither β- nor di-DNP-V, was isolated by Sanger's DNP method, and no free carboxyl group was detected by Koshland's glycine condensation method. Therefore, it was established that the primary amine in bleomycin was free, but that neither of the carboxylic acids in Fig. 5 was free. We thought that the secondary amine shown in Fig. 5 should not be free because it could neither be dinitrophenylated by Sanger's DNP method nor methylated by methyl iodide-triethylamine. Potentiometric titration of BLM B_2 showed the presence of four measurable basic functional groups at $pK_a > 11.5$, 7.4, 4.7, and 2.7. The $pK_a > 11.5$ was assigned to the guanidine group at the terminal amine, 7.4 was assigned to the primary amino group, 4.7 to the imidazole (which was confirmed by the pH dependence of the ^{1}H- and ^{13}C-NMR chemical shifts), and 2.7 to the 4-aminopyrimidine.

The low pK_a assignment was ostensibly confirmed by the fact that among II and its three derivatives, only when the aliphatic amino group in the side chain

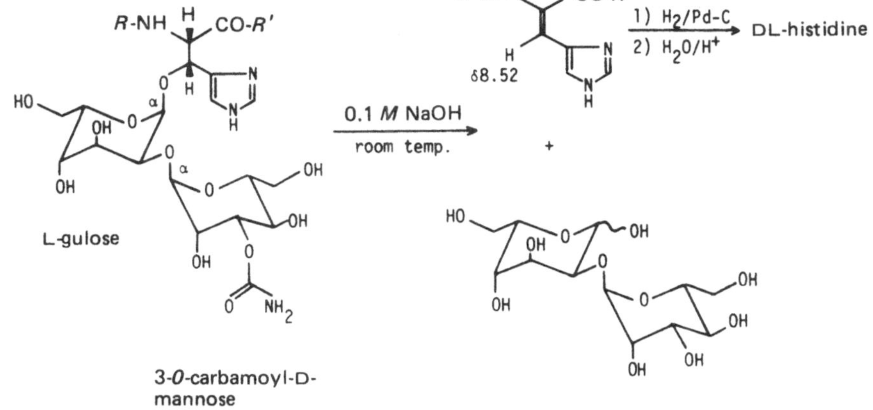

Fig. 4. Structure of the sugar moiety of bleomycin.

was masked by acetylation was the basicity of the 4-aminopyrimidine strong enough to be measurable by potentiometric titration; this measurement yielded a value of pK_a 2.7, the same as that of BLM[21]. Therefore, we assumed that the secondary amine was additionally substituted in BLM, a conclusion consistent with the resistance of this amine to dinitrophenylation and methylation. Thus, only two possible structures were left for bleomycin. In each of the carboxyl groups shown in Fig. 5 one would exist as an amide and the other as a β-lactam.

BLM was methylated with methyl iodide-triethylamine, followed by mild acid hydrolysis, to give β-aminoalanine betaine amide as a degradation product. Under the same reaction conditions, BLM gave β-aminoalanine but not β-amino-alanine amide. The steric hindrance of the trimethylamino group must function to resist acid hydrolysis of the carboxamide bond. Thus the missing nitrogen was found to be present as the carboxamide of V and a total structure was proposed for BLM (Fig. 6). After presentation of this total structure in 1972, we carried out a chemical study to provide proof for the β-lactam structure; in 1977 we isolated a diazepine derivative, which seemed to provide indirect chemical evidence for the presence of the β-lactam moiety[21].

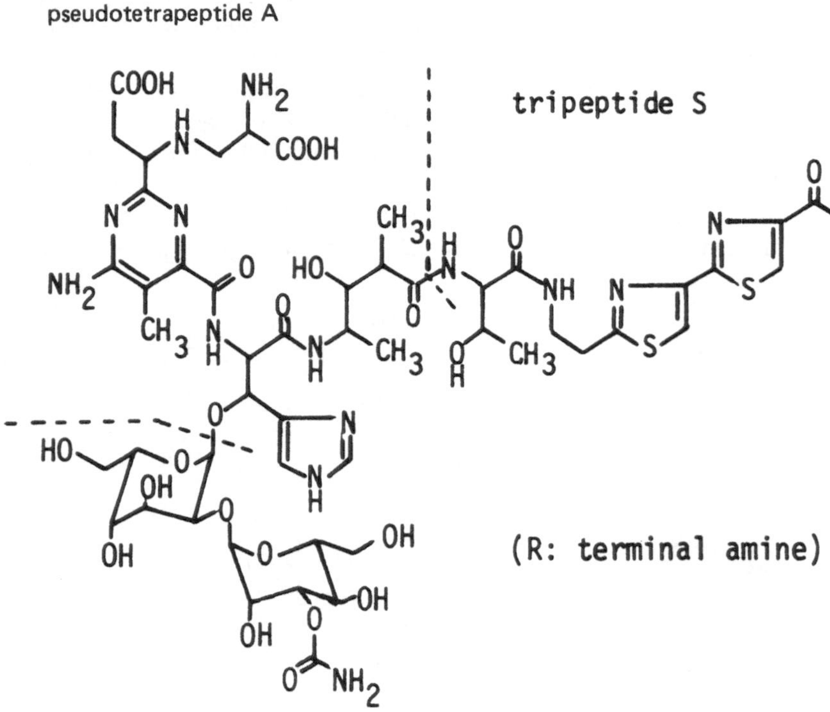

Fig. 5. An assembly of degradation products of bleomycin.

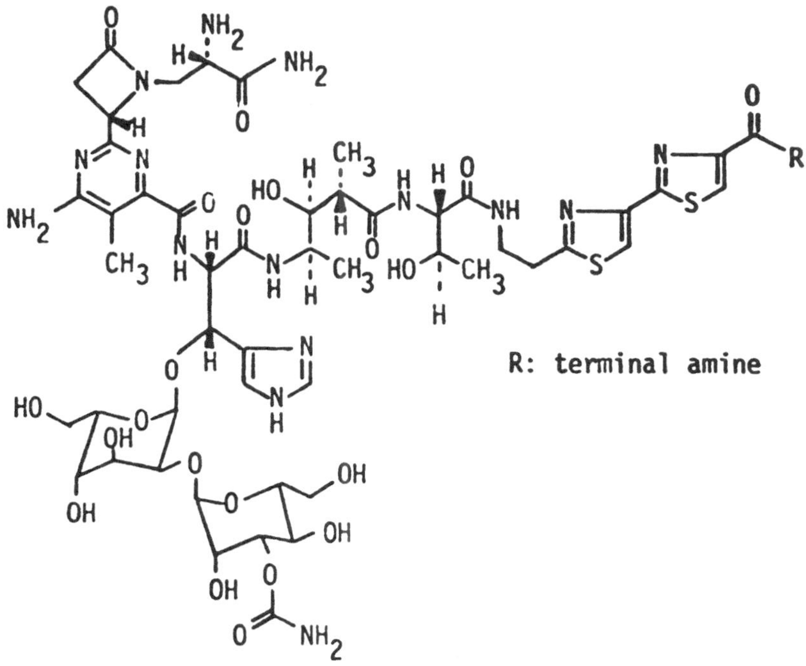

Fig. 6. Total structure of bleomycin (1972).

Structure Revision of Bleomycin

Very recently, as a result of biosynthetic studies on bleomycin, we have isolated several peptides structurally related to BLM. One of the peptides, designated P-3A, was isolated as a crystalline copper complex. Acid hydrolysis of metal-free P-3A gave amine components V, demethyl-II, histidine, and alanine. We presumed the presence of a β-lactam ring in P-3A from the ^{13}C-NMR spectroscopic comparison with BLM. The result of the X-ray crystallographic analysis supported our expected structure except for the β-lactam moiety; the analysis indicated a ring-open structure for the "β-lactam," but the exact structure could not be specified due to thermal vibration.

The most probable structure of the substituent at the 2-position of the pyrimidine ring of P-3A was assumed to be the same as that of the triamide of pseudodipeptide. It was prepared by esterification followed by ammonolysis of pseudodipeptide. The ^{13}C-NMR chemical shifts of the six carbon atoms in the substituent at the 2-position of the pyrimidine ring of pseudodipeptide triamide indicated that the structures of the substituents at the 2-positions of the pyrimidine rings of BLM and P-3A were the same because the chemical shift values were almost the same (Table 1).

Table 1: ^{13}C-NMR Chemical Shifts of BLM A$_2'$-c, P-3A, and Pseudodipeptide
Triamide

	Assignment	BLM A$_2'$-c	P-3A	Pseudodipeptide triamide
II	CO (side)	<u>177.0</u>	<u>176.9</u>	<u>176.9</u>
	CO (ring)	168.5	169.3	172.1
	2	166.1	166.0	166.5
	4 (-NH$_2$)	165.5	165.8	165.1
	6	153.0	154.9	155.8
	5	113.1	103.2	111.2
	CH	<u>53.3</u>	<u>53.2</u>	<u>53.2</u>
	CH$_2$	<u>41.0</u>	<u>40.9</u>	<u>41.1</u>
	CH$_3$	11.7		11.9
V	CO	<u>171.9</u>	<u>171.5</u>	<u>171.9</u>
	CH	<u>60.5</u>	<u>60.8</u>	<u>60.8</u>
	CH$_2$	<u>47.8</u>	<u>47.9</u>	<u>47.8</u>
His	CO		172.4	
	2		135.5	
	4		130.0	
	5		118.9	
	CH		53.5	
	CH$_2$		28.6	
Ala	CO		180.8	
	CH		52.1	
	CH$_3$		17.9	

If BLM has the structure shown in Fig. 7, as suggested by the ^{13}C-NMR study
of the triamide of pseudodipeptide, the secondary amine must be free in BLM.
The observed pK_a of 2.7 must be assigned to this secondary amine, although its
basicity would be extremely weak relative to other aliphatic amines. Therefore,
the dissociation constants of the pyrimidine chromophores were reinvestigated
by UV spectrometry. Fig. 8 shows the difference UV spectra of N-acetyl-II-
diamide, pseudodipeptide triamide, and BLM B$_2$ between 0.1 M HCl and aqueous
solution. The pK_a value of N-acetyl-II-diamide was 2.7 by potentiometric titration[21]. From the difference UV absorption at 256 nm of this compound, the
difference molecular extinction calculated as $\Delta\epsilon$ 6300. From the difference UV
absorption at 256 nm, the pK_a value of the pyrimidine chromophore of psuedo-
dipeptide triamide was calculated to be 1.2 and that of BLM B$_2$ was estimated
to less than 1.0.

These results indicated that the pK_a value of 2.7 was observed for BLM should
be assigned to the secondary amine. The pK_a value of the secondary amine of
pseudodipeptide triamide was found to be 3.4 by potentiometric titration. Both
of the pK_a values indicated that the secondary amines were extremely weak

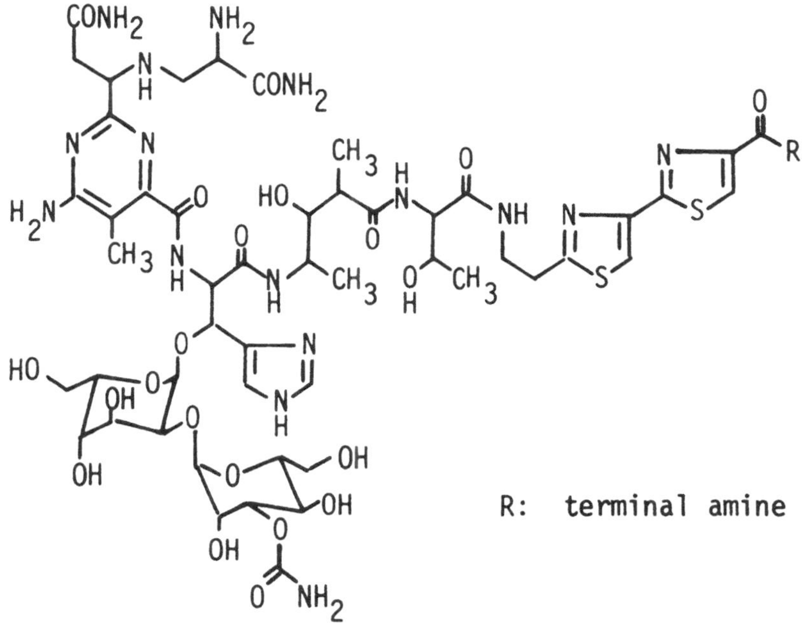

Fig. 7. Revised structure of bleomycin.

aliphatic amines. This can be explained by the effect of the neighboring primary amino group and masking of the lone pair electrons of the secondary amine by intramolecular 6-membered ring hydrogen bonding with a carboxamide proton. The pK_a value of the secondary amine of BLM B_2 was significantly lower than that of pseudodipeptide triamide. This may be due to the presence of an additional more strongly basic functional group, namely, the imidazole of the β-hydroxyhistidine moiety, which has a pK_a value of 4.7. Prevention of dinitrophenylation and methylation of the secondary amine by Sanger's DNP method and with methyl iodide-triethylamine, respectively, also can be explained by its very weak nucleophilicity and steric hindrance.

Although chemical verification of the existence of a new carboxamide group in BLM has not yet been achieved and elemental analysis has always shown a somewhat smaller nitrogen content, the ^{13}C-NMR studies of pseudodipeptide triamide, P-3A, and BLM already shown and studies of the dissociation constants of the pyrimidine and secondary amino groups by UV spectrometry and potentiometric titration indicated that the structure of BLM should be revised as shown in Fig. 7. Diazepine ring formation described in our previous paper can be explained by ring-closure rather than ring-expansion of a β-lactam ring[21].

Phleomycins have been converted to BLMs by oxidation of the thiazoline moiety[14, 22]. Thus, their structures should be revised analogously with that of

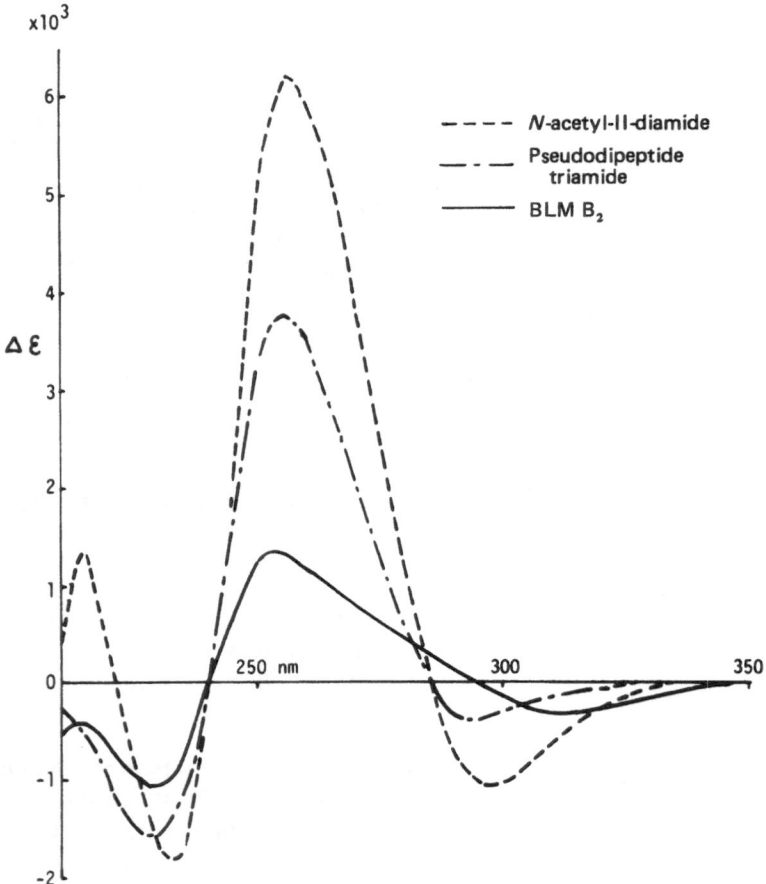

Fig. 8. Difference UV spectra of N-acetyl-II-diamide, pseudodipeptide triamide, and BLM B_2 ($\epsilon_{0.1 \, N \, HCl} - \epsilon_{H_2O}$).

BLM. The structures of the "β-lactam" moiety of other bleomycin–phleomycin group antibiotics, such as YA-56X[23] and tallysomycin[24], also should be revised, because their assigned structures were proposed analogously with our proposed BLM structure.

References

1. H. Umezawa, Y. Suhara, T. Takita, and K. Maeda, *J. Antibiot. (Tokyo)*, **19A**, 200 (1966).
2. A. Fujii, T. Takita, K. Maeda, and H. Umezawa, *J. Antibiot. (Tokyo)*, **26**, 398 (1973).

3. T. Takita, Y. Muraoka, K. Maeda, and H. Umezawa, *J. Antibiot. (Tokyo),* **21,** 79 (1968).
4. G. Koyama, H. Nakamura, Y. Muraoka, T. Takita, K. Maeda, H. Umezawa, and Y. Iitaka, *Tetrahedron Lett.,* 4635 (1968).
5. Y. Muroka, T. Takita, K. Maeda, and H. Umezawa, *J. Antibiot. (Tokyo),* **23,** 252 (1970).
6. T. Takita, T. Yoshioka, Y. Muraoka, K. Maeda, and H. Umezawa, *J. Antibiot. (Tokyo),* **24,** 795 (1971).
7. G. Koyama, H. Nakamura, Y. Muraoka, T. Takita, K. Maeda, and H. Umezawa, *J. Antibiot. (Tokyo),* **26,** 109 (1973).
8. H. Nakamura, T. Takita, H. Umezawa, Y. Muraoka, and Y. Iitaka, *J. Antibiot. (Tokyo),* **27,** 352 (1974).
9. H. Nakamura, T. Yoshioka, T. Takita, H. Umezawa, and Y. Iitaka, *J. Antibiot. (Tokyo),* **29,** 762 (1976).
10. T. Yoshioka, Y. Muraoka, T. Takita, K. Maeda, and H. Umezawa, *J. Antibiot. (Tokyo),* **25,** 625 (1972).
11. T. Yoshioka, T. Hara, T. Takita, and H. Umezawa, *J. Antibiot. (Tokyo),* **27,** 356 (1974).
12. A. Fujii, T. Takita, N. Shimada, and H. Umezawa, *J. Antibiot. (Tokyo),* **27,** 73 (1974).
13. K. Y. Zee-Cheng and C. C. Cheng, *J. Heterocycl. Chem.,* **7,** 1439 (1970).
14. D. A. McGowan, U. Jordis, D. K. Minster, and S. M. Hecht, *J. Am. Chem. Soc.,* **99,** 8078 (1977).
15. T. Takita, Y. Muraoka, K. Maeda, and H. Umezawa, *Proc. 8th Symposium on Peptide Chem.,* Osaka, 179 (1970).
16. Y. Muraoka, T. Takita, K. Maeda, and H. Umezawa, *J. Antibiot. (Tokyo),* **25,** 185 (1972).
17. T. Takita, K. Maeda, H. Umezawa, S. Omoto, and S. Umezawa, *J. Antibiot. (Tokyo),* **22,** 237 (1969).
18. S. Omoto, T. Takita, K. Maeda, H. Umezawa, and S. Umezawa, *J. Antibiot. (Tokyo),* **25,** 752 (1972).
19. T. Takita, Y. Muraoka, T. Yoshioka, A. Fujii, K. Maeda, and H. Umezawa, *J. Antibiot. (Tokyo),* **25,** 755 (1972).
20. H. Naganawa, Y. Muraoka, T. Takita, and H. Umezawa, *J. Antibiot. (Tokyo),* **30,** 388 (1977).
21. Y. Muraoka, A. Fujii, T. Yoshioka, T. Takita, and H. Umezawa, *J. Antibiot. (Tokyo),* **30,** 178 (1977).
22. T. Takita, Y. Muraoka, A. Fujii, H. Itoh, K. Maeda, and H. Umezawa, *J. Antibiot. (Tokyo),* **25,** 197 (1972).
23. Y. Ohashi, Ph.D. Dissertation, Kyoto Univ. (1974).
24. M. Konishi, K. Saito, K. Numata, T. Tsuno, K. Asama, H. Tsukiura, T. Naito, and H. Kawaguchi, *J. Antibiot. (Tokyo),* **30,** 789 (1977).

Studies on the Total Synthesis of Bleomycin

SIDNEY M. HECHT, DONALD J. BURLETT, YOSHITAKA MUSHIKA,
YASUHISA KURODA, AND MARK D. LEVIN

Several practical rewards may be expected to derive from the development of a workable total synthesis of bleomycin. In addition to permitting the direct verification of the proposed structure (which has been revised during this meeting by Dr. Umezawa and his co-workers) and hopefully extending the repertoire of available synthetic techniques, bleomycins of modified structure should facilitate the study of the mechanism of action of the drug and thus help to define the structures of analogs that may have improved anticancer activity. More extensive structural modification might also afford bleomycin analogs that are useful, e.g., as tumor radioimaging agents or for coadministration with bleomycin to reduce serious side effects of the drug such as pulmonary toxicity.

At present, the emphasis in our laboratory is on the synthesis of tetrapeptide S and on reconstruction of the carbohydrate moiety of bleomycin from its component parts. This chapter summarizes our synthetic progress toward tetrapeptide S (1)[1] and the preparation of L-gulose, which is the key building block for elaboration of the carbohydrate component of bleomycin (2).

At the time that we initiated our synthetic studies on bleomycin, a synthesis of 2'-(2-aminoethyl)-2,4'-bithiazole-4-carboxylic acid had been described by Cheng and Cheng [2]. Nevertheless, we chose to investigate an additional pathway to this species involving dehydrative cyclization of the corresponding cysteinyl peptide (i) and oxidation of the formed bithiazoline (ii) which, presumably, parallels the biosynthetic elaboration of iii [3]. In this fashion we hoped to be able to achieve an efficient synthesis of iii and render synthetically accessible a number of other peptide-derived antibiotics containing one or more thiazoles or thiazolines [3, 4].

1

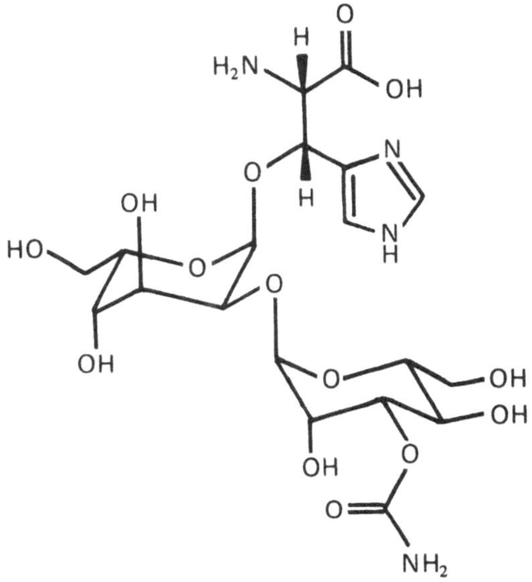

2

N-Acetyl-β-alanylcysteine ethyl ester (**3a**) was treated with each of several reagents previously utilized for the formation of thiazolines in an effort to effect the preparation of ethyl 2-(2-acetamidoethyl)-Δ^2-thiazoline-4-carboxylate. Most of these reagents had been used previously for the preparation of structurally simple thiazolines but failed to yield appreciable quantities of **4a**. However, when treated with hydrogen chloride at 0°C, a chloroform solution of **3a** deposited a clear oil which, after extractive workup, could be purified by crystallization from benzene–chloroform–petroleum ether, by column chromatography on silica gel, or by distillation (160°C/0.1 mm). Thiazoline **4a** was isolated as colorless crystals in yields up to 77%. The analogous conversion **3b** → **4b** was also carried out in the same fashion, although in somewhat lower yield. While the basis for the difference in yield is not altogether clear, it may possibly be due to the lesser solubility of **4a** in acidic chloroform as the thiazoline-forming reaction is, in principle, reversible.

The conversion of simple thiazolines to the corresponding thiazoles has been achieved with a variety of oxidants, including potassium dichromate[5], hydrogen peroxide[5], mercuric acetate[6,7], potassium ferricyanide[6,7], cupric sulfate[8], MnO_2[9], and phenanthrenequinone[9]. Of these, only activated MnO_2 was useful for the conversion of **4a** to **5a**, effecting formation of the latter in yields of about 65%. It was subsequently found that more efficient conversion could be achieved with NiO_2[10], a reagent not utilized previously for thiazoline dehydrogenations. When shaken in chloroform solution in the presence of NiO_2, **5a** was obtained in 93% yield from **4a**, workup involving simple filtration of the oxidant, concentration of the chloroform solution, and

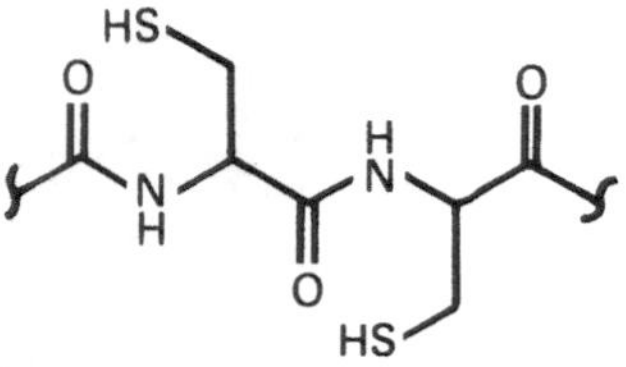

i

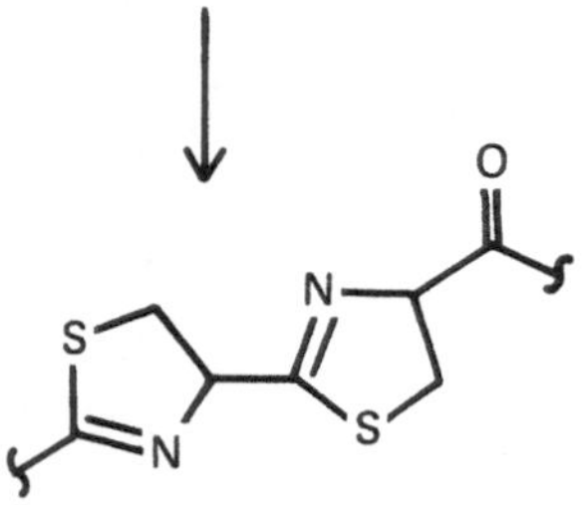

ii

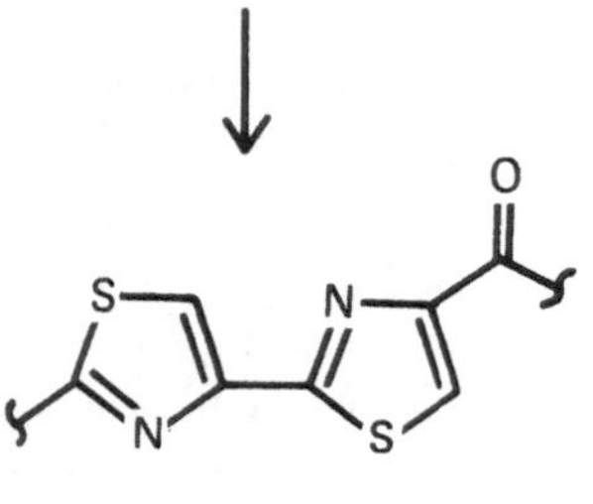

iii

crystallization of the residue from ether. The analogous transformation **4b** → **5b** could be achieved by the use of MnO$_2$ or NiO$_2$.

Completion of the synthesis of the bithiazole moiety involved saponification of **5b** and condensation of the resulting thiazole carboxylic acid with S-trityl-cysteine ethyl ester. Tripeptide analog **6** (91% overall yield from **5b**) was obtained as a white foam and converted to the respective mercaptan (**7**) in quantitative yield via the silver mercaptide. The mercaptan was dissolved in chloroform and treated with hydrogen chloride at room temperature for 36 hr; the concentrated reaction mixture was partitioned between aqueous carbonate and ethyl acetate. The presumed thiazolythiazoline (**8**), obtained as a clear oil in 90% yield from **6**, had the expected ultraviolet spectrum ($\lambda_{max}^{1:1 \ C_2H_5OH-HCl}$ 233 and 300 nm) and was employed directly for the preparation of bithiazole **9**, using either NiO$_2$ or MnO$_2$ as oxidant. Crystallization of the product from ethyl acetate–petroleum

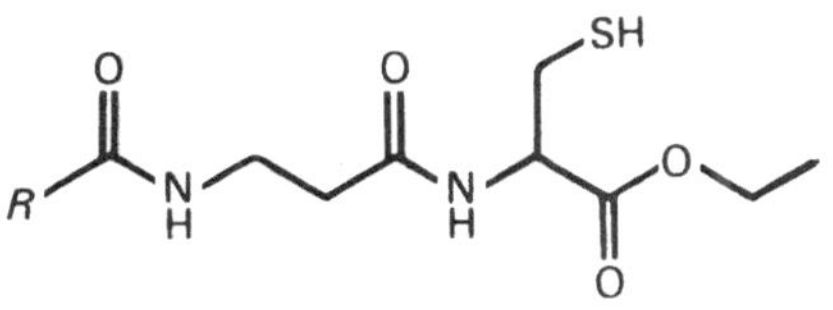

3

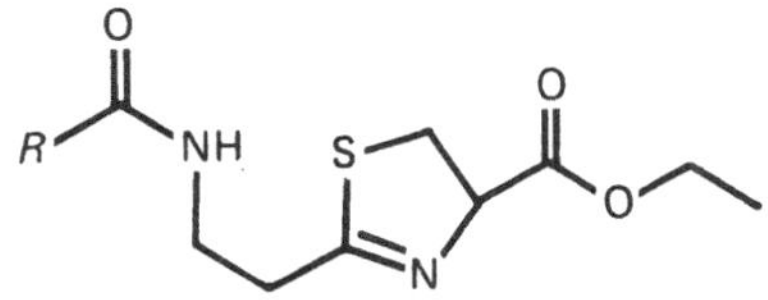

4

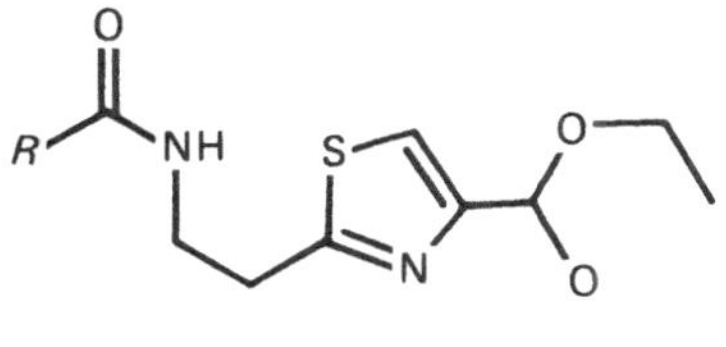

5

a, R = CH₃

b, R = C₆H₅

ether afforded ethyl 2'-(2-benzamidoethyl)-2,4'-bithiazole-4-carboxylate as colorless needles in 22–24% yield; the mother liquors contained 7 as the disulfide. The identity of bithiazole 9 was established by comparison of its physical properties (NMR, quantitative ultraviolet and mass spectra, mp and mmp, tlc mobility) with those of an authentic sample[2].

As suggested above, the biosynthetic elaboration of the bithiazole moiety of bleomycin may well involve the intermediacy of a bithiazoline (ii), formed by dehydrative cyclization of the corresponding cysteinyl peptide. Therefore, it was of interest to attempt the conversion of N-acetyl-β-alanylcysteinylcysteine ethyl ester (10) to the respective bithiazole (12) by simultaneous dehydrative cylization of both cysteinyl moieties (affording bithiazoline 11) and subsequent

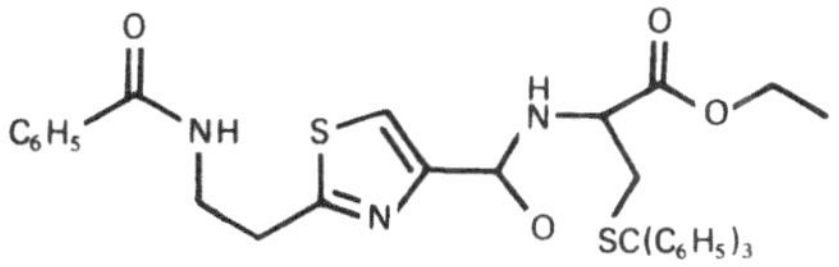

6

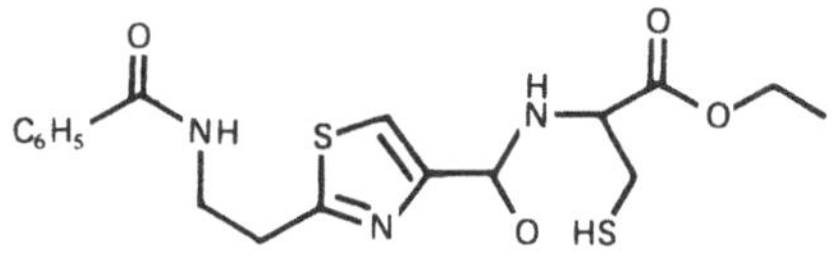

7

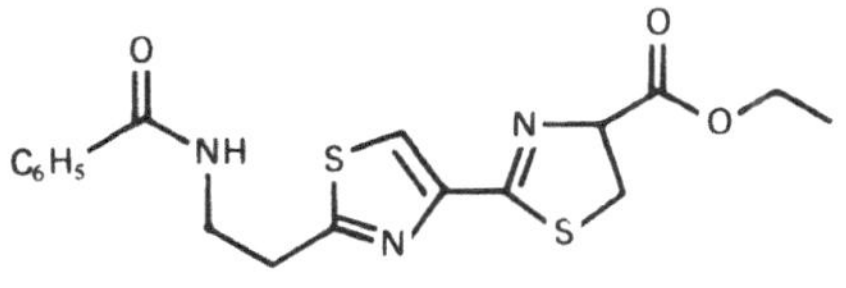

8

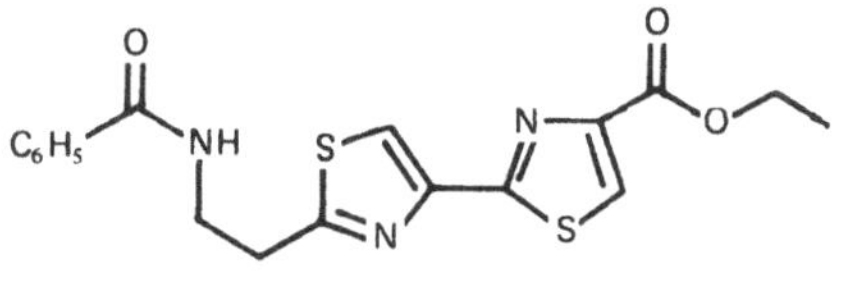

9

dehydrogenation. Treatment of **10** with hydrogen chloride in ethanol-free $CHCl_3$ (24 hr, workup by concentration *in vacuo*) gave a water-sensitive product having an ultraviolet spectrum $[\lambda_{max}^{1:1\ C_2H_5OH-HCl}$ 266 nm (ϵ 9200)] consistent with that expected for **11**. Unfortunately, treatment of the bithiazoline with NiO_2 gave disulfide **13** rather than the desired bithiazole, the formation of **13** presumably proceeding via hydrolysis of **11** by water formed during dehydrogenation or associated with the NiO_2. Further support for the intermediacy of **11** may be inferred from the results of dehydrative cyclization of **10** with hydrogen chloride in commercial reagent grade $CHCl_3$ (containing 1% ethanol as stabilizer). Isolated from this reaction in 81% yield after NiO_2 treatment were colorless crystals of thiazole **5a**, which must have arisen by ethanolysis of the S-acyl peptide[6] accessible from **11**.

The facility with which NiO_2 was found to effect thiazoline dehydrogenations suggested the potential utility of the reagent for the oxidation of other partially reduced heterocycles. As shown in Table 1, a number of such transformations were attempted and several were found to proceed in good yield[4].

Work on the synthesis of pseudotetrapeptide S has led to the successful condensation of the threonine and methyl 2′-(2-aminoethyl)-2,4′-bithiazole-4-carboxylate in high yield. o-Nitrophenylsulfenyl-L-threonine was converted to the corresponding dinitrophenyl ester (**14**) (CH_3CN; DCC) in 89% yield, mp 123.5–5.0°C. Treatment of **14** with methyl vinyl ether ($CHCl_3$, $CH_3C_6H_4SO_3H$, 25°C, 2 min) or isopropenyl methyl ether[11] (CH_3CN, $POCl_3$, 25°C, 15 hr) gave acetals **15a** and **15b** in yields of 86 and 83%, respectively; consistent with its formulation as a single optical isomer only the latter could be induced to crystallize. In the presence of a slight molar excess of threonine derivative **15b** in ethanol-free chloroform, methyl 2′-(2-aminoethyl)-2,4′-bithiazole-4-carboxylate was consumed completely with the formation of the desired coupled

10   11

12   13

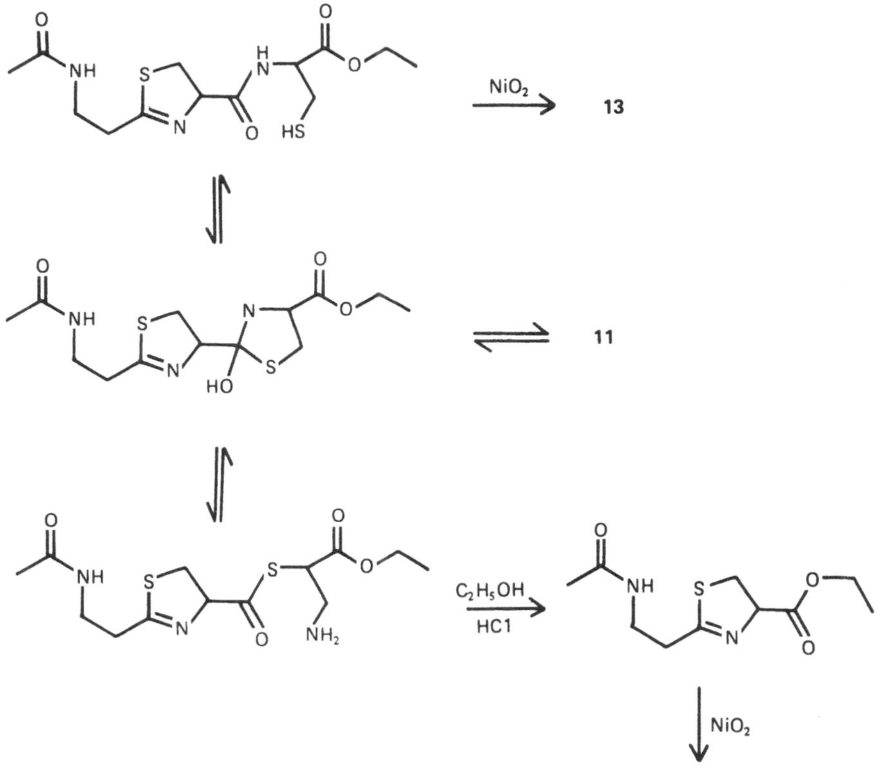

product (**16b**). The latter was isolated (97% yield) as a chromatographically homogeneous yellow glass by treatment of the reaction mixture with a small amount of hydroxylamine to consume unreacted **15b**, followed by extraction with aqueous sodium bicarbonate and concentration of the dried organic layer. Compound **16a** was obtained in 82% yield by the analogous condensation of methyl 2'-(2-aminoethyl)-2,4'-bithiazole-4-carboxylate with **15a**. Deblocking of peptides **16** could be achieved readily by vigorous agitation of chloroform solutions of these species with 1–2 equivalents of concentrated hydrochloric acid. The desired product precipitated as the hydrochloride salt within several minutes and could be isolated conveniently in approximately 70% yield by simple decantation of the chloroform layer. One may note that compound **17**, an intermediate of potential importance for the reconstruction both of tetrapeptide S and of bleomycin itself, has been obtained in good overall yield and without utilization of any chromatographic procedures, so that this species should be accessible in quantity.

Condensation of **17** with (2S,3S,4R)-4-amino-3-hydroxy-2-methylvaleric

Table 1:   Nickel Peroxide Oxidations of Partially Reduced Heterocycles

Heterocycle	NiO$_2$ (equiv O$_2$/equiv substrate)	Solvent	Conditions	Percentage yield[a]
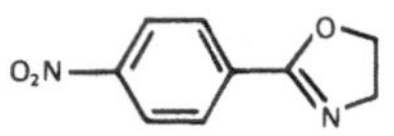	5.0	C$_6$H$_6$	7 hr, reflux	28
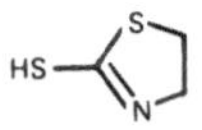	1.0	CHCl$_3$	12 hr, 25°C	88[b]
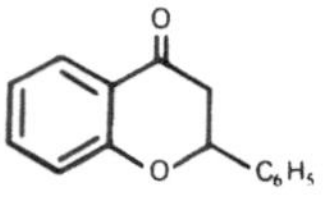	1.6	C$_6$H$_6$	4 hr, reflux	71
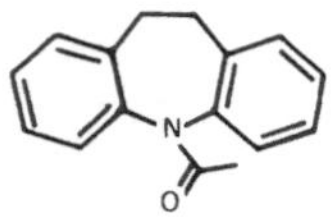	1.7	C$_6$H$_6$	4 hr, reflux	73
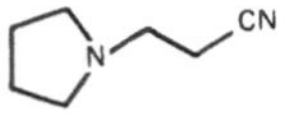	5.0	C$_6$H$_6$	7 hr, reflux	26
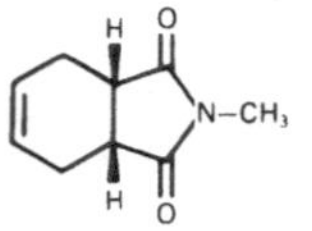	2.2	C$_6$H$_6$	1.5 hr, reflux	30[c]
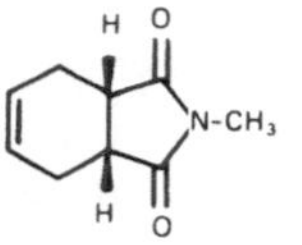	2.3	C$_6$H$_6$	7 hr, reflux	62[c]
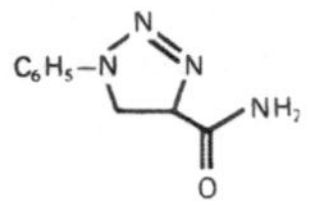	1.7	C$_6$H$_6$	4 hr, reflux	41
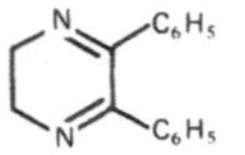	1.7	C$_6$H$_6$	4 hr, reflux	92

[a]Isolated yields. After each reaction was complete, workup involved filtration of the catalyst (Celite), concentration in vacuo, and purification where necessary (chromatography on silica gel or crystallization).
[b]The product was 2,2'-bis-(2-thiazoline) disulfide.
[c]The product was N-methylphthalimide.

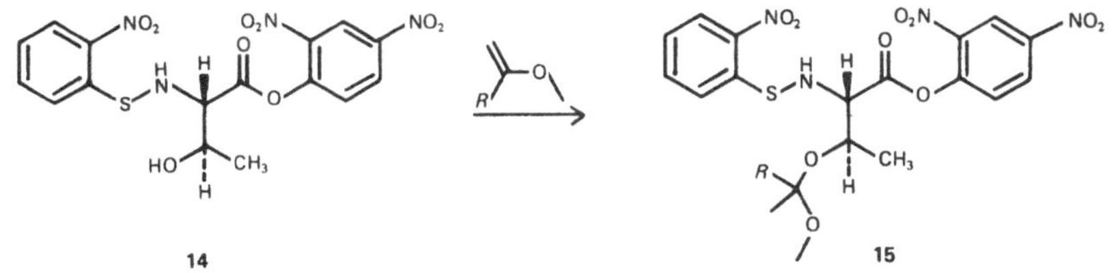

acid (20) would afford the immediate synthetic precursor to tetrapeptide S. While the synthesis of 20 as a mixture of four isomers has been reported[12], and the desired 2S,3S,4R-isomer can be isolated from the mixture by ion exchange chromatography, no preparatively useful route to (2S,3S,4R)-20 has been described. Ketone 18 has been prepared previously[12] and utilized for the synthesis of 4-amino-3-hydroxy-2-methylvaleric acid; it had the R-configuration at C-4, reflecting its derivation from D-alanine. Careful analysis of 18 revealed that it was a 2:1 mixture of configurational isomers at C-2, the major component being accessible in optically pure form via crystallization from 1:1 carbon tetrachloride-hexane. Although the borohydride reduction of 18, originally described by Yoshioka et al.[12], ostensibly establishes only a single additional asymmetric center, it was found that under any given set of experimental conditions the isomeric composition of 19 obtained was independent of the (C-2) configuration of 18 utilized for reduction.

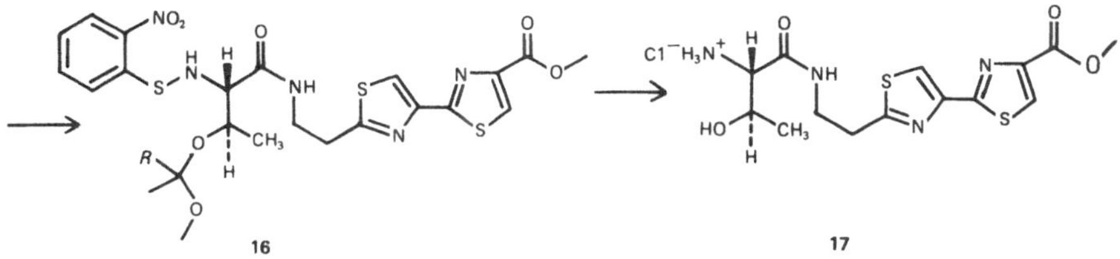

As the configurations at both C-2 and C-3 were apparently established during reduction of 18, it seemed reasonable to inquire whether the isomeric composition of 19 could be influenced substantially by the choice of reaction conditions. While subsequent experiments indicated that the composition could be so influenced, intensive efforts failed to raise the relative yield of the 2S, 3S, 4R-isomer of 19 above 23%, but did permit each of the three remaining isomers to be obtained with relative facility. The availability of these other species suggested an alternate approach to (2S,3S,4R)-20, namely, inversion of configuration of (2R,3S,4R)-20 or (2S,3R,4R)-20 at C-2 or C-3, respectively.

Given the presence of a secondary hydroxyl group at C-3, inversion of configuration at this carbon atom seemed technically feasible and efforts were
made to effect the transformation of (2S, 3R, 4R)-**20** to (2S, 3S, 4R)-**20**. The
first problem encountered in this process was conceptual in nature, namely,
that while the absolute configuration of the 2S,3S,4R-isomer of 4-amino-3-
hydroxy-2-methylvaleric acid had been established by X-ray crystallographic
analysis of material obtained by degradation of bleomycin, the absolute configurations of each of the other three species actually in hand were unknown, so
that (2S,3R,4R)-**20** could not be identified as such. The obvious solution to
this problem was the arbitrary choice of a single isomer of **20** for inversion at
C-3 since the product would be either (a) the desired (2S,3S,4R)-**20** or
(b) another of the three synthetically available isomers of **20**. In the latter case,
the starting material and product would necessarily have R-configurations at
C-2, identifying the third isomer as (2S,3R,4R)-**20**. One isomer, that was obtained in quantity by slow addition of borohydride (1.3 equivalents hydride,
ethanolic solution) to ketone **18** and that could be separated readily as an
optically pure species by fractional crystallization, was chosen as a substrate for
inversion at C-3. This amino acid was converted in quantitative yield to
lactam **22**.

A number of derivatives of optically pure lactam **22** were prepared, including
the acetate, trifluoroacetate, benzoate, methanesulfonate, and p-toluenesul-
fonate derivatives. In the presence of base (in amounts insufficient) to effect
epimerization at C-2 in **22** itself) or any of several oxygen nucleophiles, each of

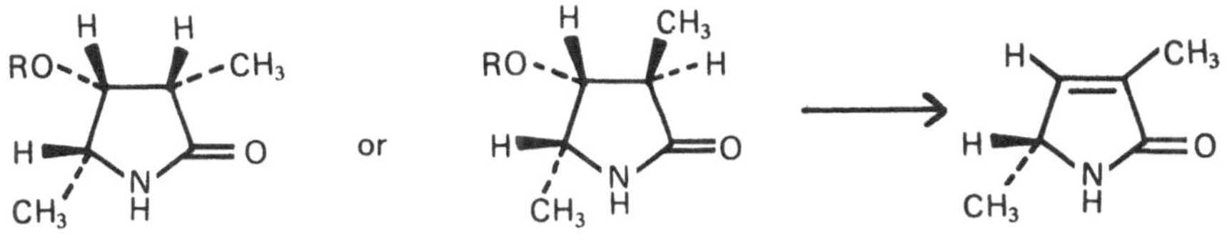

2S,3R,4R   2R,3S,4R   23

these derivatives was immediately converted to α, β-unsaturated lactam **23**. While the propensity of these derivatives to undergo elimination under very mild conditions precluded their use as intermediates for inversion of configuration at C-3, the facility with which the transformation occurred suggested the *transoid* relationship of the C-2 hydrogen atom and C-3 oxygen atom, i.e., that **22** had either the 2S,3R,4R- or 2R,3S,4R-configuration.

Actual inversion of configuration of this isomer at C-3 was achieved by N-acetylation of the ester (**21**) [CH_3CCl, (C_2H_5)$_3$N; 93% yield], and treatment of **24** with thionyl chloride at 0°C. Oxazoline **25** was isolated in 49% yield as a yellow oil unstable at room temperature; hydrolysis with 4 M hydrochloric

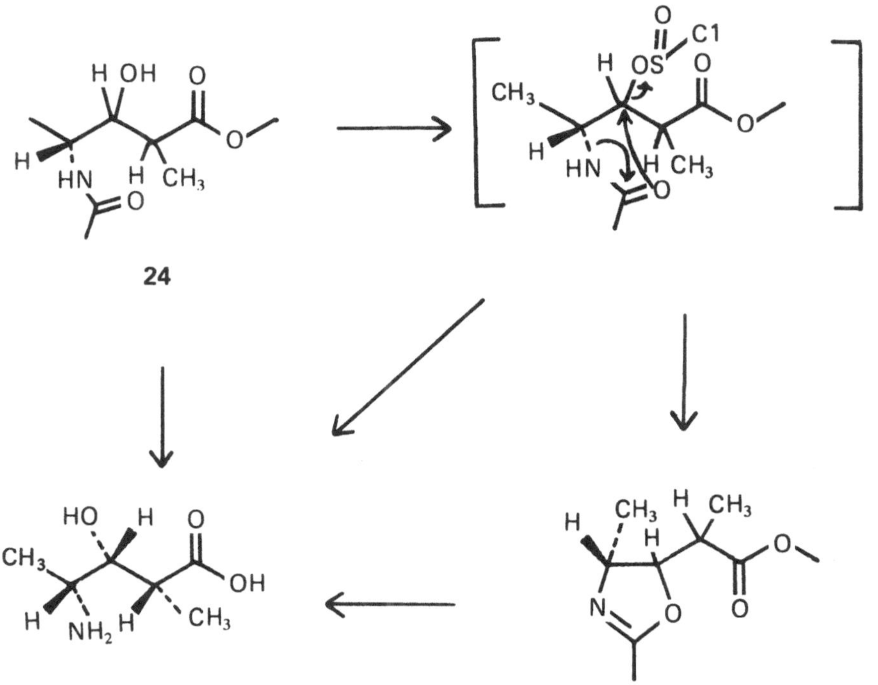

24

2R,3R,4R-**20**   **25**

acid (reflux, 4 hr) afforded a single amino acid (20) identical neither with the starting amino acid nor with the desired $2S,3S,4R$-amino acid. The same product was obtained in quantitative yield from 24 when isolation of the intermediate oxazoline was omitted. On the basis of the accumulated evidence, the amino acid employed for this transformation was assigned as the $2R,3S,4R$-isomer of 20, the product as the $2R,3R,4R$-isomer, and the remaining species as the $2S,3R,4R$-isomer. Having thus assigned absolute configurations to each of the four isomers, the putative $(2S, 3R, 4R)$-4-amino-3-hydroxy-2-methylvaleric acid (20) was esterified, N-acetylated, and treated with thionyl chloride in an effort to prepare the desired $2S,3S,4R$-isomer of 20. Remarkably, under the conditions used to effect quantitative inversion of $2R,3S,4R$-24 at C-3, the corresponding $2S,3R,4R$-isomer simply gave decomposition products. The same result was also obtained starting with $2R,3R,4R$-24, and it was not possible to devise experimental conditions useful for inversion of configuration of either of these two species.

In spite of the difficulties encountered in effecting inversion of configuration at C-3 according to the two methods described above, these experiments did suggest a solution to the problem. Specifically, the isomer of 20 originally chosen for study on the basis of its ease of isolation was shown to be $2R,3S,4R$-20, i.e., differing from the desired product only at C-2. Inspection of the structure of the corresponding lactam (22) indicated a *cisoid* arrangement of the C-2 and C-3 substituents as compared with the all-*trans* arrangement of substituents in the desired product. Therefore, we hoped that under basic conditions (in the absence of a good leaving group on C-3) it might be possible to effect epimerization at C-2. In fact, when $2R,3S,4R$-22 was heated at reflux for 3 days in methanol containing 0.3 equivalent of sodium methoxide, workup gave a 3:1 mixture (quantitative recovery) of $2S,3S,4R$-22 and $2R,3S,4R$-22. After conversion to the corresponding amino acids, pure $2S,3S,4R$-20 was obtained by trituration or fractional crystallization of $2R,3S,4R$-20. At present this amino acid is being blocked in a fashion that will facilitate condensation with 17. Synthesis of tetrapeptide S would then involve simple coupling with an appropriate amine, analogous to the derivatization of bleomycinic acid[13]

The synthesis of L-gulose has been accomplished previously[14], although not in a form directly useful for elaboration of the carbohydrate moiety of bleomycin. We have prepared L-gulose in a suitably blocked form by exploiting the obvious structural relationship between L-gulose and the readily available sugar D-glucose. Key intermediate 29 was obtained in 80% overall yield from D-glucose via the known[15,16] 3,4,6-tri-O-acetyl-1,2-O-ethylorthoacetyl-α-D-glucopyranose (26), which was deacetylated with methanolic methoxide and then tritylated (1.5 equiv $(C_6H_5)_3CCl$, DMF, Hünigs base) to afford 27 as a foam in 84% overall yield from 26. The corresponding 3,4-di-O-benzyl derivative, prepared by treatment of 26 with 10 equivalents of benzyl chloride–NaH in dry glyme at reflux, was further treated with dry acetic acid (N_2, 25°C, 1.5 hr) to afford 1,2-di-O-acetyl-3,4-di-O-benzyl-6-O-trityl-β-D-glucopyranose

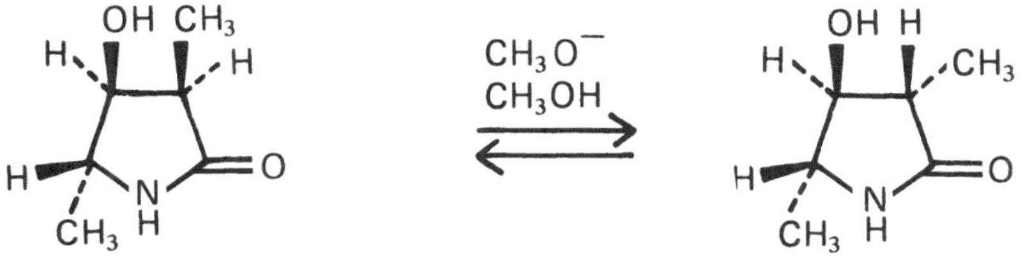

2R,3S,4R-**22**   2S,3S,4R-**22**

1   :   3

(28) in quantitative yield as a chromatographically homogeneous yellow oil having 1,2-*trans*-diequatorial geometry[14]; crystallization from ethanol afforded **28** as colorless prisms in 71% yield (based on **27**), mp 144–145°C. Oily **28** could be converted to intermediate **29** by treatment with acetic acid in 50% aqueous tetrahydrofuran at 45–50°C for 8 hr or by exposure (as a benzene solution) to acidic silica gel for 18 hr. The isolated products from these reactions were crystallized from ether-pentane, affording **29** as colorless needles in 67 and 95% yields, respectively.

D-Glucose   L-Gulose

Conversion of glucopyranosyl diacetate **29** to a suitable gulopyranose derivative involved interchange of oxidation states at C-1 and C-6. Since the most
direct approach for achieving "aldose interchange" required the preparation of
an intermediate having C-1 and C-6 at the same oxidation level, it was important
to develop some strategy for differentiating C-1 and C-6 to preclude the ultimate
formation of a mixture of D-glucose and L-gulose derivatives.

Initial attempts to oxidize compound **29** at C-6 involved the use of pyridinium
chlorochromate[17] and $SO_3 \cdot C_6H_5N$-dimethyl sulfoxide[18], the latter of
which has been employed previously for the oxidation of carbohydrates[19].
Reaction products obtained using these oxidants included benzaldehyde and
dinitrophenylhydrazine-positive species, the latter of which were converted to
more polar compounds during the course of the reaction, as judged by tlc. The
NMR spectra of the reaction products reflected possible epimerization at C-5
and loss of the well-defined acetate and anomeric proton signals, the latter suggesting Lewis acid-catalyzed conversion of the 1,2-diacetate to an acetoxonium
ion and hence the need for a less acidic oxidant. Corey and Kim[20] have reported on the use of N-chlorosuccinimide-dimethylsulfide-triethylamine for the
oxidation of alcohols; when applied to the oxidation of **29** this procedure
produced a chromatographically homogeneous product having spectral data
[IR, 1710 cm^{-1}; NMR, δ1.93 (s,3), 2.05 (s,3), and 9.59 (s,1)] expected of the
desired 6-aldehydo sugar. As this material was found to be unstable it was
immediately treated with 1,1-dimethylhydrazine (2 equiv $(C_2H_5)_3N$, 25°C),
effecting conversion to the respective N,N-dimethylhydrazone (**30**) which was
isolated as colorless needles in 66% yield, mp 146–147°C.

Compound **30**, having C-1 and C-6 at the same oxidation level, was treated
with catalytic sodium methoxide to unmask the C-1 aldehyde and then with

NaBH$_4$ in aqueous ethanol to effect selective reduction of the aldehyde. The desired product (31), which was isolated in quantitative yield as a white foam, was shown by NMR to consist of a 70:30 equilibrium mixture of hydrazino-glycoside - acyclic hydrazone. While compound 30 itself represented a reasonable intermediate for continuation of our synthetic studies on bleomycin, and did not require further modification for this purpose, it seemed prudent to verify the structure of this species before proceeding. Therefore, gulose derivative 30 was methylated (CH$_3$I, tetrahydrofuran, 25°C, 12 hr) and converted to the tosylate salt (32) by further treatment with 2 equivalents of silver tosylate in methanol. When 32 was dissolved in glyme and heated in the presence of an additional equivalent of p-toluenesulfonic acid, workup afforded 1,6-anhydro-L-gulose derivative 33 as a pale yellow oil in 74% yield (overall from 30). Acetyla-

tion [(CH$_3$$\overset{\overset{\text{O}}{\|}}{\text{C}}$)$_2$O, pyridine] and debenzylation (Pd(OH)$_2$/C, 3 atm H$_2$, 40 hr) of 33 gave 2-O-acetyl-1,6-anhydro-β-L-gulopyranose which was further acetylated to afford 2,3,4-triacetyl-1,6-anhydro-β-L-gulopyranose (34), isolable as colorless needles from chloroform-pentane, mp 114°C; $[\alpha]_D^{25}$ −22°C (c 2.4, CHCl$_3$). While 34 was a new compound, its enantiomer has been reported[21] and had the expected physical properties [mp 114–115°C; $[\alpha]_D^{25}$ + 22.1°C (c 1.5, CHCl$_3$)]. Deacetylation of 34 gave 1,6-anhydro-β-L-gulopyranose (35) as colorless prisms [mp 153°C; $[\alpha]_D^{25}$ − 50°C (c 2.23, H$_2$O)], the optical antipode of which is also known[21,22] [mp 154–155°C; $[\alpha]_D^{25}$ + 50.4°C (c 2.8, H$_2$O)].

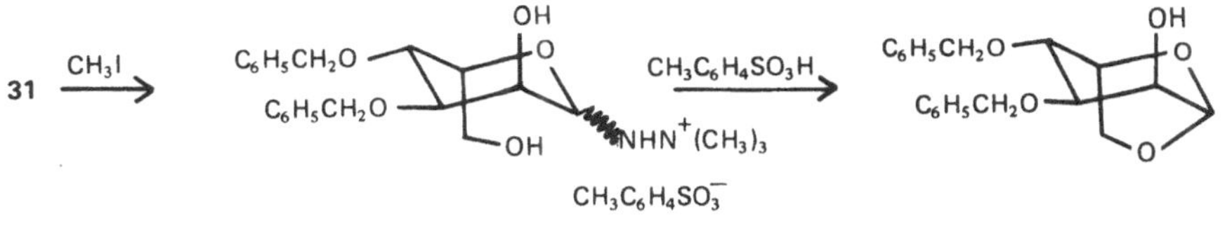

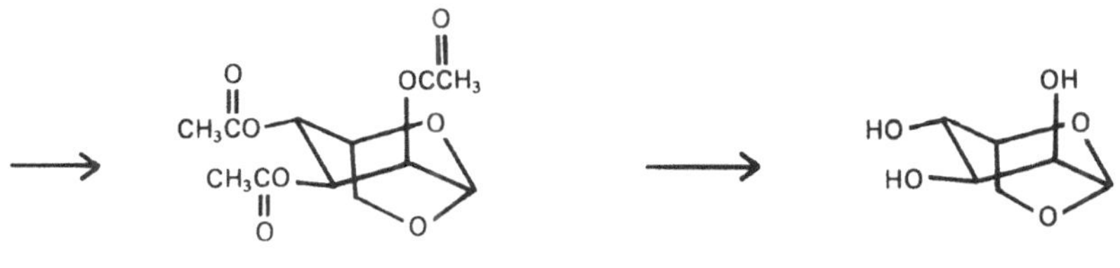

Acknowledgement

This investigation was supported by contract NO1-CM-43712 (from the Division of Cancer Treatment) and research grant CA 22614, National Cancer Institute, National Institutes of Health, Department of Health, Education and Welfare.

References

1. Y. Muraoka, T. Takita, K. Maeda, and H. Umezawa, *J. Antibiot. (Tokyo)*, **25**, 185 (1972).
2. K. Y. Zee-Cheng and C. C. Cheng, *J. Heterocycl. Chem.*, **7**, 1439 (1970).
3. See, e.g., D. A. McGowan, U. Jordis, D. K. Minster, and S. M. Hecht, *J. Am. Chem. Soc.*, **99**, 8078 (1977), and references therein.
4. D. K. Minster, U. Jordis, D. L. Evans, and S. M. Hecht, *J. Org. Chem.*, **43**, 1624 (1978), and references therein.
5. F. Asinger, M. Thiel, and L. Schroder, *Justus Liebigs Ann. Chem.*, **610**, 49 (1957).
6. J. Walker, *J. Chem. Soc., C*, 1522 (1968).
7. N. A. Fuller, *J. Chem. Soc., C*, 1526 (1968).
8. J. C. Vederas, Ph.D. Thesis, Massachusetts Institute of Technology (1973).
9. M. A. Barton, G. W. Kenner, and R. C. Sheppard, *J. Chem. Soc., C*, 1061 (1968).
10. M. V. George and K. S. Balachandran, *Chem. Rev.*, **75**, 491 (1975).
11. A. F. Kluge, K. G. Untch, and J. H. Fried, *J. Am. Chem. Soc.*, **94**, 7827 (1972).
12. T. Yoshioka, T. Hara, T. Takita, and H. Umezawa, *J. Antibiot. (Tokyo)*, **27**, 356 (1974).
13. H. Umezawa, Y. Takahashi, A. Fujii, T. Saino, T. Shirai, and T. Takita, *J. Antibiot. (Tokyo)*, **26**, 117 (1973).
14. D. K. Minster and S. M. Hecht, *J. Org. Chem.*, **43**, 3987 (1978), and references therein.
15. R. U. Lemieux, *Methods Carbohyd. Chem.*, **2**, 221 (1963).
16. R. U. Lemieux and A. R. Morgan, *Can. J. Chem.*, **43**, 2199 (1965).
17. E. J. Corey and J. W. Suggs, *Tetrahedron Lett.*, 2647 (1975).
18. W. von E. Doering and J. R. Parikh, *J. Am. Chem. Soc.*, **89**, 5505 (1967).
19. G. M. Cree, D. W. Mackie, and A. S. Perlin, *Can. J. Chem.*, **47**, 511 (1969).
20. E. J. Corey and C. U. Kim, *J. Am. Chem. Soc.*, **94**, 7586 (1972).
21. L. C. Steward and N. K. Richtmeyer, *J. Am. Chem. Soc.*, **77**, 1021 (1955).
22. M. Prystus, H. Gustafsson, and F. Sorm, *Collect. Czech. Chem. Commun.*, **36**, 1487 (1971).

A Synthetic Approach to the Pyrimidine Moiety of Bleomycin

Yoji Umezama, Hajime Morishima, Takeo Yoshioka, Masami Otsuka and Masaji Ohno

The original structure of bleomycin (BLM) was proposed on the basis of a combination of direct and indirect evidence, including total and stepwise hydrolysis, proton and ^{13}C-NMR, infrared spectra and pK_a' values (1)[1]. A revised structure (2) containing an open chain β-aminopropionamide group instead of the β-lactam ring has been proposed at this symposium by Dr. Takita. It should be pointed out that the new structure (2) was considered to be one of the most probable candidates for the structure of BLM from the early stages of the structure determination, but it was extremely difficult to establish an unambiguous structure solely on the basis of the degradation study.

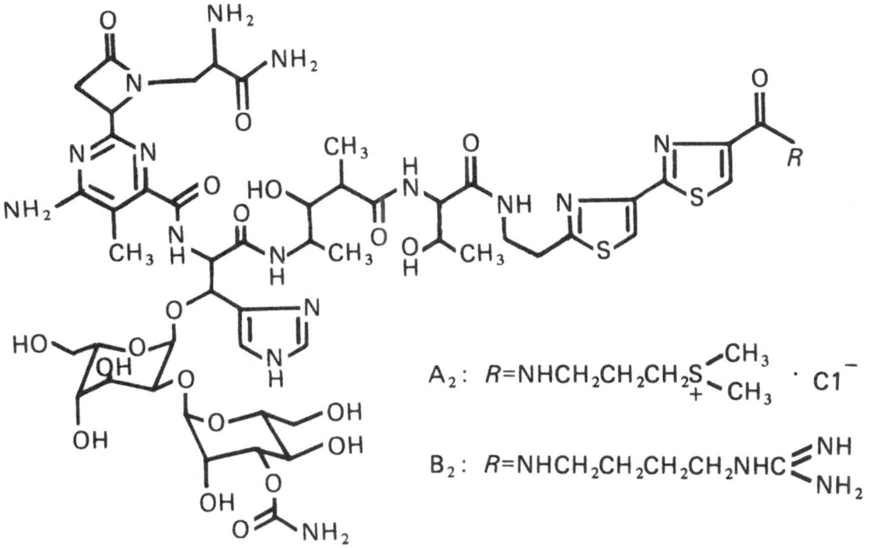

A_2: $R=NHCH_2CH_2CH_2\overset{+}{S}\underset{CH_3}{\overset{CH_3}{<}}$ · Cl$^-$

B_2: $R=NHCH_2CH_2CH_2CH_2NHC\underset{NH_2}{\overset{NH}{<}}$

1

Therefore, we became interested in a synthetic approach to the pyrimidine moiety, which would help to clarify the structure of BLM by preparing the most probable structural candidates corresponding to the pyrimidine moiety.

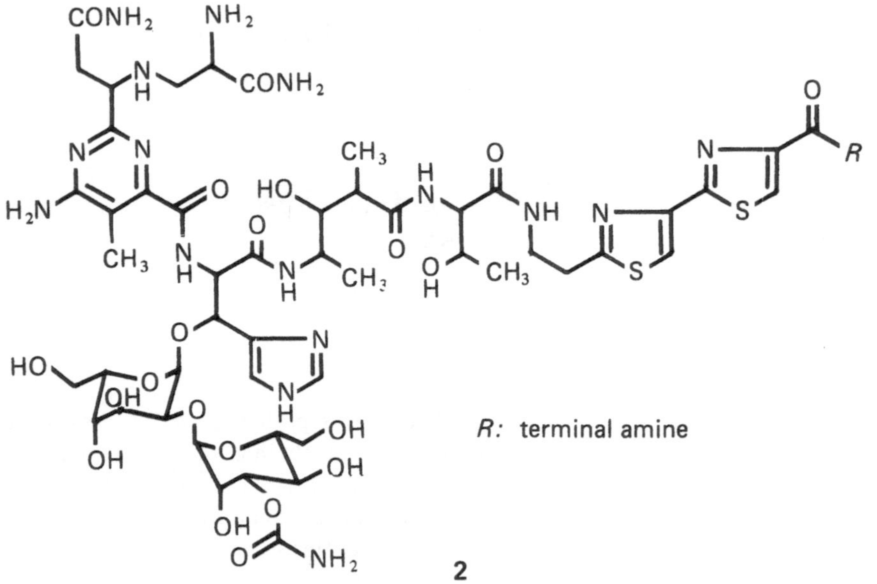

R: terminal amine

2

In our opinion, a synthetic approach of this type may constitute one of the best methods for structural verification in the case of BLM. Additionally, we believed that there might be other significant reasons for the synthesis of the pyrimidine moiety of BLM.

These are (i) BLM binds and reacts with DNA, causing single-strand scission[2]. This reaction, which occurs even at 0°C, is enhanced by radical-producing compounds and inhibited by Cu^{2+}. The β-aminoalaninamide and carbamoyl moieties of BLM are involved in its action on DNA and in the chelation of Cu^{2+}, so it is not unlikely that the "active site" of BLM includes the pyrimidine moiety. Therefore, it seemed of interest to test synthetically prepared model compounds for their biological activity and chelation chemistry.

(ii) A synthesis of the pyrimidine moiety could be useful as part of a total synthesis of BLM.

(iii) Correlation of structure and physical data such as ^{13}C-NMR chemical shifts, IR spectra, and pK_a' values is not only useful but also necessary for the study of a glycopeptide as complex as BLM. This is especially true since Dr. Takita and his co-workers have already found that the IR and pK_a' values of BLM are not easily understood in terms of known data from more usual peptides[3].

Our synthetic strategy for the elaboration of the pyrimidine moiety is depicted in Scheme 1.

The key compound is pyrimidine aldehyde 3. This aldehyde could be used for the introduction of the β-lactam ring (6) by the reaction of an appropriate ketene and the Schiff base (5) containing the desired amino acid or β-amino-

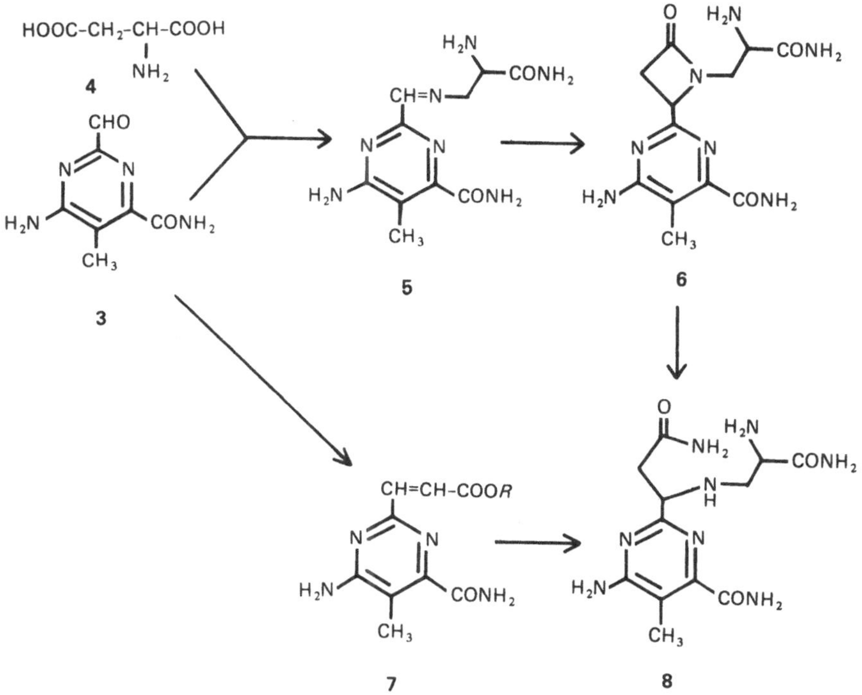

Scheme 1. Synthetic strategy for the pyrimidine moiety.

propionamide, the latter of which should be accessible from L-aspartic acid (4). Alternatively, this aldehyde could be converted directly into a β-aminopropion-amide derivative (8) using Knoevenagel reaction product 7. In this synthetic scheme, there are at least three main problems. They are the effective synthesis of the pyrimidine aldehyde, the yield of formation of the β-lactam moiety, and functional group protection during the transformation of synthetic intermediates.

We first thought that a more simplified β-lactam-containing compound might be useful for diagnostic purposes. The infrared spectra and pK_a' values of such simplified compounds might suggest the similarity or differences between such

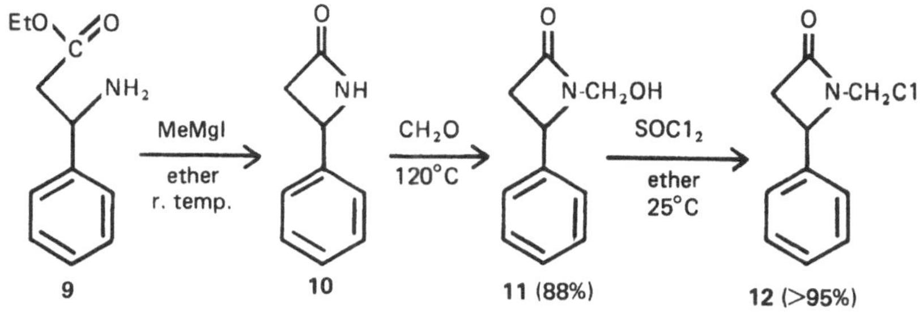

species and the pyrimidine moiety of BLM, and therefore the probable presence
or absence of a β-lactam in BLM. A 4-phenyl-2-azetidinone derivative was considered to be a good candidate for a model compound, since the starting material
(10) is known[4] and easily accessible.

Compound 9 was prepared according to Rodionow and Malewinskaja[5].
They found that the reaction between benzaldehyde and malonic acid in
alcoholic ammonia gives not only cinnamic acid but also a β-amino acid in
practically equal amount. Ethyl β-amino-β-phenylpropionate (9) thus obtained
was treated with methyl magnesium iodide at room temperature, affording
4-phenyl-2-azetidinone (10) in 20% yield. Although the yield was low, this compound could be converted into the desired structure, and so was adequate for
our purposes. Direct introduction of the β-aminoalaninamide moiety was attempted. A dehydroalanine derivative was prepared by treatment of the corresponding β-chloroalanine derivative with triethylamine[6] and this species
was combined with 4-phenyl-2-azetidinone in the presence of sodium hydride.
After preparative thin-layer chromatography, a Michael reaction product of this
desired structure was obtained, but the yield was disappointing, only ca. 1%.
Therefore, another route was investigated. Addition of a C_1-unit to the amino
group of the β-lactam was achieved by treatment with paraformaldehyde at
120°C in a sealed tube for 4 hr, affording hydroxymethyl compound 11 in 88%
yield.

The hydroxymethyl compound was quantitatively converted into the chloride
(12) with thionyl chloride in ether at room temperature. The chloromethyl
compound (12) was treated with dibenzyl benzoxycarbonylaminomalonate in
the presence of sodium hydride in ether solution, affording the expected condensation product in 70% yield. However, selective removal of the protecting
groups was not successful. Hydrogenolysis of compound 13 afforded compounds
14 and 16, showing that hydrogenolysis had occurred at the C–N bond of the
β-lactam ring. Therefore, the N-protecting group was changed to a t-butyloxycar-
bonyl (Boc) group and the O-protecting groups to t-butyl groups. Compound
15 was obtained in 66% yield by the reaction of compound 12 and di-t-butyl
t-butyloxycarbonylaminomalonate in the presence of sodium hydride in ether
at room temperature. In this case the free amino acids were obtained quantitatively as a mixture of two diastereomers. The yield was excellent, but the
carboxyl group would have to be converted into a carboxamide at the end of
our synthesis.

It should be mentioned here that such nonfused β-lactam rings were found
rather stable in dilute basic media. Treatment of compound 13 with 0.1 M
NaOH in methanol at 0°C, for example, simply afforded a product in which
one of the benzyl esters had undergone methanolysis, while treatment of compound 13 with 0.1 M NaOH in acetone followed by reflux in chloroform
effected smooth decarboxylation. In neither case was the β-lactam ring attacked
by base. Therefore, similar reaction conditions can presumably be used for
further elaboration of such β-lactams.

13 (70%) → Pd-C, H$_2$ / MeOH → **14 (50%)** + 16 (>50%)

15 (66%) → CF$_3$COOH / 0°C → **16 (>95%)**

two diastereomers

Chloromethyl derivative (**12**) was treated with dimethyl *t*-butyloxycarbonyl-aminomalonate in a fashion similar to that indicated above, affording the expected compound (**17**) in excellent yield. One of the methoxycarbonyl groups was removed under alkaline conditions, using sodium hydroxide in acetone at 0–10°C, followed by reflux in chloroform. Compound **18** was then treated with ammonia to afford the carboxamide, and the *t*-butyloxycarbonyl group at the α-position was removed smoothly with trifluoroacetic acid affording the desired β-lactam (**20**).

At this stage, the infrared spectra of several 2-amino-3-(2-oxo-4-phenyl-azetidinyl)propionamide derivatives were carefully examined (Table 1). As mentioned previously, two diastereomers were obtained. They were separated by thin-layer chromatography on silica gel using 3:1 benzene-ethyl acetate for development. Both of the β-lactam compounds protected with the *t*-butyloxy-carbonyl group had strong absorptions at 1760 cm^{-1}. One diastereomer had an absorption at 1741 cm^{-1}. However, one of the hydrochlorides absorbed at 1710 cm^{-1}, a much lower frequency than is usual for a β-lactam. The other diastereomer had an absorption at 1735 cm^{-1}. Both of the trifluoroacetic acid salts had absorptions around 1740 cm^{-1}. We wondered why such a large change in the IR absorption should occur and which of the diastereomers had the unusual spectrum. There is an absorption shoulder at 1720–1735 cm^{-1} for BLM·HCl, which was originally assigned to the β-lactam ring. We first thought that the hydrochloride of the diastereomer with the absorption at 1710 cm^{-1} represented a product cleaved at the β-lactam ring; therefore, we checked the structure by X-ray crystallography. Amazingly, it was found to have an intact β-lactam ring. Therefore, while the existence of the β-lactam moiety originally proposed as part of the BLM structure may be in doubt, the results obtained with these model compounds were not inconsistent with the proposed structure. It seems afe to say that the infrared spectrum alone cannot be used as a criterion for the presence or absence of the β-lactam.

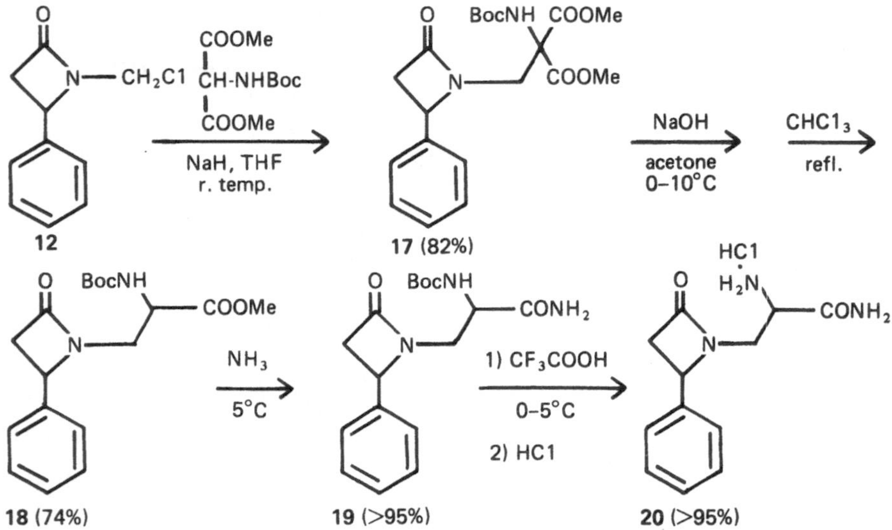

The molecular structure of the hydrochloride of 2-amino-3-(2-oxo-4-phenyl-1-azetidinyl)propionamide (20) as determined by X-ray diffraction is shown in Fig. 1. The stereochemistry of the two asymmetric carbons was S,S or R,R, since the sample was used in racemic form. The stereochemistry of the pyrimidine moiety of BLM is also S,S. Strong hydrogen bonding between the carbonyl group of the β-lactam ring and one of the hydrogens of the protonated amino group at the position α to the carboxamido group was clearly demonstrated, so the low energy absorption in the IR spectrum can probably be attributed to such

Table 1:   Infrared Spectra of 4-Phenyl-2-azetidinones

R		IR (cm^{-1})	
H		1735	
CH$_2$OH		1735	
CH$_2$–C(COOMe)$_2$–NHBoc		1775 (s), 1750 (sh)	
CH$_2$–CH(COOMe)–NHBoc	1755 (R_f=0.38)	1746 (R_f=0.22)	
CH$_2$–CH(CONH$_2$)–NHBoc	1760	1760 (sh), 1730	
CH$_2$–CH(CONH$_2$)–NH$_2$·TFA	1738	1740	
CH$_2$–CH(CONH$_2$)–NH$_2$	1746	1741	
CH$_2$–CH(CONH$_2$)–NH$_2$·HCl	1710	1735	

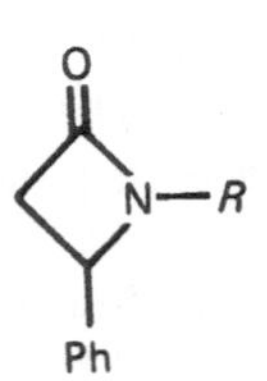

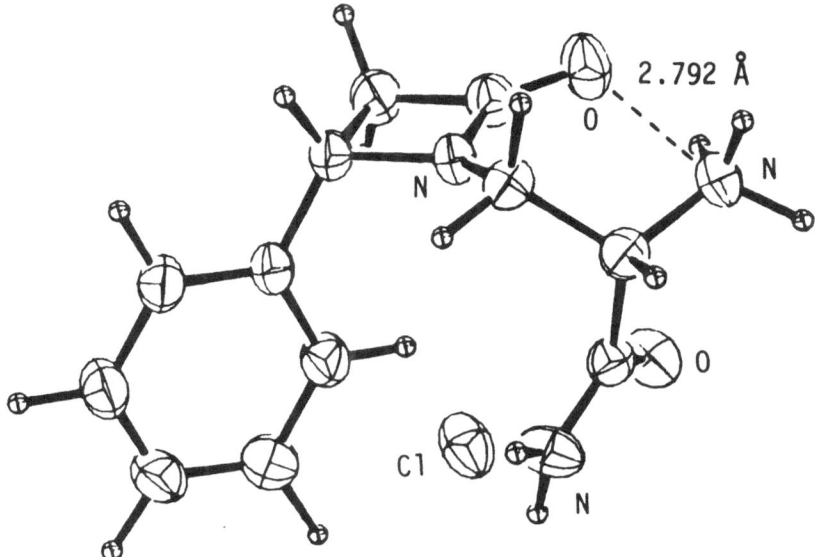

Fig. 1. Molecular structure of hydrochloride of 2-amino-3-(2-oxo-4-phenyl-1-azetidinyl)propionamide determined by X-ray diffraction.

hydrogen bonding. To the best of our knowledge, this is the lowest IR absorption frequency ever definitively demonstrated for any β-lactam. Furthermore, it should be pointed out that in this compound the β-lactam ring is perpendicular to the benzene ring.

We have now made another model compound which has the same side chain as the pyrimidine moiety of the revised structure of bleomycin. Methyl β-amino-β-phenylpropionate (21) was used as a starting material for this model compound. A Michael reaction of 21 and a dehydroalanine derivative afforded the desired compound (22) in 30% yield. The ester was converted into a carboxamido group in excellent yield, and 23 was hydrogenolyzed over palladium–carbon in ethanol, affording the target molecule, 3-N-(2-amino-2-carboxamidoethyl)amino-3-phenylpropionamide (24).

The ^{13}C-NMR spectra of several substituted β-aminopropionamides were compared (Table 2) [compound (25) was prepared from a degradation product of BLM[7]]. Since it was shown that each of the BLM analogs had very similar ^{13}C-NMR spectra[8], it seemed reasonable to ask which of the model compounds (20 or 24) had a spectrum closer to that of BLM itself. As shown in Table 2, compound 24 had a spectrum very similar to that of BLM, suggesting that BLM does not contain a β-lactam.

As shown in Table 3, the $pK_a{}'$ values of various β-aminopropionamides were compared with those of BLM. In this case, dramatic agreement was obtained between the dicarboxamido compound (24) and BLM. The carboxamido compound with the pyrimidine nucleus (25) had a pK_a of 3.4 for the secondary amino group, which was very low but not exactly the same as that of BLM. Various factors can influence pK_a values including hydrogen bonding, conformational stability, and electronic effects, but in any case the $pK_a{}'$ values of these model compounds clearly indicate the absence of a β-lactam in BLM.

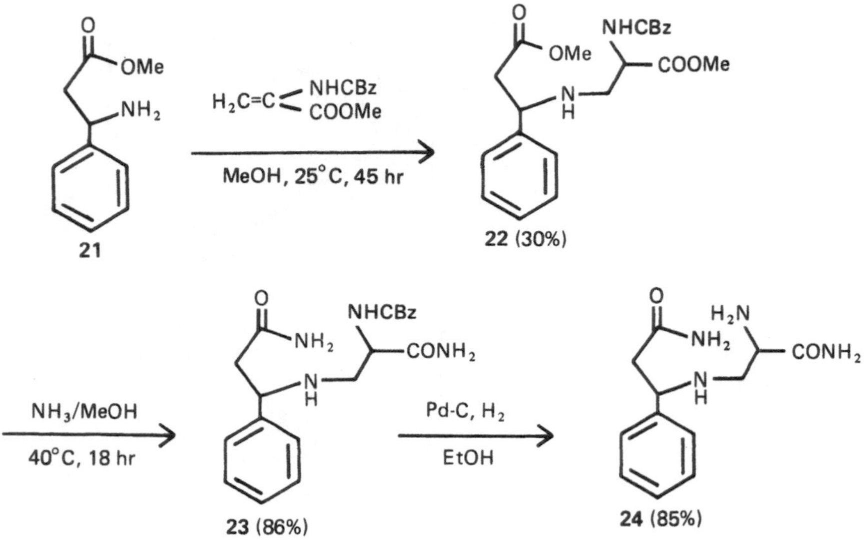

We have now prepared some pyrimidines having a β-lactam ring. As noted above, pyrimidine aldehyde 3 was considered to be a key compound for syn-

Table 2: ^{13}C-NMR of Substituted β-Aminopropionamides

	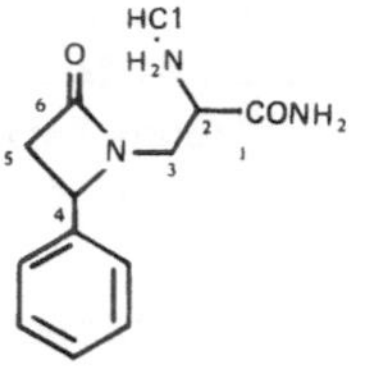 20	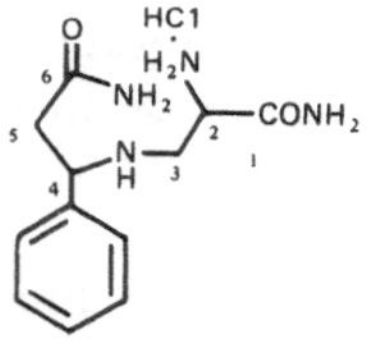 24	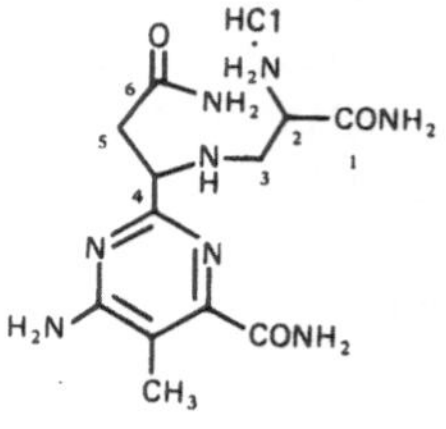 25	BLM
6	173.1 (s)	175.4 (s)	176.9 (s)	176.7 (s)
1	169.9 (s)	170.2 (s)	171.9 (s)	171.6 (s)
Aromatic ring	137.0 (s)	136.2 (s)		
	130.1 (d)	130.4 (d)		
	130.0 (d) X 2	130.3 (d) X 2		
	128.0 (d) X 2	128.7 (d) X 2		
2	56.3 (d)	61.0 (d)	60.8 (d)	60.5 (d)
4	52.5 (d)	51.7 (d)	53.2 (d)	53.3 (d)
3	45.8 (t)	46.6 (t)	47.8 (t)	47.8 (t)
5	43.1 (t)	40.0 (t)	41.1 (t)	41.0 (t)

Table 3: pK_a' Values of Substituted β-Aminopropionamides

	20	24	25	BLM
Primary amino	5.7	7.4	7.5	7.5
Secondary amino	–	2.7	3.4	2.7

thesis of the pyrimidine moiety. At first, we planned to obtain the aldehyde from dichloromethylpyrimidine **26** or monochloromethyl analog **27** by hydrolysis followed by oxidation. The dichloromethylpyrimidine (**26**) and monochloromethylpyrimidine (**27**) were prepared by a known procedure[9], but the yields of these compounds were generally low (10 to 30%). The hydrolysis of **26** under various conditions was attempted without any success. An ethyl acetal (**29**) was obtained from **28**, but the overall yield was low.

Therefore, another approach was attempted, as shown below. The known 2,5-dimethyl-4-ethoxycarbonyl-6-oxopyrimidine (**30**) was prepared from acetamidine hydrochloride and ethyl 2-ethoxalylpropionate in the presence of sodium ethoxide in ethanol[9]. This pyrimidine derivative (**30**) was treated with benzaldehyde in the presence of zinc chloride in DMF at reflux, affording a condensation product (**31**) in 45% yield which involved selective condensation at the 2-methyl group[10]. After chlorination of the hydroxyl group of the pyrimidine with phosphorus oxychloride, 6-chloro-4-ethoxycarbonyl-2-formyl-5-methylpyrimidine (**33**) was obtained by ozonolysis of **32** in methylene chloride at –60°C[11]. This aldehyde is rather stable, but was still obtained in low yield.

The third approach successfully afforded the pyrimidine aldehyde (**33**) in two steps. The reaction of diethoxyacetamidine and ethyl 2-ethoxalylpropionate in the presence of triethylamine in ethanol afforded compound **29** in 46% yield and **29** was converted into **33** with phosphorous oxychloride in excellent yield. The chloro and ethoxycarbonyl groups of (**33**) may be replaced easily by amino groups at a later stage in the synthesis, so now we have a key synthetic intermediate.

The synthesis of the diaminopropionic acid derivative which is required for combination with the pyrimidine aldehyde (**33**) has also been investigated. Diaminopropionic acid (**34**) was prepared from L-aspartic acid by the Schmidt

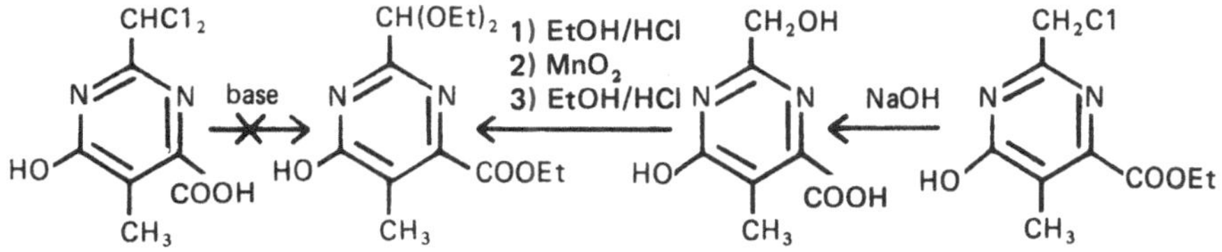

26   29 (70%)   28 (80%)   27

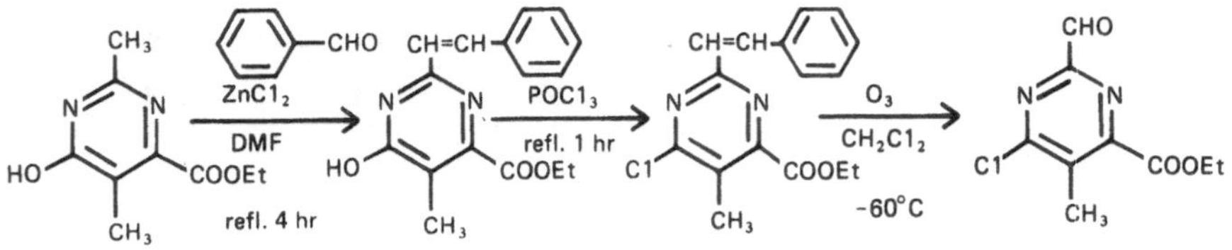

reaction[12]. The β-amino group was initially protected with *t*-butyloxycarbonyl azide (pH 9.0)[13] and then the α-amino group of **35** was protected with *N*-carboethoxyphthalimide to afford the corresponding phthaloyl derivative (**36**). The β-amino protecting group was removed with hydrochloric acid and the amino acid was isolated as the hydrochloride (**37**). Treatment of compound **37** with hydrogen chloride in methanol afforded the desired phthaloyl amino acid ester (**38**).

Protection of the α-amino group with *N*-carboethoxyphthalimide was shown to be necessary due to the propensity of such functionalities to react with Schiff bases. For example, compounds **39**[14] and **40** were formed by addition of tosylamino and carboxamido groups, respectively, to intermediate Schiff bases.

Pyrimidine aldehyde **33** was treated with methyl β-amino-α-phthalimidopropio-nate (**38**) to afford the corresponding Schiff base, which was treated with a substituted acetyl chloride or bromide in the presence of triethylamine at 0–25°C. β-Lactams were obtained, although the yields were very low (**42**, X=Cl 15%, X=Br 1%, X=SCH$_2$Ph 18%).

The reduction, stereochemical study, and nucleophilic reactions of β-lactams **42** and **43** and a direct conversion of **33** to pyrimidine moiety **25** of BLM are under investigation.

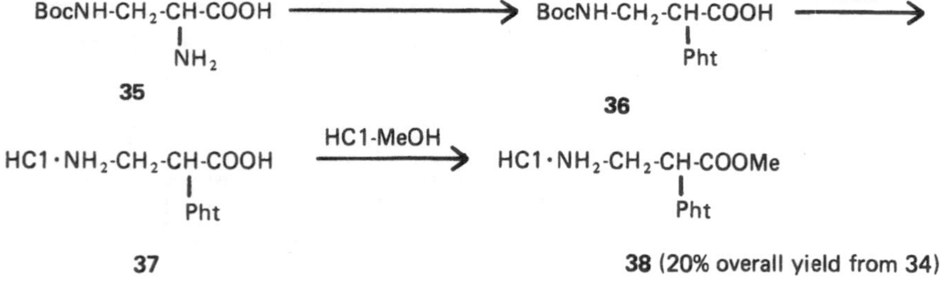

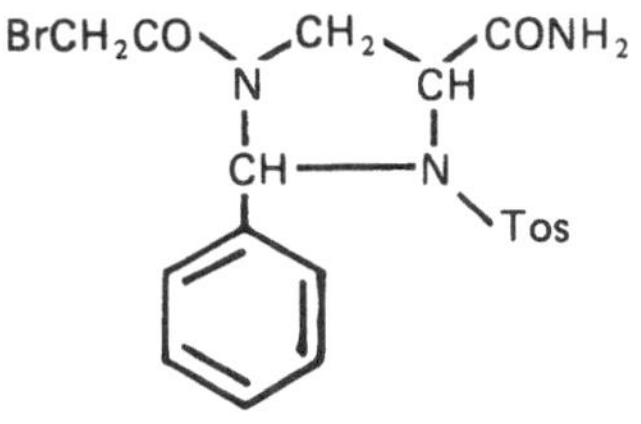

39

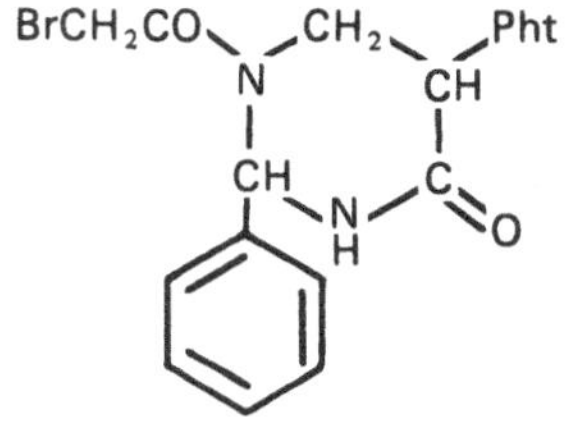

40

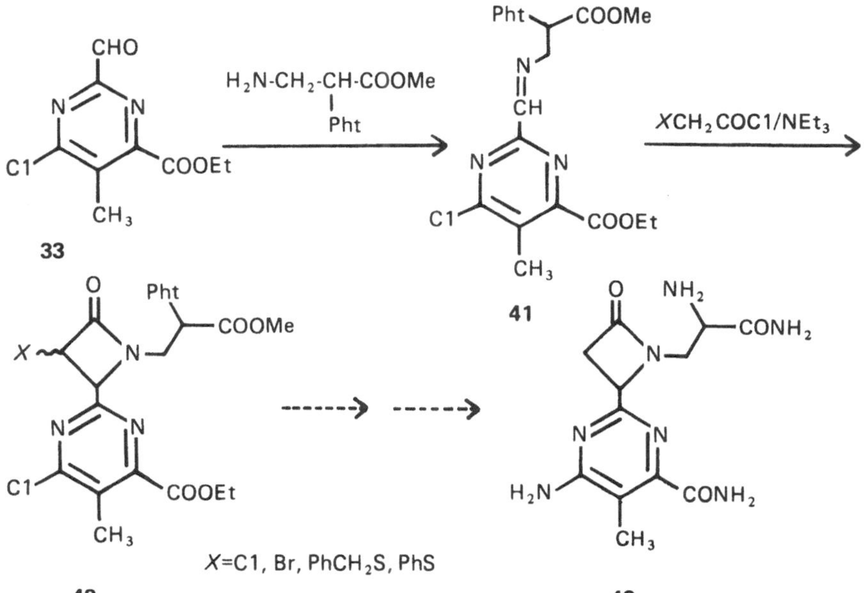

33

41

42

X=C1, Br, PhCH₂S, PhS

43

Acknowledgments

We would like to express our cordial thanks to Prof. H. Umezawa for his guidance and encouragement throughout this work, and to Dr. T. Takita for discussing many problems involved in this synthetic work. Also, we would like to express our deep thanks to Dr. Y. Muraoka for the preparation of some samples used in the present synthesis. We are also indebted to Dr. H. Naganawa for his kind help in measuring proton and ^{13}C-NMR spectra, and mass spectra, and to Dr. Y. Iitaka and Dr. H. Nakamura of the Faculty of Pharmaceutical Sciences of the University of Tokyo for the X-ray crystallographic analyses.

References

1. T. Takita, Y. Muraoka, T. Yoshioka, A. Fujii, K. Maeda, and H. Umezawa, *J. Antibiot. (Tokyo)*, **25**, 755 (1972).
2. H. Umezawa, *Gann Monograph Cancer Res.*, **19**, 3 (1976).
3. Y. Muraoka, A. Fujii, T. Yoshioka, T. Takita, and H. Umezawa, *J. Antibiot. (Tokyo)*, **30**, 178 (1977).
4. E. Testa, L. Fontanela, and V. Aresi, *Justus Liebig's Ann. Chem.*, **656**, 114 (1962).
5. W. M. Rodionow and E. R. Malewinskaja, *Chem. Ber.*, **59**, 2952 (1926).
6. L. Benoiton, *Can. J. Chem.*, **46**, 1549 (1968).
7. T. Takita, Y. Muraoka, K. Maeda, and H. Umezawa, *Proc. 8th Symp. Peptide Chem. Osaka*, 179, (1970).
8. H. Naganawa, Y. Muraoka, T. Takita, and H. Umezawa, *J. Antibiot. (Tokyo)*, **30**, 388 (1977).
9. Y. Muraoka, T. Takita, K. Maeda, and H. Umezawa, *J. Antibiot. (Tokyo)*, **23**, 252 (1970).
10. H. Kondo and M. Yanai, *Yakugaku Zasshi*, **57**, 747 (1937).
11. J. J. Pappas, W. P. Keaveney, E. Gancher, and M. Berger, *Tetrahedron Lett.*, **36**, 4273 (1966).
12. T. Kitagawa, T. Ozawa, and H. Taniyama, *Yakugaku Zasshi*, **89**, 285 (1969).
13. J. Leclerc and L. Benoiton, *Can. J. Chem.*, **46**, 1047 (1968).
14. J. C. Sheehan and E. J. Corey, *Org. React.*, **9**, 388 (1957).

Biogenetic Aspects of Bleomycin–Phleomycin Group Antibiotics

Akio Fujii

Bleomycin–phleomycin group antibiotics have attracted great interest in a variety of fields. Their biogenesis is no exception, because they are very sophisticated natural products containing unusual amino acids as their building blocks[1]; they also have interesting glycosidic moieties[2] and are efficient chelating agents. In addition, new members of this family have recently been isolated[3-7]. The production of various bleomycins[8,9] and phleomycins[10,11] and their unique structures prompted us to study their biosynthesis. When we started our study, we had several problems which we wished to solve quickly, namely (i) the possible incorporation at the C-terminus of amines added to the culture media, (ii) the biogenetic relationship between bleomycin and phleomycin, (iii) the biosynthetic mechanisms utilized for elaboration of the unusual amino acids, (iv) the mechanism of hydroxylation of L-histidine, and (v) the mechanism of glycoside formation. The first problem has already been solved and reported[9]. As an approach to solution of the other problems, we chose to isolate intermediate peptide fragments from the culture medium. The exciting results obtained, which exceeded our expectations, are reported in this paper.

According to Lipmann's hypothesis[12], proposed for the biosynthesis of gramicidin S, peptide elongation is initiated at the N-terminus and progresses toward the C-terminus. In analogy, the chain growth of bleomycin would be initiated at the pseudodipeptide moiety and conclude with introduction of the terminal amine. Therefore, intermediate N-terminal peptide fragments at various stages of modification and of various sizes would be expected to be present in mycelia producing bleomycin.

Isolation of Intermediate Peptides and Their Biogenetic Implications

A typical fermentation process for the production of bleomycin by *Streptomyces verticillus*[13] is shown in Fig. 1. When bleomycin potency increases maximally at the last stage of the process, the pH of the medium tends to increase persistently, almost to pH 8. Therefore the pH must be controlled frequently to avoid damage to the mycelia. Ammonia production also increases rapidly, while the number of mycelia decreases. This pattern seems to indicate that antibiotic

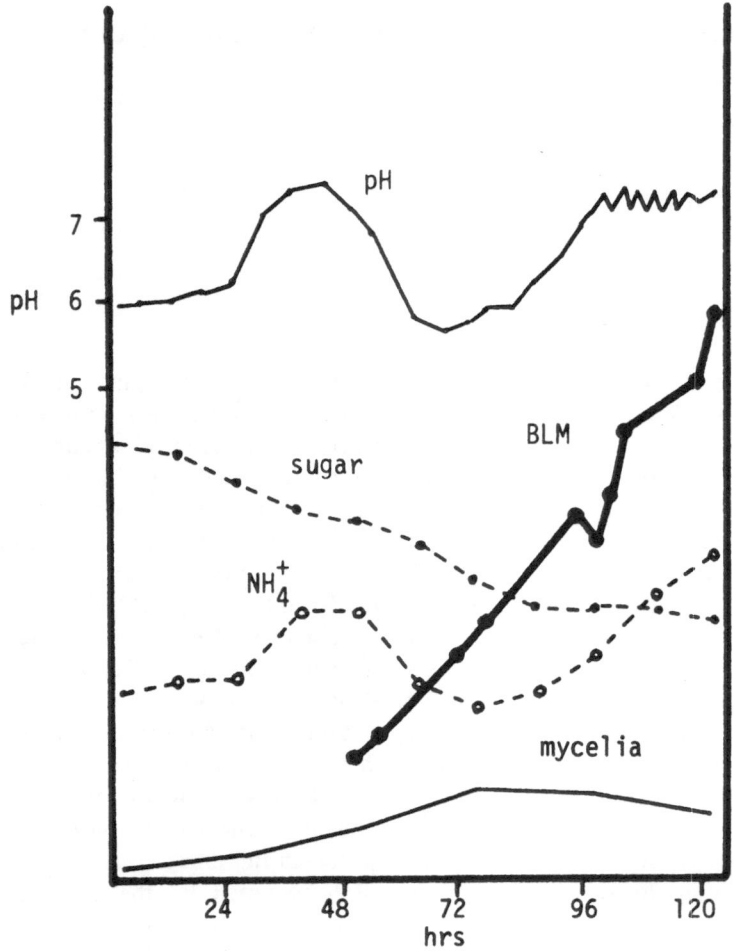

Fig. 1. Fermentative production of BLM.

production and partial autolysis of mycelia occur simultaneously. It was thought that some intermediate peptide fragments might be liberated under such conditions, and an attempt was made to isolate these peptide fragments from the culture broth.

A culture was made by using a high-potency strain and a high-productivity medium. When antibiotic production was maximal, the broth was harvested. The broth filtrate was processed as summarized in Fig. 2. It was first passed through a column of Amberlite IRC-50 (H^+ form). The adsorbed peptides were eluted with dilute hydrochloric acid, then adsorbed on an activated charcoal column; the eluate resulting from washing with acidic 80% methanol was applied to a column of neutral alumina packed with methanol. After bleomycin

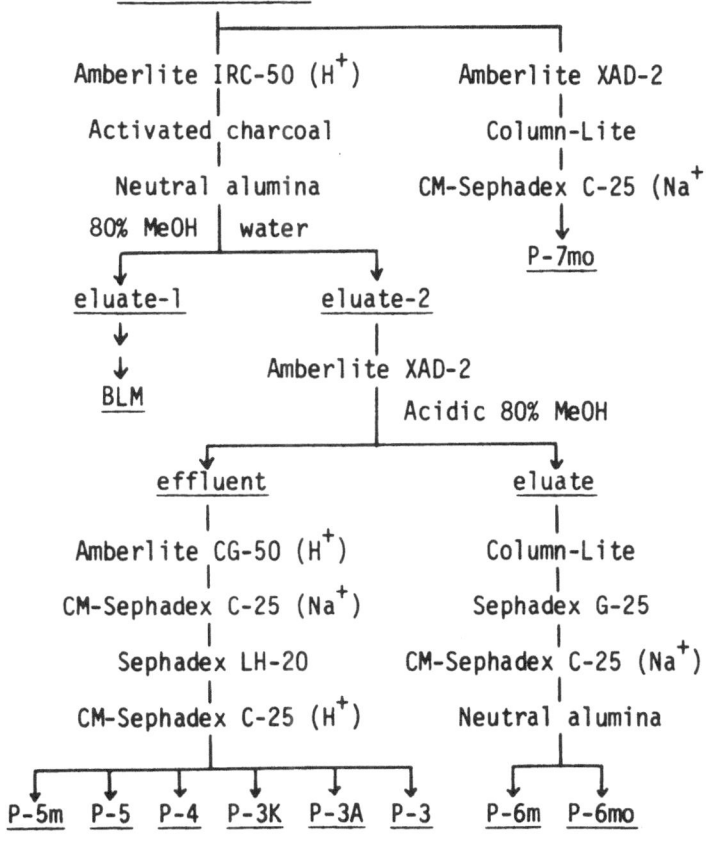

Fig. 2. Isolation of intermediate peptides of bleomycin.

was washed from the column with 80% aqueous methanol, water was used to effect elution of the intermediate peptides. The peptide mixture obtained was passed through a column of Amberlite XAD-2. The less hydrophobic fraction was adsorbed on a column of Amberlite CG-50 and eluted with acidic 80% methanol. After removal of the methanol, successive column chromatographies using CM-Sephadex C-25 and Sephadex LH-20 afforded six peptides. The more hydrophobic fraction on the Amberlite XAD-2 column was eluted with acidic 80% methanol. The eluate was passed through a column of Column-Lite, which is composed of aluminum, magnesium, and silicic acid (Fuji-Kagaku Industry Co., Japan). The eluted peptides were further purified by chromatography on Sephadex G-25, CM-Sephadex C-25, and alumina. Two peptides, P-6m and P-6mo, were separated by elution of the alumina with distilled water.

An independent purification procedure was carried out to isolate the most basic peptide, P-7mo. After chromatography on Amberlite XAD-2, the broth filtrate was adsorbed on Column-Lite and eluted with dilute hydrochloric acid. All peptides were isolated as blue-colored copper complexes[14].

The peptides P-3 and P-3A, having the pseudodipeptide moiety in common with bleomycin, were found to have L-histidine and L-alanine as their C-terminal amino acids, respectively. The latter was crystallized from hot water and subjected to X-ray crystallographic analysis. As reported by Dr. Takita, the structures in Fig. 3 were proposed for P-3 and P-3A on the basis of chemical, NMR spectral, and X-ray crystallographic studies[15], and this structure suggested the need for revision of the structure of bleomycin[16]. The X-ray crystallographic analysis of P-3A also contributed to an understanding of the three-dimensional structure of the P-3A copper complex[15]. From a biogenetic point of view, the finding of amino acid II*, the demethyl form of II, as an intermediate was unexpected and very suggestive of the biosynthesis of the methylpyrimidine ring. The same was also true for the finding of L-histidine in place of β-hydroxyhistidine.

4-Aminopentan-3-one III* was detected in the hydrolyzate of a peptide denoted P-3K. The aminoketone was identical to an authentic sample synthesized by the Dakin–West reaction[17], which involved a condensation of alanine and propionic anhydride (Fig. 4). It is obvious that part of the aminoketone ori-

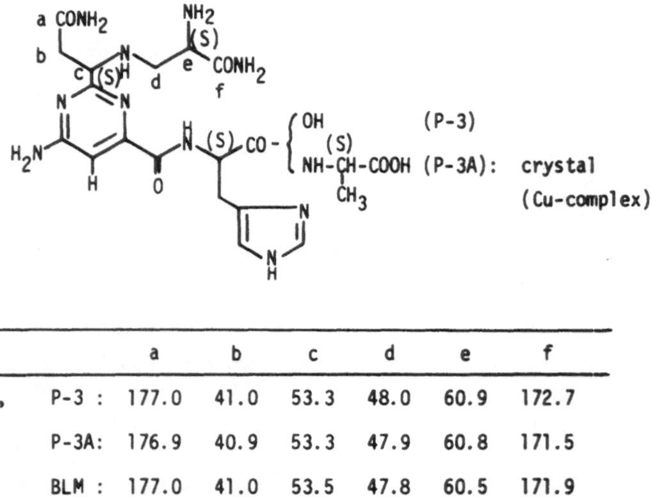

	a	b	c	d	e	f
^{13}C-NMR, P-3 :	177.0	41.0	53.3	48.0	60.9	172.7
P-3A:	176.9	40.9	53.3	47.9	60.8	171.5
BLM :	177.0	41.0	53.5	47.8	60.5	171.9

Fig. 3. Structures of P-3 and P-3A.

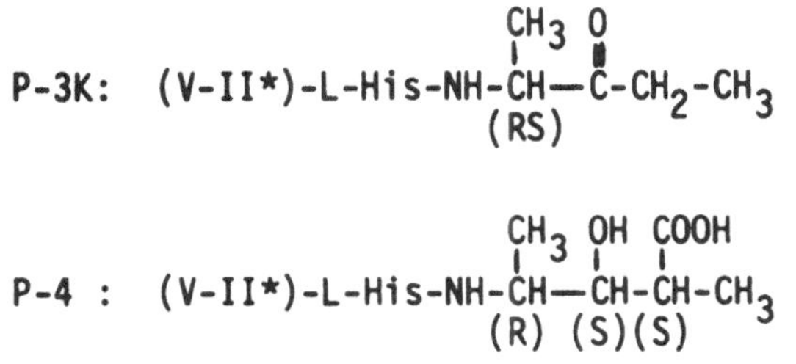

P-3K: (V-II*)-L-His-NH-$\overset{\overset{\displaystyle CH_3}{|}}{CH}$—$\overset{\overset{\displaystyle O}{\|}}{C}$-CH$_2$-CH$_3$
                         (RS)   III*

P-4 :   (V-II*)-L-His-NH-$\overset{\overset{\displaystyle CH_3}{|}}{CH}$—$\overset{\overset{\displaystyle OH}{|}}{CH}$-$\overset{\overset{\displaystyle COOH}{|}}{CH}$-CH$_3$
                         (R) (S)(S)   III

---

^{1}H-NMR of P-3K   δ1.5 (d-t, 3H),   1.9 (d-d, 3H)
(for III* part)

                          3.1 (d-q, 2H),   5.1 (d-q, 1H)

NH$_2$-$\overset{\overset{\displaystyle CH_3}{|}}{CH}$-COOH + 3(CH$_3CH_2$CO)$_2$O $\xrightarrow[\text{reaction}]{\text{Dakin-West}}$ NH$_2$-$\overset{\overset{\displaystyle CH_3}{|}}{CH}$—$\overset{\overset{\displaystyle O}{\|}}{C}$-CH$_2$-CH$_3$
   (S)   (RS)

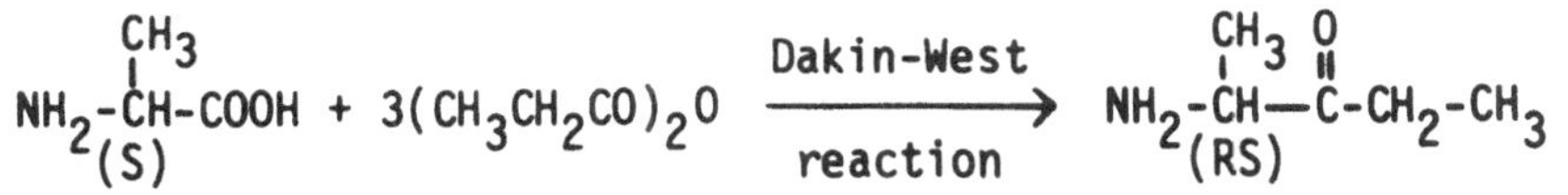

Fig. 4. Structures of P-3K and P-4.

ginated from the L-alanine moiety of P-3A. The absence of a C-terminus in P-3K suggests that it may be an artifact resulting from decarboxylation at position 2, especially in light of the structure of P-4, which has amino acid III in intact form. In the proton NMR spectrum, two sets of signals were detected at δ1.5 (assigned to the terminal methyl group), at δ1.9 (assigned to the other methyl group), at δ3.1 (assigned to the methylene group), and at δ5.1 (assigned to the methine hydrogen). This phenomenon can be explained by racemization at the methine carbon. Another interesting fact is that the S-configuration of L-alanine in P-3A was changed to R in the course of the biosynthesis of P-4.

A possible biosynthetic route from P-3A to P-4, that is a biosynthetic mechanism for the elaboration of amino acid III, is shown in Fig. 5. P-3A bound to the enzyme is probably elongated by a C$_3$ unit to give P-3AP, which was postulated on the basis of the isolation of P-3K, since β-ketocarboxylic acids are easily decarboxylated. The P-3AP is thought to be hydrogenated in stereospecific fashion to give P-4. This process is probably analogous to that believed to be utilized for fatty acid synthesis[18]. The origin of the C$_3$ unit is usually believed to be propionate, as in the case of the macrolide antibiotics[19-22] of Streptomyces origin. But in this case, propionate was not the precursor, because the incorporation of [^{14}C] propionate was too low (Table 1). Since propionate is known to be transformed to alanine or β-alanine via aspartic acid[23], the incorporation into III actually observed can be explained by labeling of the

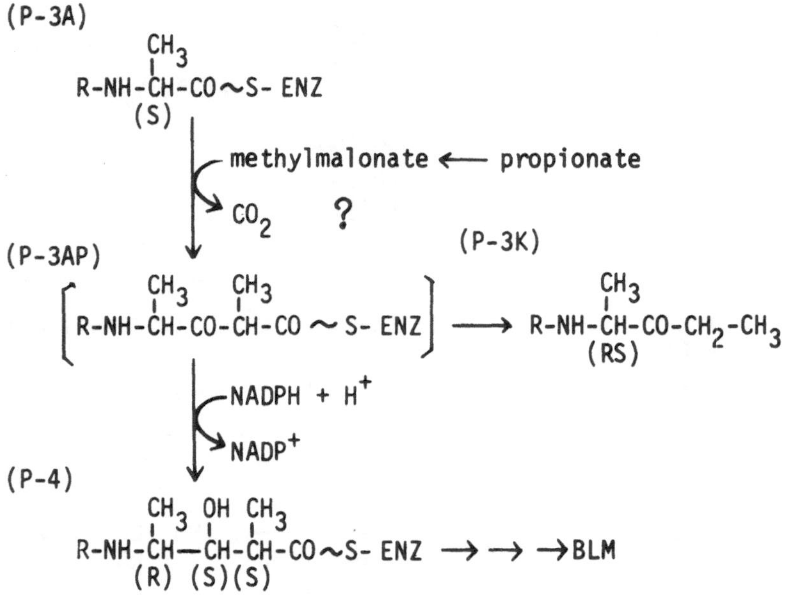

(P-3A)

$$CH_3$$
$$R-NH-CH-CO\sim S- ENZ$$
(S)

methylmalonate ⟵ propionate

$$CO_2 \qquad ?$$

(P-3AP)   (P-3K)

$$\left[R-NH-\overset{CH_3}{\underset{|}{CH}}-CO-\overset{CH_3}{\underset{|}{CH}}-CO \sim S- ENZ \right] \longrightarrow R-NH-\overset{CH_3}{\underset{|}{CH}}-CO-CH_2-CH_3$$
(RS)

NADPH + H⁺

NADP⁺

(P-4)

$$CH_3 \quad OH \quad CH_3$$
$$R-NH-CH—CH-CH-CO\sim S- ENZ \rightarrow \rightarrow \rightarrow BLM$$
(R) (S)(S)

Fig. 5. Biosynthesis of amine component III.

alanine moiety of III. Therefore we tested the possibility that the branched methyl group originated from some other C_1 unit. L-[$CH_3-^{14}C$] methionine or [^{14}C] formate were added in portions to a culture medium actively producing bleomycin. Formate showed very poor incorporation due to its wide distribution, while L-[$CH_3-^{14}C$] methionine gave remarkably good incorporation. An acid hydrolyzate of the labeled bleomycin was shown to have 62% of the radio-label associated with amino acid III[14]. This result supports a mechanism involving methylation after elongation of P-3A by a C_2 unit (Fig. 6). S-Adenosyl-methionine is thought to act as the methyl donor.

Table 1: Origins of C_3 Unit in Amine Component III

Labeled compound (dpm)	dpm in BLM (%)	Distribution (%)	
L-[$CH_3-^{14}C$]Methionine (9.99×10^7)	1.578×10^6 (1.58)	III	62
		II	28
		others	10
[^{14}C] Formic acid (9.99×10^7)	2.460×10^4 (0.025)	II, III	
[$U-^{14}C$] Propionic acid (1.07×10^8)	7.310×10^4 (0.068)	II, III, VI	

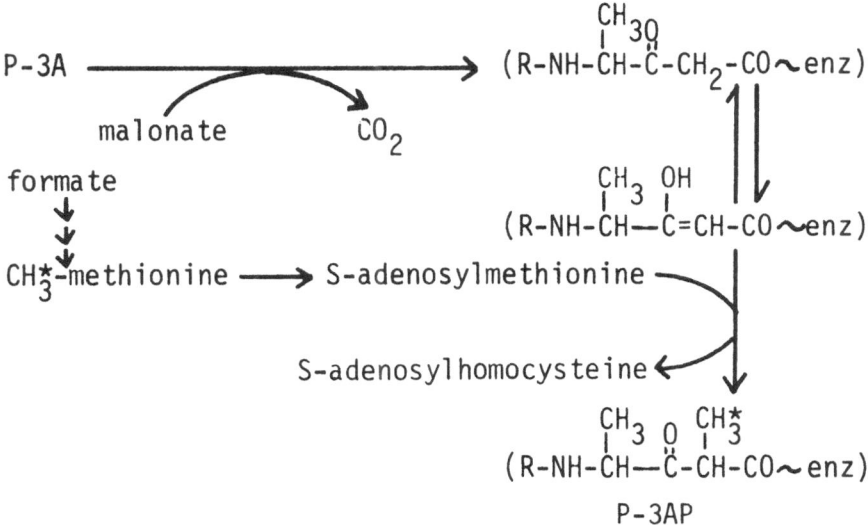

Fig. 6. A mechanism of formation of the C_3 unit.

γ-Amino-β-hydroxycarboxylic acids found as natural products of *Streptomyces* origin are shown in Table 2. The constituents on the left side of the dotted lines can originate from amino acids, and those on the right side from fatty acids. The top three belong to the bleomycin–phleomycin family. In tallysomycin[7], the combination is alanine and acetic acid, but the resulting product is not methylated. This difference from bleomycin seems very important biogenetically. In YA-56[3] or zorbamycin[4], biosynthesis would seem to involve homoserine, acetate, and a C_1 unit. Pepstatin[24], an inhibitor of

Table 2: γ-Amino-β-hydroxy Acids Produced by *Streptomyces*

Peptide	Amino acid
Bleomycin (BLM) and phleomycin (PHM)	$\underset{(R)\ \ (S)\ \ \ \ (S)}{CH_3\text{-}CH\text{-}CH\text{-}\!\!\vdots\!\!\text{-}CH\text{-}COOH}$ with NH_2, OH, CH_3 substituents
Tallysomycin (BLM-like antibiotic)	$CH_3\text{-}CH\text{-}CH\text{-}\!\!\vdots\!\!\text{-}CH_2\text{-}COOH$ with NH_2, OH substituents
YA-56 (PHM-like antitiotic)	$\underset{}{HO\text{-}CH_2\text{-}CH_2\text{-}CH\text{-}CH\text{-}\!\!\vdots\!\!\text{-}CH\text{-}COOH}$ with NH_2, OH, CH_3 substituents
Pepstatin	$\underset{(S)\ \ \ \ \ (S)}{CH_3\text{-}CH\text{-}CH_2\text{-}CH\text{-}CH\text{-}\!\!\vdots\!\!\text{-}CH_2\text{-}COOH}$ with CH_3, NH_2, OH substituents

pepsin, would be biosynthesized from L-leucine and acetic acid. Morishima *et al.* have already reported that L-leucine and malonate were incorporated into this amino acid in pepstatin[25]. It is interesting that all such types of amino acids have been found in natural products of *Streptomyces* origin.

All of the peptides isolated were arranged by increasing length (Table 3). The numbering for each component was reported previously[1]. A pair of closely related peptides, P-5 and P-5m, have L-threonine (I) at their C-termini. P-5 has demethyl II, while P-5m has the same amino acid II as does bleomycin. The evidence for these assignments included the ^{13}C-NMR spectrum of P-5m which had a signal at δ 11.7 corresponding to this methyl group, while that of P-5 did not. Another prominent difference was seen in the signals assigned to C-5 of the pyrimidine, which was methylated in P-5m (Table 4). Isolation of this pair of compounds has two biogenetic implications. The first is that the methyl group is introduced after the formation of the pyrimidine ring; the other is that methylation must occur immediately after condensation with I. It was shown in Table 1 that L-$[CH_3-^{14}C]$ methionine was incorporated into amino acid II effectively. This fact indicates clearly that the origin of the methyl group is methionine and that the more direct methyl donor is S-adenosylmethionine. An analogous reaction is the methylation of uridine in tRNA to afford ribothymidine[26]. It is catalyzed by ribothymidine methylase, which requires S-adenosylmethionine.

The biosynthetic components of the pseudodipeptide moiety are thought to be those shown in Fig. 7. An analogous structure is lycomarasmin, which is a natural product produced by a *Fusarium*. According to the study of Popplestone and Urau[27], it is biosynthesized from glycine, phosphoenolpyruvate,

Table 3:   Peptide Chain Formation of Bleomycin[a]

Peptide	Peptide chain
P-3	(V–II*)–His
P-3A	(V–II*)–His–Ala
(P-3AP)	(V–II*)–His–III** $\xrightarrow{\quad CO_2 \quad}$ P-3K
	(V–II*)–His–III*
P-4	(V–II*)–His–III
P-5	(V–II*)–His–III–I
P-5m	(V–II)–His–III–I
P-6m	(V–II)–His–III–I–VI
P-6mo	(V–II)–IV–III–I–VI
P-7mo	(V–II)–IV–III–I–VI–VII

[a]II*, demethyl-II; III**, NH–CH(CH$_3$)–CO–CH(CH$_3$)–COOH; III*, NH–CH(CH$_3$)–CO–CH$_2$–CH$_3$.

Table 4: ^{13}C-Chemical Shift Map of Intermediate Peptides

Assignment		P-3	P-3A	P-4	P-5	P-5m	P-6m	P-6mo
I	CO				177.4	177.5	172.7	172.7
	β-CH				68.8	68.8	67.9	67.9
	α-CH				60.9	60.9	59.9	59.9
	CH$_3$				20.1	20.0	19.7	19.7
II	S-CO	177.0	176.9	177.0	176.0	176.9	176.9	176.8
	R-CO	169.3	169.3	169.4	169.4	169.1	168.5	168.2
	2	166.0	166.0	166.0	166.0	166.6	166.2	166.0
	4	165.4	165.8	166.0	166.0	165.1	165.0	165.0
	6	155.3	154.9	155.0	155.0	155.1	154.3	153.6
	5	103.1	103.2	103.3	103.3	111.8	112.0	112.3
	CH	53.3	53.2	53.2	53.3	53.4	53.3	53.2
	CH$_2$	41.0	40.9	41.0	41.0	41.1	41.0	40.9
	CH$_3$	··	—	—	—	11.7	11.7	11.6
III	CO		180.8*	184.0	177.9	177.9	178.3	178.2
(Ala)*	β-CH		75.9	75.1	75.2	75.1	75.1	
	γ-CH			48.6	48.3	48.2	48.2	48.3
	α-CH		52.1*	46.1	43.7	43.5	43.6	43.6
	α-CH$_3$		17.9*	14.9	15.5	15.8	14.9	15.1
	γ-CH$_3$			14.0	12.9	12.3	13.3	13.4
IV	CO	117.2	172.4	172.4	172.4	173.2	172.4	170.1
(His)	2	135.0	135.5	135.8	136.0	136.6	136.0	136.9
	4	131.7	130.0	131.7	131.9	132.8	131.9	136.9
	5	118.0	118.9	118.0	118.0	118.0	118.1	117.5
	β-CH	—	—	—	—	—	—	67.4
	(β-CH$_2$)	29.0	28.6	29.0	29.0	29.2	28.9	—
	α-CH	55.4	53.5	54.4	54.6	55.2	54.7	58.9
V	CO	172.7	171.5	171.5	171.8	172.2	171.7	172.0
	CH	60.9	60.8	60.9	60.9	60.8	60.6	60.4
	CH$_2$	48.0	47.9	48.0	48.0	48.2	48.0	47.8
VI	CO						171.1	171.0
	2						169.2	169.1
	2'						162.4	162.3
	4						154.2	154.2
	4'						148.1	148.0
	5'						125.6	125.5
	5						118.9	118.9
	β-CH$_2$						39.9	39.9
	α-CH$_2$						32.8	32.8

and aspartic acid. In the case of the pseudodipeptide, phosphoenolpyruvate, aspartic acid, aspargine, and methionine may be involved in the biosynthesis. We are now carrying out some experiments in an effort to verify this hypothesis.

Another pair of closely related peptides, P-6m and P-6mo, have 2-amino-ethylbithiazolecarboxylic acid (VI) at their C-termini, as judged by chemical

pseudodipeptide   lycomarasmin

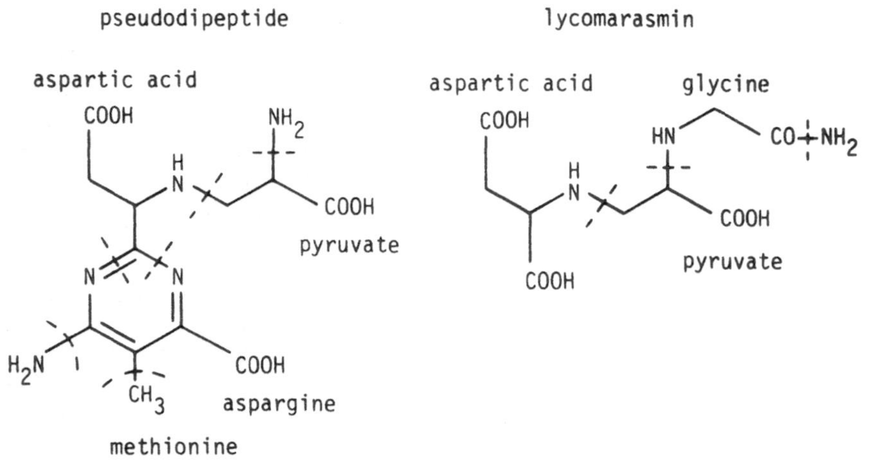

Fig. 7. Biosynthetic origins of pseudodipeptide (postulated) and lycomarasmin.

and NMR studies. The ^{13}C-NMR chemical shift map described by Dr. Naganawa[28] was especially useful in making the assignments. P-6m has L-histidine, while P-6mo has β-hydroxy-L-histidine (IV) instead. Therefore P-6mo is more closely analogous to bleomycin. The existence of this pair of intermediates suggests that IV is formed by hydroxylation of L-histidine at this stage of structural elaboration. It may be noted that an analogous reaction, catalyzed by proline hydroxylase[29], is known. This hydroxylation occurs in the peptide in presence of oxygen.

The last peptide, P-7mo, is a basic peptide having a terminal amine. The yield was the lowest of all the peptides. This intermediate was found to have the complete peptide chain of bleomycin, but no sugar moiety. Therefore it is the aglycone of bleomycin. All of the peptide intermediates were isolated as copper complexes. It is interesting that even the smallest intermediate, P-3, forms a stable copper complex. It is known that bleomycin production proceeds poorly when cultured without enrichment of cupric ion, and that equimolar cupric ion blocks perfectly the active center of bleomycin and protects it from degradation. These facts suggest the possibility that equimolar cupric ion may be one of the essential factors in bleomycin biosynthesis.

Biogenetic Relationships between Bleomycin and Phleomycin

The biogenetic origin of the bithiazole moiety of bleomycin from β-alanine and two moles of cysteine is suggested by its structure, as shown in Fig. 8. After the formation of the β-alanylcysteinylcysteine chain, cyclizations take place and are followed by dehydrations to give thiazoline rings; dehydrogenation then occurs to give phleomycin or bleomycin. This hypothesis can be supported experimentally. β-[1-^{14}C] Alanine and [3-^{14}C] cysteine were added to culture media;

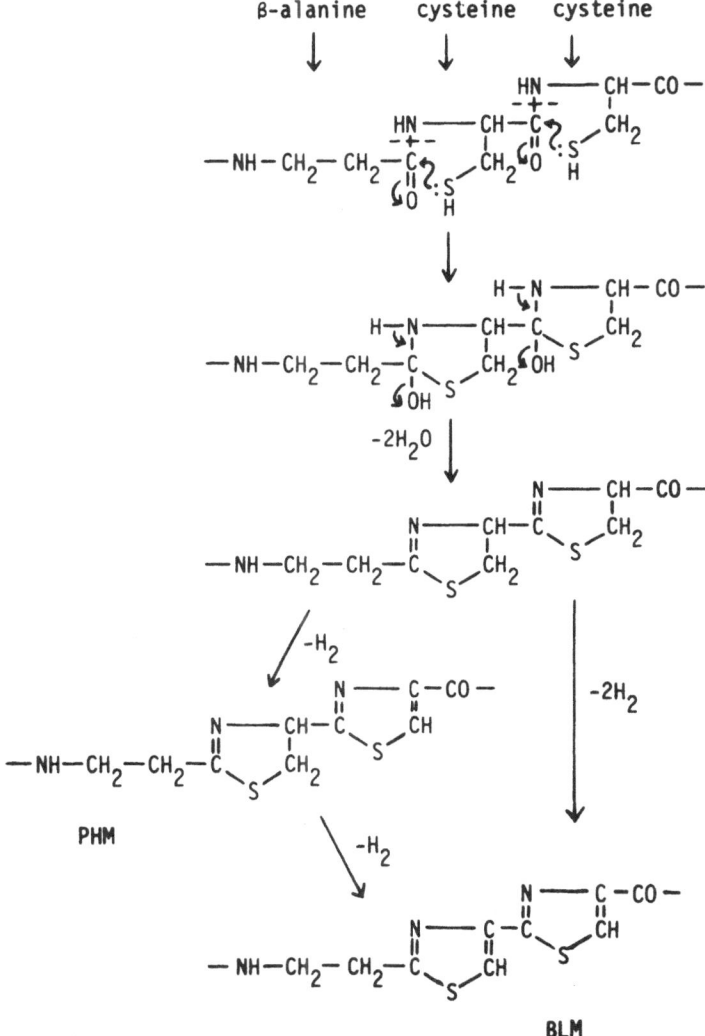

Fig. 8. Biosynthesis of VI moieties of BLM and PHM.

as two types of *Streptomyces verticillus*, a bleomycin-producing strain and a phleomycin-producing strain[11] were available, both were tested in comparison. β-Alanine and cysteine were found to be incorporated exclusively into amino acid VI[30]. The phleomycin-producing strain seemed to have a bigger endogenous β-alanine pool, supplied by decarboxylation of aspartate[31]. Tritiated VI was added into a culture medium under the same conditions, but showed no incorporation. Since a new intermediate peptide having β-alanine at its C-terminus was isolated very recently, the possibility of the direct incorporation of VI could be excluded. We are still attempting to isolate a peptide having cysteine or a related amino acid (Table 5).

The chemical transformation from phleomycin to bleomycin by dehydrogenation has been reported[1,32]. As described above, an enzymatic thiazoline dehydrogenation reaction is thought to occur naturally in the intermediate peptide. Becuase no suitable intermediate peptide has ever been available to verify the enzymatic reaction, phleomycin was utilized as a substrate.

A cell-free system was prepared from the mycelia of *Streptomyces verticillus* as summarized in Fig. 9. Mycelia were disrupted with a French press. After ultracentrifugation, the supernatant was treated with protamine sulfate to remove nucleic acids, followed by ultrafiltration (Diaflo UM-20 membrane) The eluate was subjected to gel filtration on Sephadex G-25; elution was effected as shown in Fig. 10 (left panel). The F-I fraction was found to contain a relatively stable essential cofactor and could not be replaced with other cofactors or metals. The higher molecular-weight fraction retained by the ultrafiltration membrane was purified by chromatography on Sephadex G-200 as shown (right panel). The F-II fraction was found to contain the enzyme required for dehydrogenation.

The protocol for the enzymatic dehydrogenation of phleomycin is shown in Fig. 11. Mixed together were the lower molecular-weight cofactor F-I, enzyme F-II, NAD as a hydrogen accepter, and [^{14}C] phleomycin as a substrate. After an incubation period of 1 hr at 30°C, cold carriers were added to the mixture, which was subjected to an isolation procedure utilized for bleomycin. The result of the reaction is expressed as an elution profile from a column of CM-Sephadex C-25 (Fig. 12). Peak IV is the bleomycin produced in the reaction, V is the residual phleomycin, and the other three peaks were found to be degradation products having no bithiazole ring. Therefore our enzyme system was still contaminated with some other enzymes. Nevertheless, we confirmed the enzymatic dehydrogenation of a thiazoline ring in a cell-free system for the

Table 5: Incorporation of β-[1-^{14}C] Alanine or [3-^{14}C] Cysteine into Bleomycin and Phleomycin[a]

Strain	Added compounds (dpm × 10^8)	Produced BLM or PHM (mg)	Radioactivity in BLM or PHM (dpm × 10^6)	Incorporation ratio (%)
BLM	β-[1-^{14}C] Alanine (2.136)	13.4	5.38	2.52
	L-[3-^{14}C] Cysteine (1.707)	13.1	3.32	1.94
PHM	β-[1-^{14}C] Alanine (2.136)	15.5	1.33	0.62
	DL-[3-^{14}C] Cysteine (1.100)	10.9	2.40	2.18

[a]Terminal amine of BLM and PHM; 3-aminopropylmorpholine.

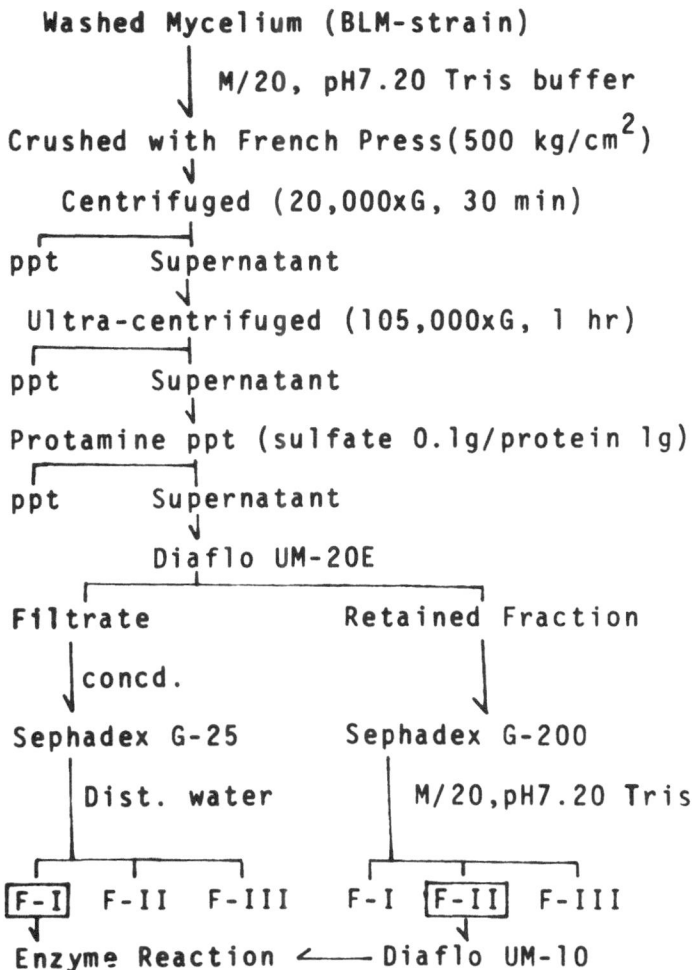

Fig. 9. Preparation of enzyme system.

first time. A cell-free enzyme extract was prepared also from the phleomycin-producing strain, which produces no bleomycin. The dehydrogenase level was compared with that of the bleomycin-producing strain under the same conditions. The upper panel of Fig. 13 shows the result from the bleomycin-producing strain. The bottom panel is the result from the phleomycin-producing strain. The activity in the latter was found to be much lower, around 15% of the former. This result seems to suggest one of the features of the difference between the two strains[30].

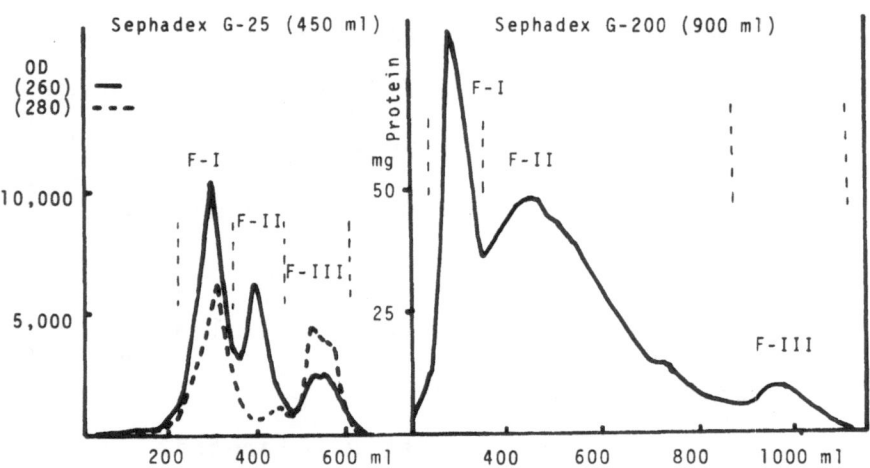

Fig. 10. Sephadex G-25 (450 ml) (left); Sephadex G-200 (900 ml) (right).

(a) Reaction Mixture

Sephadex G-25 F-I: 1 ml
Sephadex G-200 F-II: 2 ml(84 mg protein)
NAD: 30 mg/1.5 ml dist. water
^{14}C-Phleomycin: 2 mg(1.209×10^5 dpm)
Filled up to 5.0 ml, pH 7.20

(b) Incubation   at 30°C for 1 hr

(c) Isolation of Products

Diluted with cold PHM(0.8 mg) & BLM(0.8 mg)
↓
Precipitate protein with 50% MeOH & acidic 50% MeOH
↓
Column chromatographies
    Amberlite XAD-2, Alumina, Sephadex LH-20
↓
CM-Sephadex C-25: 0.05, 0.1, 0.2, 0.3 M NaCl

Fig. 11. Enzyme reaction and isolation of products.

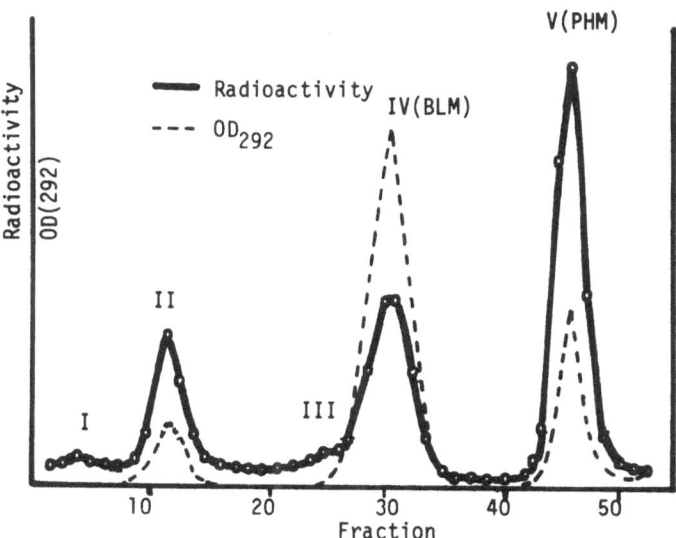

Fig. 12. Enzymatic transformation of PHM to BLM.

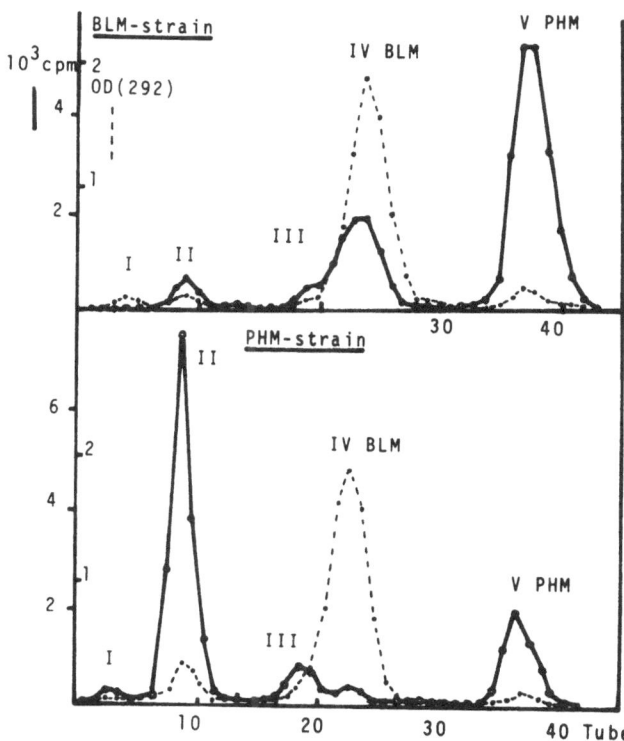

Fig. 13. Transformation of phleomycin to bleomycin. CM-Sephadex C-25 column.

References

1. T. Takita, Y. Muraoka, A. Fujii, H. Itoh, K. Maeda, and H. Umezawa, *J. Antibiot. (Tokyo)*, **25**, 197 (1972).
2. T. Takita, K. Maeda, H. Umezawa, S. Omoto, and S. Umezawa, *J. Antibiot. (Tokyo)*, **22**, 237 (1969).
3. Y. Ito, Y. Ohashi, Y. Egawa, T. Yamaguchi, T. Furumai, K. Enomoto, and T. Okuda, *J. Antibiot. (Tokyo)*, **24**, 727 (1971).
4. A. D. Argoudelis, M. E. Bergy and T. R. Pyke, *J. Antibiot. (Tokyo)*, **24**, 543 (1971).
5. I. Kawamoto, S. Takasawa, R. Okachi, M. Kohakura, I. Takahashi, and T. Nara, *J. Antibiot. (Tokyo)*, **28**, 358 (1975).
6. S. Takasawa, I. Kawamoto, I. Takahashi, M. Kohakura, R. Okachi, S. Sato, M. Yamamoto, T. Sato, and T. Nara, *J. Antibiot. (Tokyo)*, **28**, 656 (1975).
7. H. Kawaguchi, H. Tsukiura, K. Tomita, M. Konishi, K. Saito, S. Kobaru, K. Numata, K. Fujisawa, T. Miyaki, M. Hatori, and H. Koshiyama, *J. Antibiot. (Tokyo)*, **30**, 779 (1977).
8. H. Umezawa, H. Suhara, T. Takita and K. Maeda, *J. Antibiot. (Tokyo)*, **19**, 210 (1966).
9. A. Fujii, T. Takita, N. Shimada, and H. Umezawa, *J. Antibiot. (Tokyo)*, **27**, 73 (1974).
10. T. Ikekawa, F. Iwami, H. Hiranaka, and H. Umezawa, *J. Antibiot. (Tokyo)*, **17**, 194 (1964).
11. H. Umezawa, A. Fujii, T. Takita, and N. Shimada, U.S. Patent 3,984,390, 1976.
12. F. Lipman, *Science*, **173**, 875 (1971).
13. H. Umezawa, K. Maeda, T. Takeuchi, and Y. Okami, *J. Antibiot. (Tokyo)*, **19**, 200 (1966).
14. T. Nakatani, K. Ishikawa, Y. Muraoka, A. Fujii, H. Naganawa, T. Takita, and H. Umezawa, Meeting of the Agricultural Chemical Society of Japan, 1978, Abstracts.
15. Y. Iitaka, H. Nakamura, T. Nakatani, Y. Muraoka, A. Fujii, T. Takita, and H. Umezawa, *J. Antibiot. (Tokyo)*, **31**, 1070 (1978).
16. T. Takita, Y. Muraoka, T. Nakatani, A. Fujii, Y. Umezawa, H. Naganawa, and H. Umezawa, *J. Antibiot. (Tokyo)*, **31**, 801 (1978).
17. J. D. Hepworth, *Org. Syn.*, **45**, 1 (1965).
18. F. Lynen, D. Oesterhelt, E. Schweizer, and K. Willeke, In *Cellular Compartmentalization and Control of Fatty Acid Metabolism*, (F. C. Gran, Ed.). Academic Press, Inc., London and New York (1968).
19. T. Kaneda, J. C. Butte, S. B. Taubmann, and J. W. Corcoran, *J. Biol. Chem.*, **237**, 322 (1962).
20. A. J. Birch, C. Djerassi, J. D. Dutcher, J. Mayer, D. Perlman, E. Pride, R. W. Rickards, and P. J. Thomson, *J. Chem. Soc.*, 5274 (1964).
21. J. W. Corcoran and M. Chick, *Biosynthesis of Antibiotics*, Vol I (J. F. Snell, Ed.). Academic Press, New York (1966).
22. S. Omura, H. Takeshima, A. Nakagawa, and J. Miyazawa, *J. Antibiot. (Tokyo)*, **29**, 316 (1976).
23. A Meister (Ed.), *Biochemistry of the Amino Acids*, Vol. II. Academic Press, London and New York (1965).
24. H. Umezawa, T. Aoyagi, H. Morishima, M. Matsuzaki, M. Hamada, and T. Takeuchi, *J. Antibiot. (Tokyo)*, **23**, 259 (1970).

25. H. Morishima, T. Sawa, T. Takita, T. Aoyagi, T. Takeuchi, and H. Umezawa, *J. Antibiot. (Tokyo)*, **27**, 267 (1974).
26. J. Hurwitz, M. Gold, and M. Auders, *J. Biol. Chem.*, **239**, 3462 (1964).
27. C. R. Popplestone and A. M. Urau, *Can. J. Chem.*, **51**, 3943 (1973).
28. H. Naganawa, Y. Muraoka, T. Takita, and H. Umezawa, *J. Antibiot. (Tokyo)*, **30**, 388 (1977).
29. S. Udenfriend, *Science*, **152**, 1335 (1966).
30. A. Fujii, Y. Muraoka, T. Nakatani, T. Fukuoka, H. Ito, T. Takita, and H. Umezawa, Meeting of the Agricultural Chemical Society of Japan, 1973, abstracts.
31. W. Shive and J. Macow, *J. Biol. Chem.*, **162**, 451 (1946).
32. D. K. Minster, U. Jordis, D. L. Evans, and S. M. Hecht, *J. Org. Chem.*, **43**, 1624 (1978).

Liquid Chromatography of Bleomycin

Yasuhiko Muraoka

Liquid chromatography (LC) has contributed remarkably to the progress of natural products chemistry. It played an especially important role in the chemical study of bleomycin (BLM) for the following reasons: (i) BLMs are produced as a mixture by fermentation together with their biosynthetic intermediates. To facilitate their chemical study, they have been separated into individual species by LC, (ii) BLM has many susceptible functional groups. Therefore, even under mild reaction conditions, it generally gives several reaction products, and their separation is generally possible only by LC, (iii) The molecular weight of BLM is about 1400. It is a large molecule and has many polar functional groups. Therefore, minor structural modifications do not greatly affect its physico-chemical properties, and rather precise liquid chromatographic separation techniques are required to separate such similar products. Moreover, the retention time of a given compound on LC is sometimes useful for predicting what modification has occurred, (iv) The purity of BLM and its modification products can be verified by high-performance liquid chromatography (HPLC).

This chapter describes the liquid chromatographic systems shown in Table 1, which we established for analysis and purification of BLMs.

Ion Exchange Chromatography

The most useful, and most frequently used chromatographic system for the separation of bleomycins is ion exchange chromatography using CM-Sephadex C-25. This chromatographic method was first applied at the Institute of Microbial Chemistry to the separation of a mixture of phleomycins[1]. We improved this chromatographic technique by the use of a buffered solution for elution. In this case, pH control is very important for reproducibility.

Figure 1 shows the separation of a natural BLM mixture obtained by the use of CM-Sephadex C-25. These BLMs are copper complexes and differ from each other only in the terminal amine moiety[2]. The n at the bottom of Fig. 1 denotes the number of charges at the terminal amine moiety of the individual bleomycins at pH 6.8 and indicates that a major factor in determining the retention time of a given bleomycin on CM-Sephadex C-25 is the amount of positive charge. Therefore, we can predict from the retention time how many charges the solute substance has. In this chromatogram, BLM A_2'-a and -b could not be

Table 1: Liquid Chromatographic Systems for the Analysis and the Preparation of BLMs

Mode	Functional group	Column		Mobile phase	Structure–retention time relationship
		Analytical	Preparative		
Ion exchange	Carboxylic acid	CM-Sephadex		Buffer; $4 \leq pH \leq 9$ citrate, acetate, phosphate, bicarbonate	Clear
	Sulfonic acid	SP-Sephadex		Buffer; $pH < 4$ citrate	
Normal phase	Silica gel	Lichrosorb SI-60	Lichroprep SI-60	Buffer-MeOH-EtOH (MeCN)	Not clear
Reverse phase	Octadesyl silane	μ-Bondapak C18 Lichrosorb RP-18	Bondapak C18 porasil B	Buffer-MeOH (Buffer-MeCN)	Intermediate
	Octyl silane	Lichrosorb RP-8	Lichroprep RP-8		

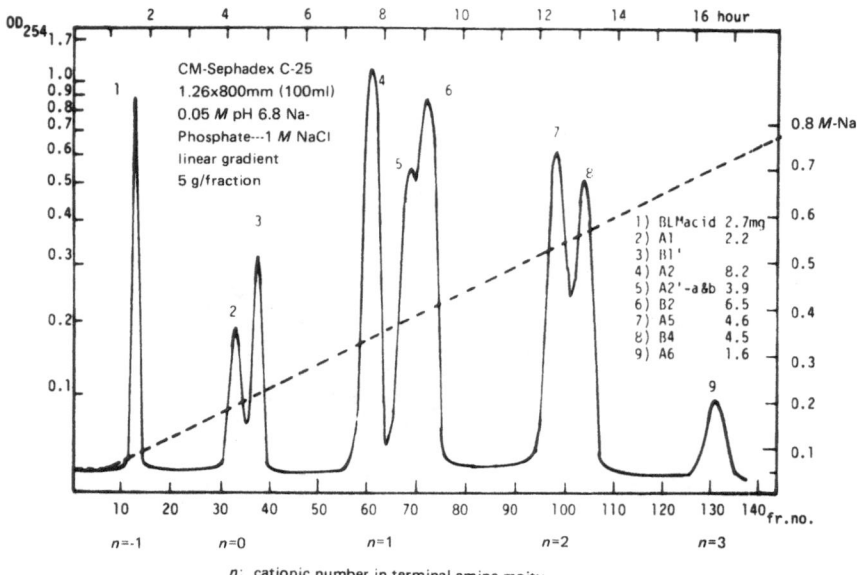

n: cationic number in terminal amine moiety

Fig. 1. CM-Sephadex chromatography of natural BLM.

separated, because the pK_a values of their terminal amines are almost identical[3].

One of the advantages of CM-Sephadex chromatography is that metal-free BLM can be fractionated as well as the copper complex. A metal-free BLM has three measurable basic functions by potentiometric titration in addition to the C-terminal amine substituent. They are the primary amino group (pK_a 7.4), the secondary amino group (pK_a 2.7), and the imidazole (pK_a 4.7). Therefore, the total number of positive charges in whole molecule varies according to the pH of the mobile phase (Table 2). Thus, for the separation of metal-free BLMs,

Table 2: pH Dependence of Number of Plus Charge of Metal-Free and Cu-Chelated BLMs

Metal-free BLMs[a]		Cu-chelated BLMs[b]	
pH	N^c	pH	N
2.7	$2.5+n^d$		
4.7	$1.5+n$	5.5	
6.05	$1\ +n$	∿	$1+n$
7.4	$0.5+n$	9.0	

[a]pK_a: 1, 2.7, 4.7, 7.4.
[b]$pK_a < 3.5$ (×2), 10.9.
[c]N=number of total plus charge.
[d]N=number of plus charge in terminal amine moiety,
    0(DM-A_2, A_1, B_1'), 1(A_2, B_2, A_2'-a and -b, pH≤7.7),
    2(A_5, B_4), 3(A_6, B_5).

the use of a buffer of well-defined pH as the mobile phase is essential. BLM–copper complex has no ionizable group between pK_a 5 and 9 except for the terminal amine; if metal-free BLMs are separated with a pH 6.0 buffer, their retention times are almost the same as those of the corresponding copper complexes. This fact provided important structural information regarding the copper complex of BLM, as described by Dr. Takita.

In the case of metal-free BLM, when the pH of the developing solvent is lowered, the number of positive charges is increased, as in the retention time. When the pH of the solvent is raised, the number of positive charges is decreased and the retention time is shortened. In the case of the copper complexes of BLMs such as BLM B_2, which have a strongly basic group at the C-terminal, the retention time is almost independent of the pH of the developing solvent between 6 and 8. Thus, Takahashi et al. succeeded in separating BLM A_2'-c copper complex and metal-free BLM A_2'-c in cell homogenates using a pH 7.7 phosphate buffer as the developing solvent[4].

Another interesting example is the separation of BLM, iso-BLM, and epi-BLM. Iso-BLM is formed when metal-free BLM is maintained in weakly alkaline aqueous solution[5], while epi-BLM is formed when the copper complex of BLM is maintained under the same conditions[6]. Iso-BLM is different from BLM only in the position of the O-carbamoyl group, while epi-BLM is different from BLM only in the stereochemistry at the α-methine carbon of the pyrimidine substituent (Fig. 2). Therefore, separation of these three compounds would seem potentially to be very difficult. In fact, a mixture of these three compounds in metal-free form could not be separated by CM-Sephadex chromatography. However, a mixture of these compounds as the respective copper complexes

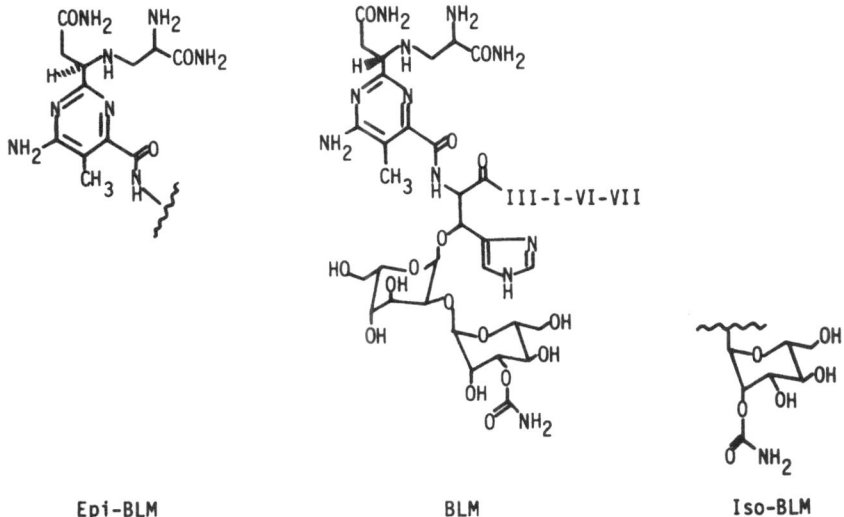

Fig. 2. Partial structures of BLM, epi-BLM, and iso-BLM.

could be separated by CM-Sephadex column chromatography, as shown in Fig. 3. Effective separation was achieved by using a pH 4.5 buffer as the developing solvent. When a pH 6.0 buffer was used, the attempted separation was not successful. This gave additional useful information regarding the structures of the metal complexes of these compounds, because their retention times reflected the ease of protonation of the ligand atoms in these three compounds.

When a developing solvent with a pH lower than 4 is required. SP-Sephadex C-25, having sulfonic acid functional groups, should be used. BLM has two basic groups (pK_a 7.4 and 2.7) in the pyrimidine substituent, while its depyruvamide derivative obtained by refluxing in 10% acetic acid[7] has only one basic group (pK_a 7.6). Therefore, it was anticipated that when an acidic buffer solution was used as the developing solvent, the separation of the two compounds would be possible. In fact, BLM and its depyruvamide derivative were separated on SP-Sephadex C-25 using a pH 2.5 sodium citrate buffer as the developing solvent (Fig. 4).

A disadvantage of chromatographic analysis by CM-Sephadex is the time required for assay. For the separation of a mixture of natural BLMs, 1 day or more is needed because high pressure cannot be applied to the soft matrix employed. Cation exchangers used in modern HPLC are sulfonated polystyrene resins or modified silica gel[8]. Even without the functional groups these matrices strongly absorbed BLM. Therefore, elution of a BLM from these exchangers is very difficult when the net charge on that BLM is more than two. Thus, HPLC analysis of BLM must be done by some other mode of separation.

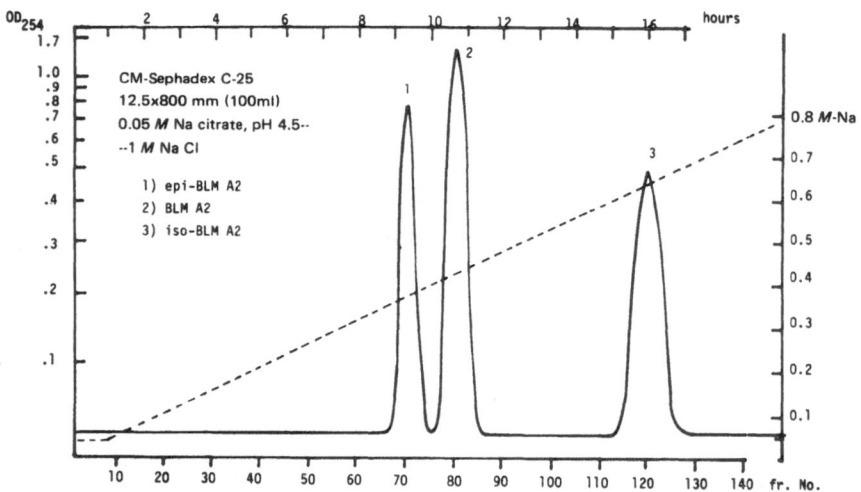

Fig. 3. Separation of copper complexes of BLM A_2, epi-BLM A_2, and iso-BLM A_2.

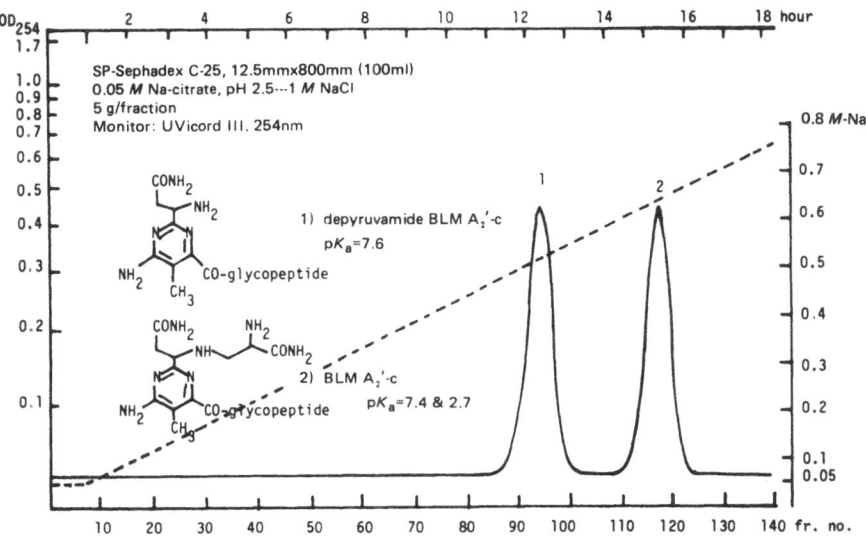

Fig. 4. Separation of BLM A_2'-C and depyruvamide BLM A_2'-C.

Normal Phase Chromatography

High-performance normal phase chromatography can be used as an extension of thin-layer chromatography (tlc) on silica gel. This technique has been employed in chemical studies of BLM. In tlc, the mobile phase consists of an equal volume of methanol and 10% aqueous ammonium acetate. In special cases, instead of ammonium acetate solution, ammonium acetate–ammonia buffer was used to obtain a satisfactory separation.

For HPLC, the following modifications were needed[6]: (i) It is necessary to fix the conformation of BLM by copper complex formation. BLM and epi-BLM cannot be separated conveniently in their metal-free forms (Fig. 5), (ii) to obtain sharp separation, the chromatographic column should be warmed to 60°C, and (iii) to adjust the retention time, a third solvent such as ethanol[6] or acetonitrile is added. Retention time is controlled by selection of the ratio of methanol and acetonitrile (Fig. 6). Addition of acetic acid or ammonia to this solution affects the retention time of some specific BLMs. For example, by the addition of ammonia, the retention time of BLMs with primary amino groups at the C-terminal, such as A_5, A_2'-a and -b, is increased, and by addition of acetic acid the retention time of bleomycinic acid is increased.

All epi-BLMs can be separated from their corresponding BLMs by silica gel chromatography. Therefore, successful kinetic study of the epimerization[6] could be achieved by the use of this system. The time required for this analysis was only 5 to 10 min, one-fiftieth of that of CM-Sephadex chromatography. It should be noted, however, that the relationship between structure and retention

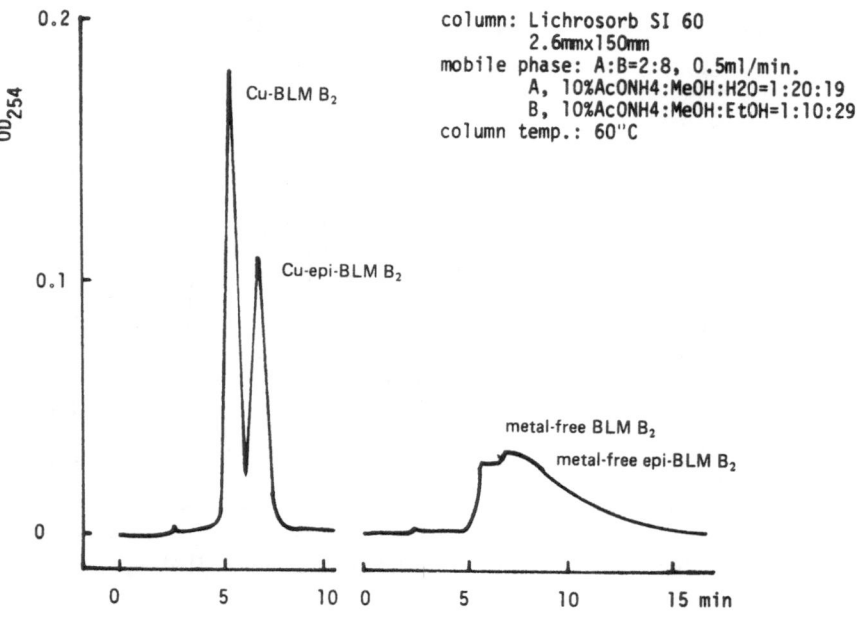

Fig. 5. Normal phase separation of BLM B_2 and epi-BLM B_2.

time in silica gel chromatography is generally very complicated, so it is recommended that reverse-phase chromatography be used first and thereafter silica gel chromatography applied if the reverse-phase chromatography does not give satisfactory results. Preparative scale application of this system using a commercially available, prepacked column has been described[7].

Reverse Phase Chromatography

Recently, reverse-phase chromatography has become popular for the analysis of water-soluble compounds. BLMs can also be successfully separated on such columns (e.g., metal-free bleomycins can be separated on μ-Bondapak C-18) (Fig. 7). The resolution of metal-free BLMs, however, was not as good as that obtained using bleomycin copper complexes. In reverse-phase chromatography, if the capacity factor of BLM is greater than 1, the elution profile suffers severe tailing, in some cases giving two peaks, even though a single substance is being analyzed. Therefore, copper complex formation of BLM is recommended for HPLC analysis.

The relationship between the retention times of various BLMs on a μ-Bondapak C-18 column and the constitution of the mobile phase is summarized in three figures (Figs. 8–10). Figure 8 shows the relationship between the retention times and the pH value of the buffer solution used. Retention time is increased by a basic group in the terminal amine moiety if the pH of the buffer is increased. On the other hand, the retention times of BLMs which have no basic group at

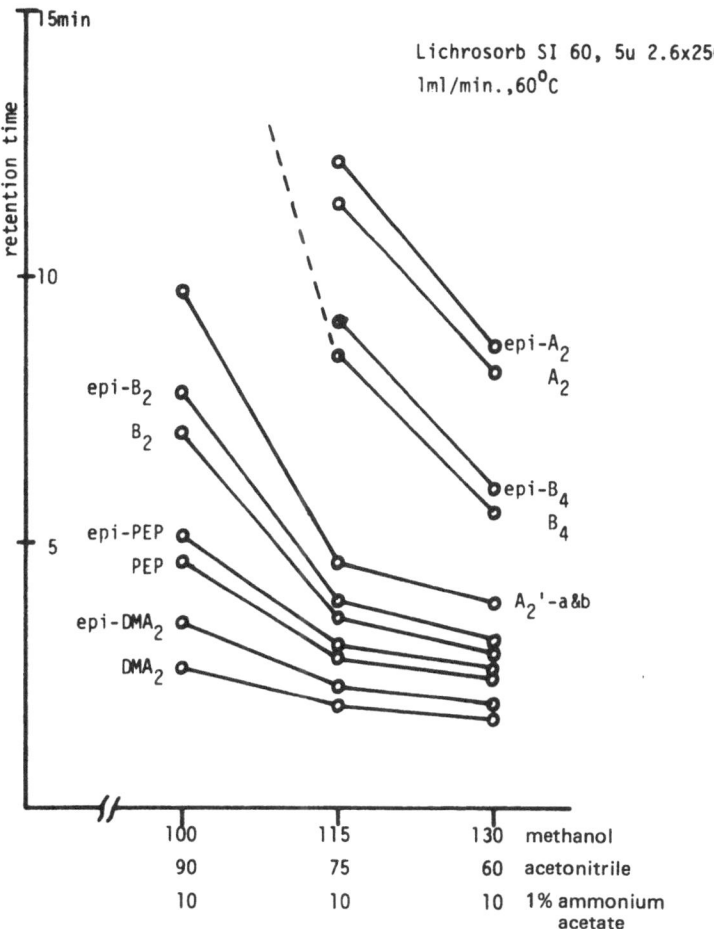

Fig. 6. Retention time versus solvent composition.

the C-terminus, such as bleomycinic acid, BLM B_1', and demethyl A_2, are almost independent of the pH of the buffer employed. In this fashion, the retention times of BLMs can be controlled by pH change of the buffer solution. The effect of the concentration of the buffer solution on the retention time is shown in Fig. 9, and the effect of methanol content is shown in Fig. 10. Generally, when the concentration of the buffer or content of methanol is increased, the retention time is shortened. Thus, the retention time can be controlled by pH, by the concentration of the buffer, and by methanol content. Our standard chromatogram of natural BLMs is shown in Fig. 11. If we fail to obtain satisfactory results under these conditions, the mobile phase is modified by change of pH or buffer concentration and methanol content, with reference to Figs. 8–10.

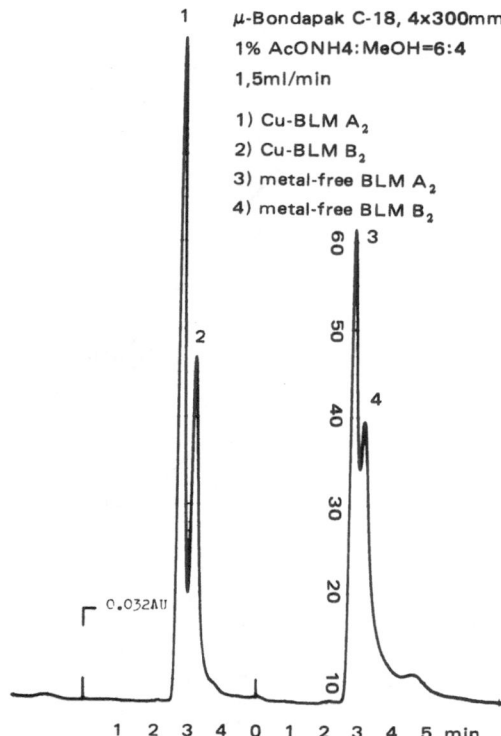

Fig. 7. Comparison of metal-free form and copper complex.

Large scale separation of natural BLMs is possible by CM-Sephadex chromatography with the exception of BLM A_2'-a and -b, which differ only in the length of the alkyl chain in the terminal amine moiety[2]. The terminal amine of the former is 1,4-diaminobutane and that of the latter is 1,3-diaminopropane. The pK_a values of the terminal amines are almost identical (~9.7). Thus, separation of these species by CM-Sephadex column is very difficult, though it can be achieved by repeated CM-Sephadex chromatography[3].

BLM A_2'-a and -b were effectively separated by reverse-phase chromatography on μ-Bondapak C-18, as shown in Fig. 12 (right). Thus, separation of A_2'-a and -b on a large scale was achieved using a Lobar column packed with Lichroprep RP-8 (size B), obtained from E. Merck Co., West Germany (Fig. 12). To obtain better resolution, the methanol content was decreased to 12.5%. One hundred milligrams of a mixture of BLM A_2'-a and -b was separated with a 25-mm-diameter and 310-mm-length column within 7 hr, giving 40 mg of A_2'-a (97% purity) and about 40 mg of pure A_2'-b. Contamination of A_2'-b in the A_2'-a fraction was due to tailing of the former. This separation was superior to that obtained by CM-Sephadex chromatography.

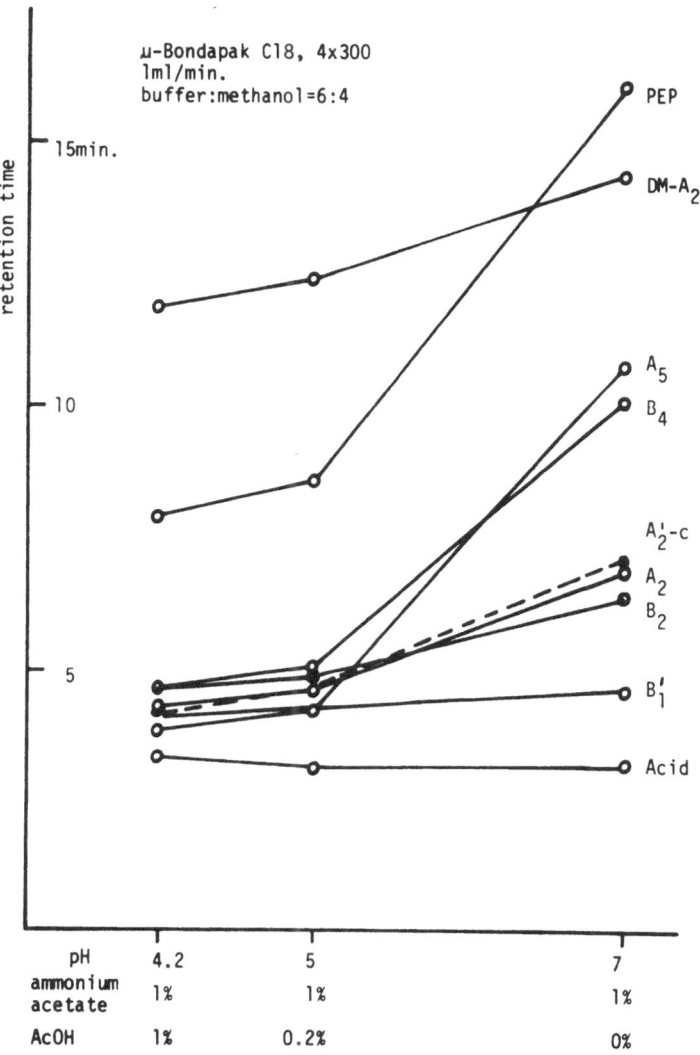

Fig. 8. Retention time versus pH.

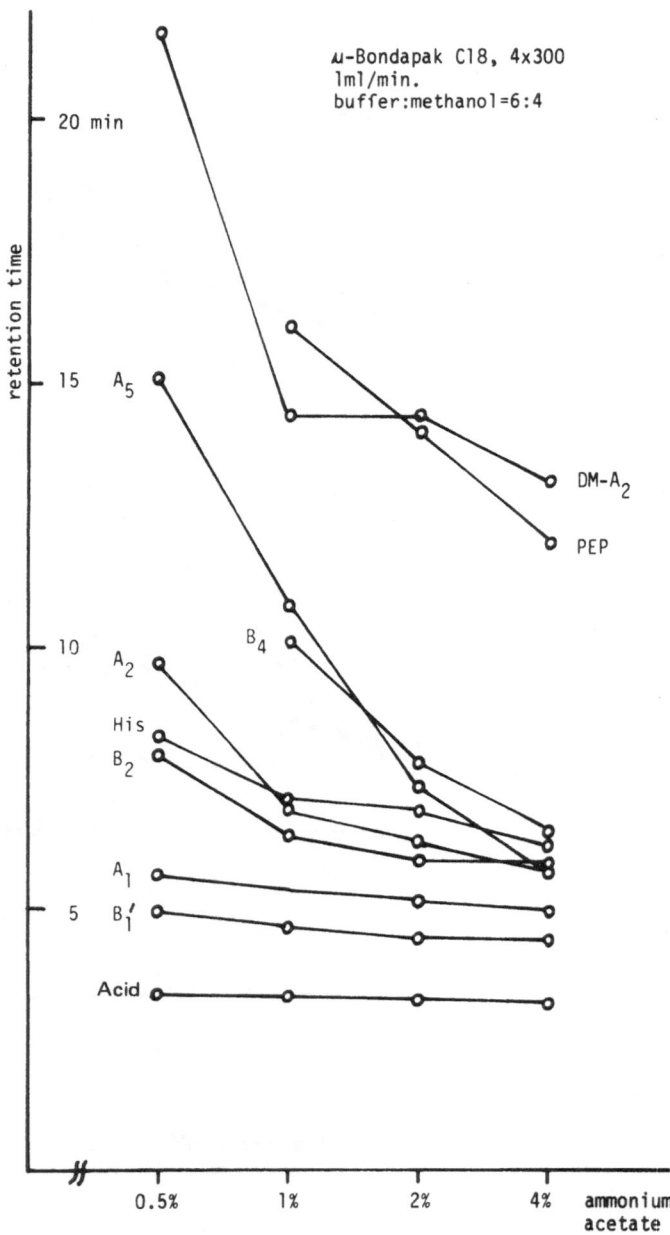

Fig. 9. Retention time versus concentration of ammonium acetate.

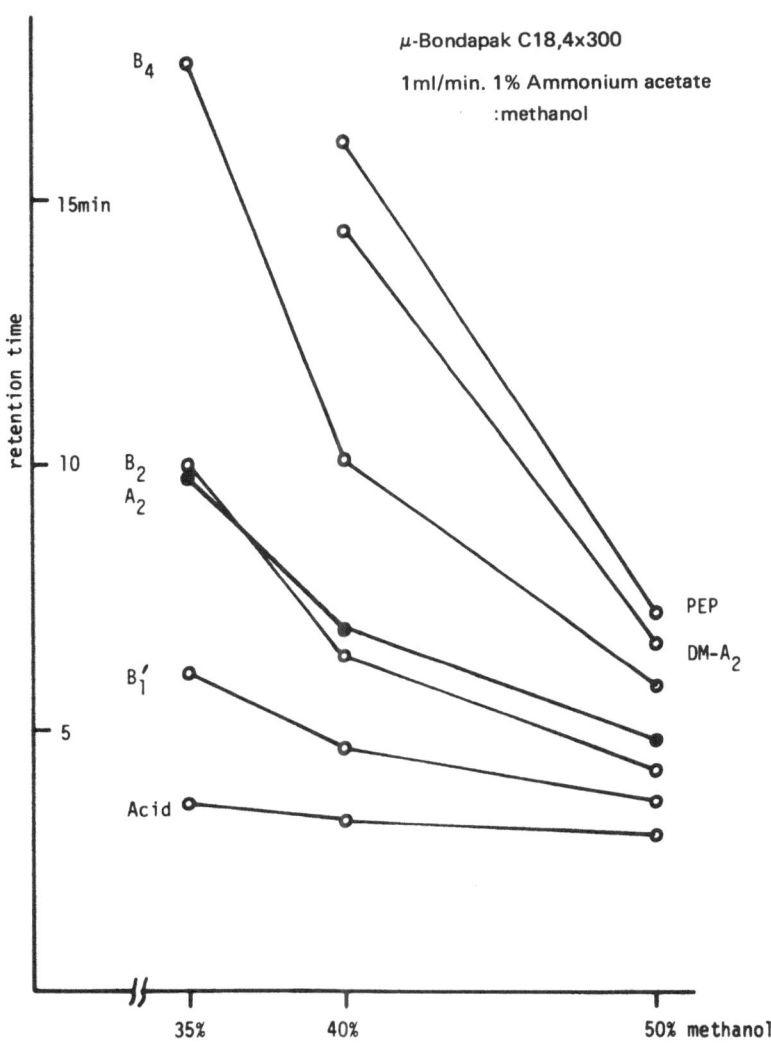

Fig. 10. Retention time versus methanol content.

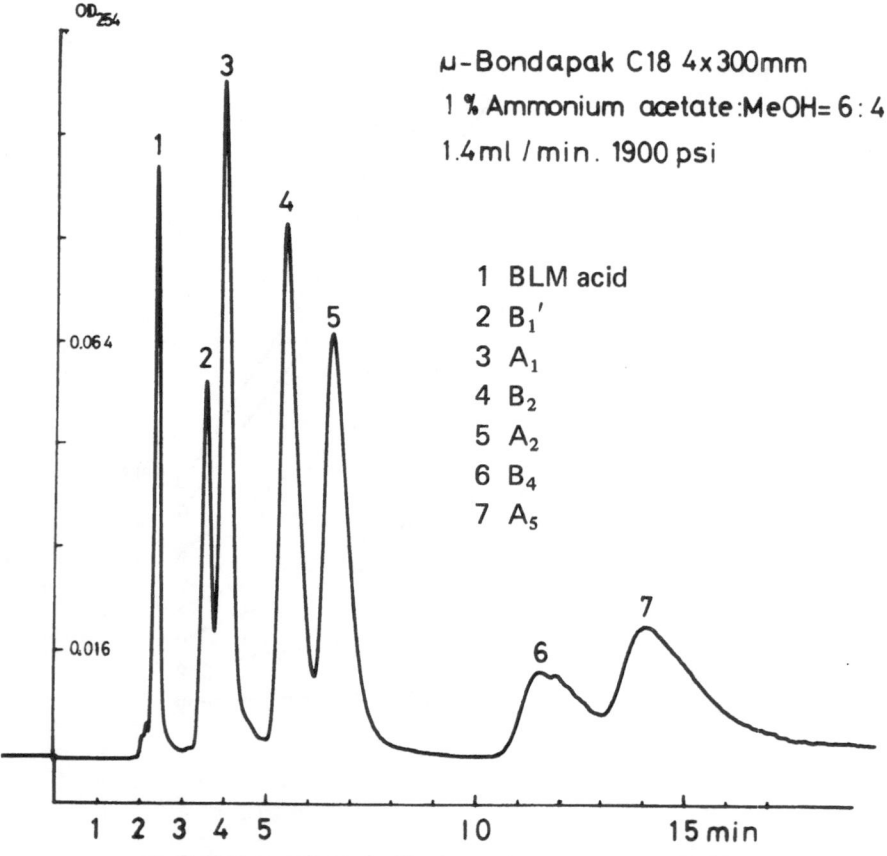

Fig. 11. Standard chromatogram of natural Cu-BLMs.

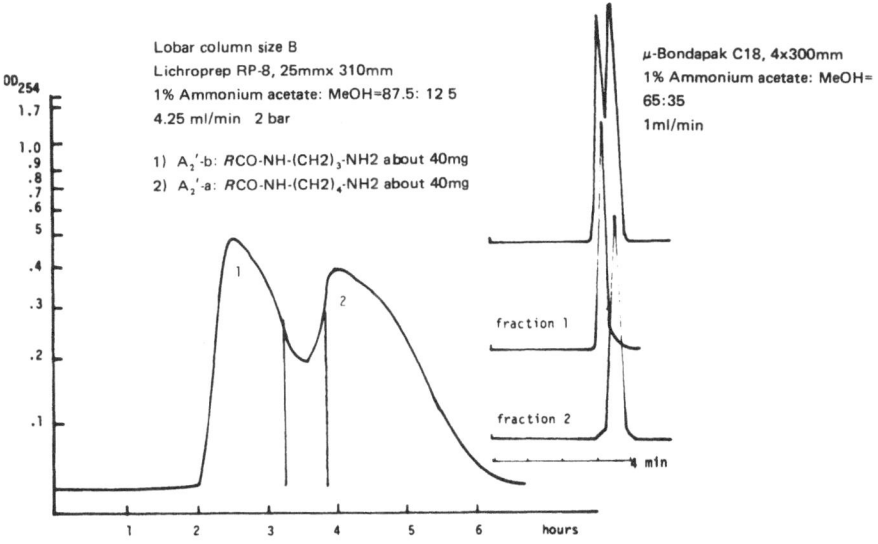

Fig. 12. Preparative separation of BLM A_2'-a and -b by reverse-phase chromatography.

References

1. T. Ikekawa, F. Iwami, H. Hironaka, and H. Umezawa, *J. Antibiot. (Tokyo)*, Ser. A **17**, 194 (1964).
2. A. Fujii, T. Takita, K. Maeda, and H. Umezawa, *J. Antibiot. (Tokyo)*, **26**, 398 (1973).
3. A. Fujii, T. Takita, K. Maeda, and H. Umezawa, *J. Antibiot. (Tokyo)*, **26**, 396 (1973).
4. K. Takahashi, O. Yoshioka, A. Matsuda, and H. Umezawa, *J. Antibiot. (Tokyo)*, **30**, 861 (1977).
5. Y. Nakayama, M. Kunishima, S. Omoto, T. Takita, and H. Umezawa, *J. Antibiot. (Tokyo)*, **26**, 400 (1973).
6. Y. Muraoka, H. Kobayashi, A. Fujii, M. Kunishima, T. Fujii, Y. Nakayama T. Takita, and H. Umezawa, *J. Antibiot. (Tokyo)*, **29**, 853 (1976).
7. Y. Muraoka, A. Fujii, T. Yoshioka, T. Takita, and H. Umezawa, *J. Antibiot. (Tokyo)*, **30**, 178 (1977).
8. R. E. Majors, *J. Chromatogr. Sci.*, **15**, 334 (1977).

NMR Study of Bleomycin

Hiroshi Naganawa

During chemical studies of bleomycin (BLM), many fragments and derivatives have been isolated and characterized. The NMR spectra of these compounds were taken and analyzed for structural information.

The ¹H-NMR of BLM and its derivatives are generally complicated with many overlapping signals; to analyze the ¹H-NMR spectra, spin decoupling studies were essential. The ¹³C-NMR spectra were taken as proton noise decoupled, off-resonance, or selective proton decoupled spectra. This chapter presents a few of our experimental results, which contributed substantially to the structural study of BLM.

Conditions Utilized for NMR Measurements

BLM and its derivatives and degradation products are generally soluble in water, methanol, and dimethylsulfoxide. Therefore, almost all of the spectra were taken in D_2O solution. The aqueous solutions of the samples were adjusted to pH 6 before dissolving them in D_2O. In the ¹³C-NMR spectrum, at ∼ pH 6.0, the overlap of signals was minimized and the chemical shifts of the signals were the least pH sensitive. The ¹H-NMR spectra were generally taken in CW-mode with a Varian HA-100D or XL-100 spectrometer. When samples were available only in small quantities or when it was important to remove the water signal, the spectra were taken in FT-mode. The ¹H-chemical shift values are expressed relative to external $(CH_3)_4Si$. All of the proton noise decoupled ¹³C-NMR spectra were taken in FT-mode. Pulse Fourier transform parameters for ¹³C-observation are shown in Table 1. The ¹³C-NMR chemical shift value is expressed relative to internal dioxane, the signal from which is assigned a value of δ67.40.

Last year a paper giving the ¹³C-NMR assignments of BLM was published[1]. The chemical shift values in that paper are slightly different from the present values due to adjustment of the data processing system. The error is only 0.2 ppm at 100 ppm downfield from internal dioxane. Therefore, the chemical shift map could be used without correction.

¹H- and ¹³C-NMR Spectra of BLM

The ¹H- and ¹³C-NMR spectra of BLM are shown in Fig. 1. The 100-MHz ¹H-NMR spectrum and ¹³C-NMR spectrum were taken in the same magnetic field.

Table 1:   Conditions of NMR Measurements

D$_2$O solution	(pH6)

Ambient temperature (~30°C)
Spectrometer: Varian HA 100-D
                        XL-100-DISK FT system
PFT parameter for ^{13}C observation:
                Sweep width: 5120 Hz; acquisition time: 0.8–1.6 sec.
                Flip angle: 30–45°; number of transients: 1–8 × 10^4
Reference: ^{1}H: External TMS   δ=0
                ^{13}C: Internal dioxane δ=67.40

In the ^{13}C-NMR, each signal was well separated except for two signals, while the ^{1}H signals between δ3.5 and 4.7 could not be resolved. Even in the 300-MHz ^{1}H-NMR spectrum, which was kindly taken in the Application Laboratory of Varian Associates at Palo Alto, the signals between δ4.0 and 4.7, which included most of the sugar moiety, could not be resolved. In the structural study of small degradation products, ^{1}H-NMR spectrometric analysis was one of the most powerful tools, because analysis of spin–spin coupling indicated not only primary structure but also configuration and conformation.

The Structure of the Sugar Moiety of Bleomycin A$_2$

Peracetylated disaccharide was obtained from bleomycin A$_2$ under strictly defined conditions of acid hydrolysis, namely 0.3 M H$_2$SO$_4$ at 80°C for 6 hr, followed by acetylation. A part of the ^{1}H-NMR spectrum of the peracetylated disaccharide is shown in Fig. 2. From the results of analysis of spin decoupling experiments, it was determined that the mannopyranose moiety was connected to the 2-O-position of gulopyranose by a glycosidic linkage.

The lowest field signal at δ5.9 was assumed to correspond to the anomeric proton of gulose. The coupling constant was 8.5 Hz, which indicated that this peracetate, the major anomer of the hydrolyzate, was the β-anomer. Irradiation at δ5.9 indicated that the C-2 proton of gulose resonated at δ4.0. The signal at δ4.0 was the highest methine resonance. Irradiation at δ4.0 showed that the C-3 proton of gulose resonated at δ5.45. Irradiation at δ5.45 caused the C-2 proton signal to collapse to a doublet; the coupling constant was 8.5 Hz. Double irradiation at δ5.90 and 5.45 caused coalescence of the C-2 proton signal to a singlet.

From a consideration of chemical shifts, coupling constants, and spin coupling patterns, the C-1, C-2, and C-3 proton signals were definitely assigned to δ5.90, 4.00, and 5.45, respectively. Appearance of the C-2 proton signals at the highest field indicated that the C-2 hydroxyl group of gulose was involved in a glycosidic linkage.

The configuration of the glycopeptide linkage between the disaccharide and the peptide was also determined by ^{1}H-NMR analysis. A part of the ^{1}H-NMR spectrum of BLM-A$_2$, which was taken by FT-mode to eliminate the water

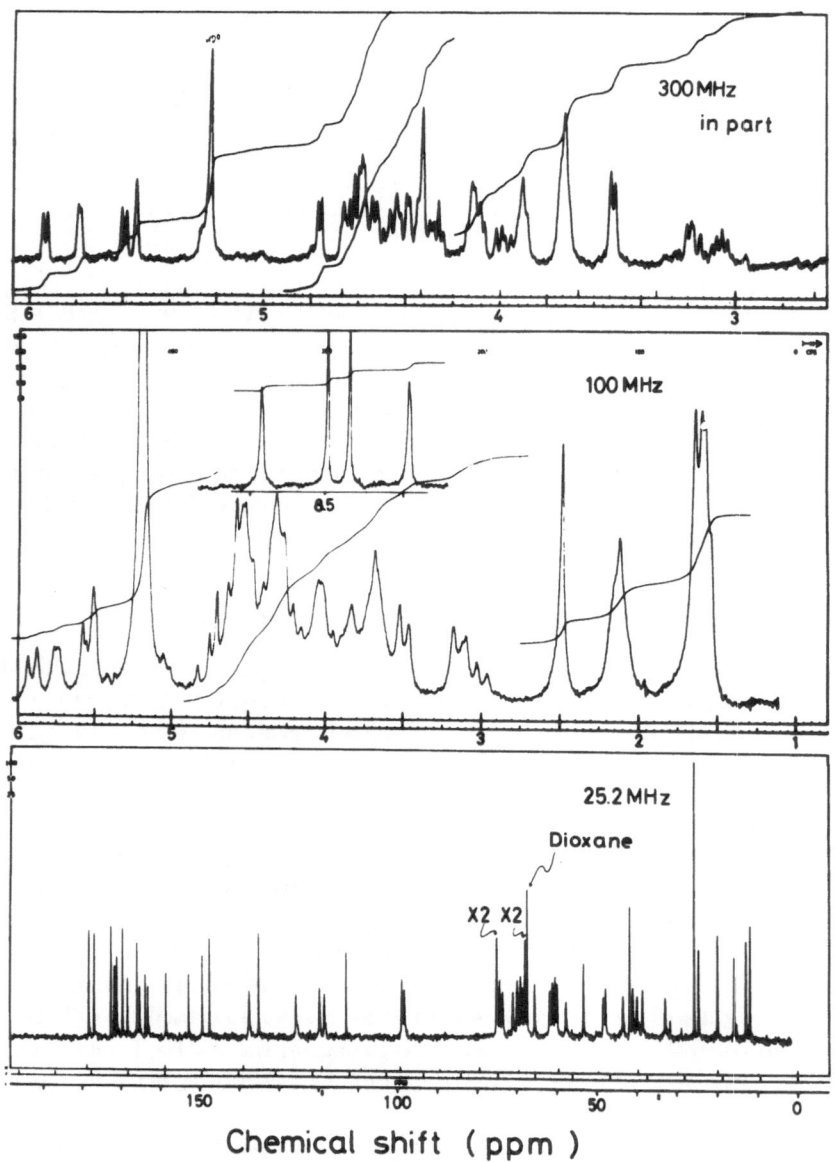

Fig. 1. ^{1}H and ^{13}C-NMR spectra of bleomycin.

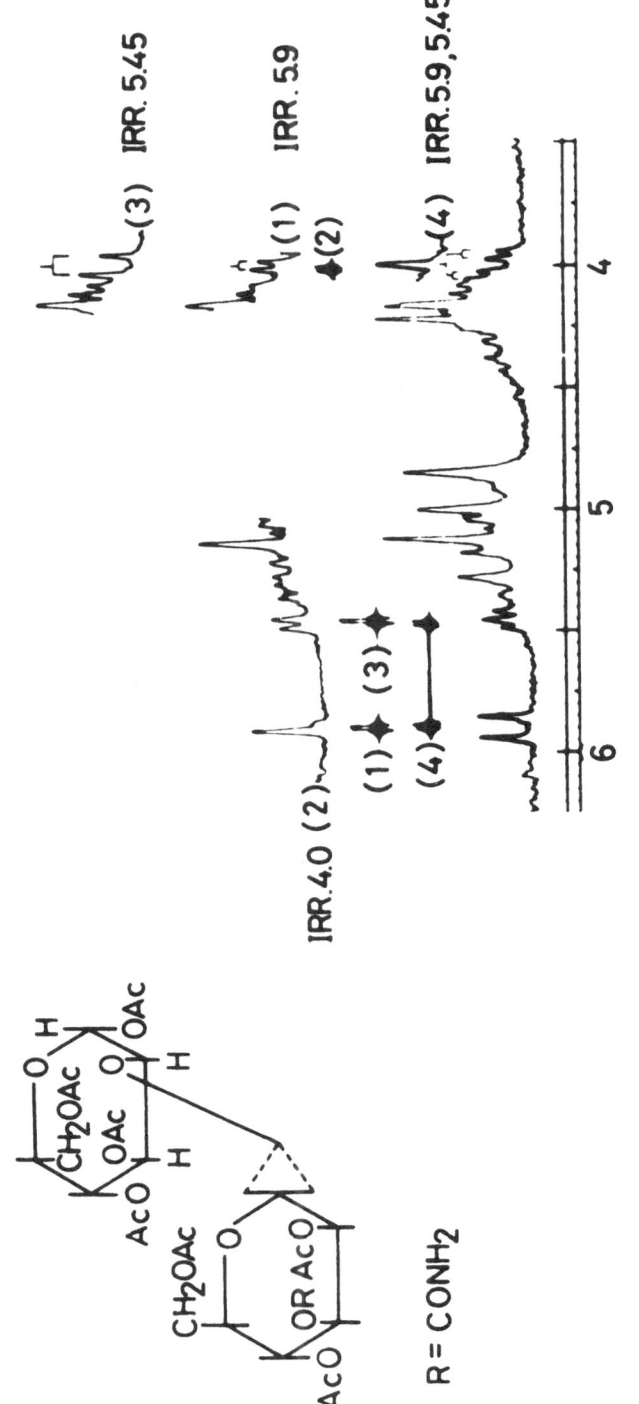

Fig. 2. Partial ^{1}H-NMR spectrum of peracetylated disaccharide (CDCl$_3$, TMS).

signal, is shown in Fig. 3. The spectrum was taken at pH 3.8 and 55°C. There are five proton signals in the region from δ5 to 6. The signal at δ5.96 was sensitive to pH change. It can be assigned to the β-methine proton of β-hydroxyhistidine. This proton coupled with the protons resonating at δ5.6. Therefore, the latter signal is assigned to the α-methine proton of β-hydroxyhistidine. The other three signals should be the anomeric protons of gulose and 3-O-carbamoyl mannose and the C-3 proton of 3-O-carbamoyl mannose. The coupling constant of the anomeric proton of α-D-mannopyranoside is characteristically small. Thus, the signal at δ5.55 is assigned to the anomeric proton of the 3-O-carbamoyl mannose. The C-3 proton of 3-O-carbamoyl mannose is axial, the C-4 proton is axial, and the C-2 proton is equatorial. Therefore, the coupling pattern of the C-3 proton of the mannose was expected to be doublet of doublets with large and small coupling constants. The signal at δ5.3 was assigned to the C-3 proton of mannose on the basis of the following decoupling experiments. Irradiation of the signal at δ4.67, which was assigned to C-2 of mannose because it resulted in sharpening of the C-1 proton signal of mannose, indicated coupling to the signal at δ5.3. Irradiation at δ4.38 caused the signal at δ5.3 to collapse to a doublet of small coupling constant.

The remaining signal at δ5.75 can be assigned to the anomeric proton of gulose. It is a broad doublet. This splitting pattern contains about 3.5 Hz and long range spin–spin coupling; it indicates that the anomeric proton is equatorial. The configuration of the glycosidic linkage of carbamoyl mannose was also determined to be α by application of Hudson's rule[2]. The coupling constant between ^{13}C and ^{1}H of the anomeric carbon was observed by using a gated decoupling method. The observed coupling constants of the gulose and mannose moieties were 170.1 and 172.9 Hz, respectively. These results support the assignment of α-configurations to both glycosidic linkages[3]. Thus, the sugar moiety of bleomycin A_2 is 2-O-(3-O-carbamoyl-α-D-mannopyranosyl)-α-L-gulopyranose.

The ^{1}H-NMR Spectrum of Deglyco-Bleomycinic Acid (P-6mo)

As described above, even in the 300-MHz ^{1}H-NMR spectrum, the signals in the δ4.0 to 4.7 region could not be resolved due to overalpping of the sugar and peptide signals. There is an interesting peptide designated as P-6mo, which was isolated during the biosynthetic study of BLM. P-6mo is deglyco-bleomycinic acid, that is, a peptide lacking the sugar moiety and the terminal amine of BLM. The ^{1}H-NMR spectra of this compound were taken at 100 and 300 MHz, the latter of which was also recorded by Dr. Schoolery of Varian Associates at Palo Alto. All of the signals in the 100-MHz spectrum were assigned by spin decoupling experiments and with reference to the accumulated data from the spectra of many smaller peptide fragments.

The 300-MHz spectrum was well-resolved, which also served to verify the assignments that we made on the basis of our 100-MHz spectrum. A part of the

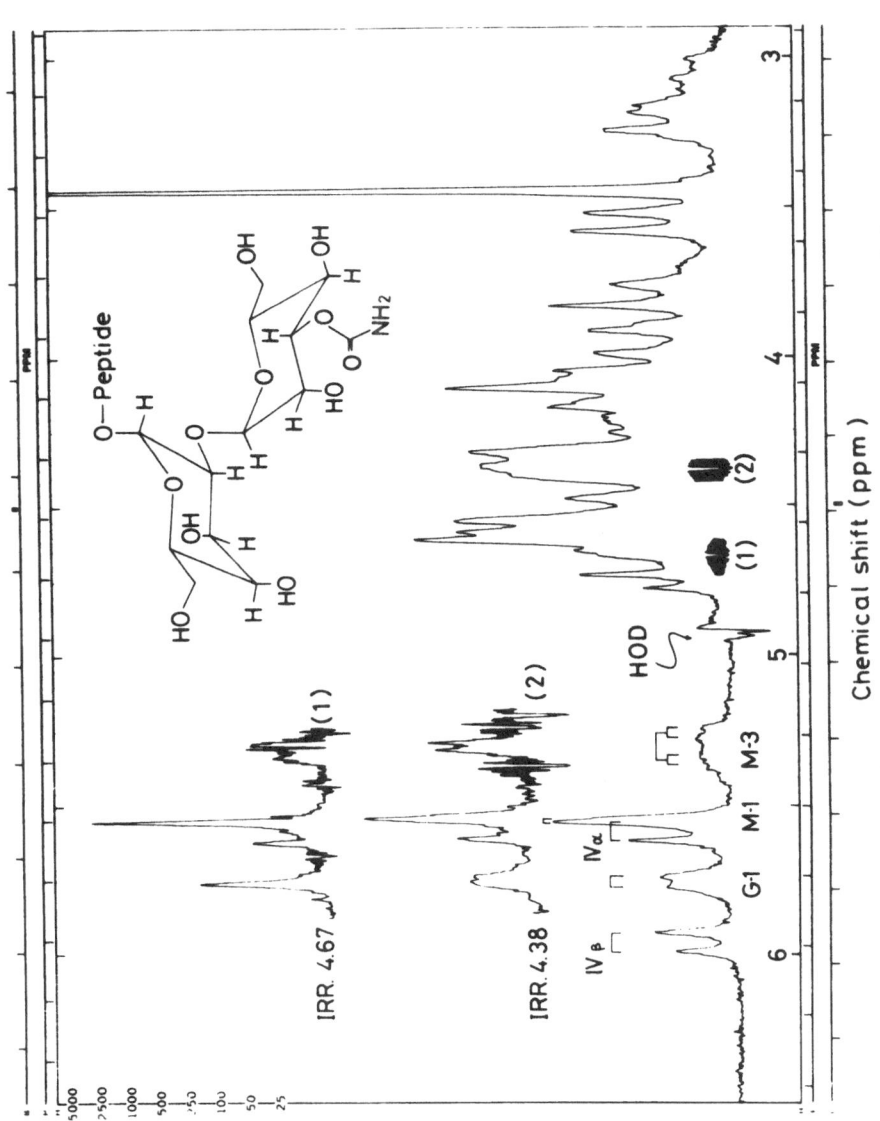

Fig. 3. Partial ^{1}H-NMR spectrum of bleomycin A$_2$ (pH 3.8, 55°C).

spectrum, the δ3 to 5 region, is shown in Fig. 4. On the basis of the 270-MHz spectra of the clinically used bleomycin mixture, and with reference to assignments made at 100 MHz[2, 4-8], Chen *et al.*[9] proposed assignments for the bleomycin proton resonances; most of their assignments were the same as ours.

The ^{1}H-NMR chemical shifts of peptide fragments were assigned by application of spin decoupling experiments. The results are shown in Table 2. These four peptides, tripeptide S, tetrapeptide S, pseudotripeptide and pseudotetrapeptide A, were described by Dr. Takita, and P-6mo and deglyco-BLM acid were described by Dr. Fujii. This ^{1}H-NMR chemical shift table was useful for the assignment of the ^{13}C-NMR spectrum.

^{13}C-NMR Studies

The ^{13}C-NMR spectroscopic study of BLM has played an important role in the final stage of determination of the total structure of BLM. That is, the ^{13}C-NMR

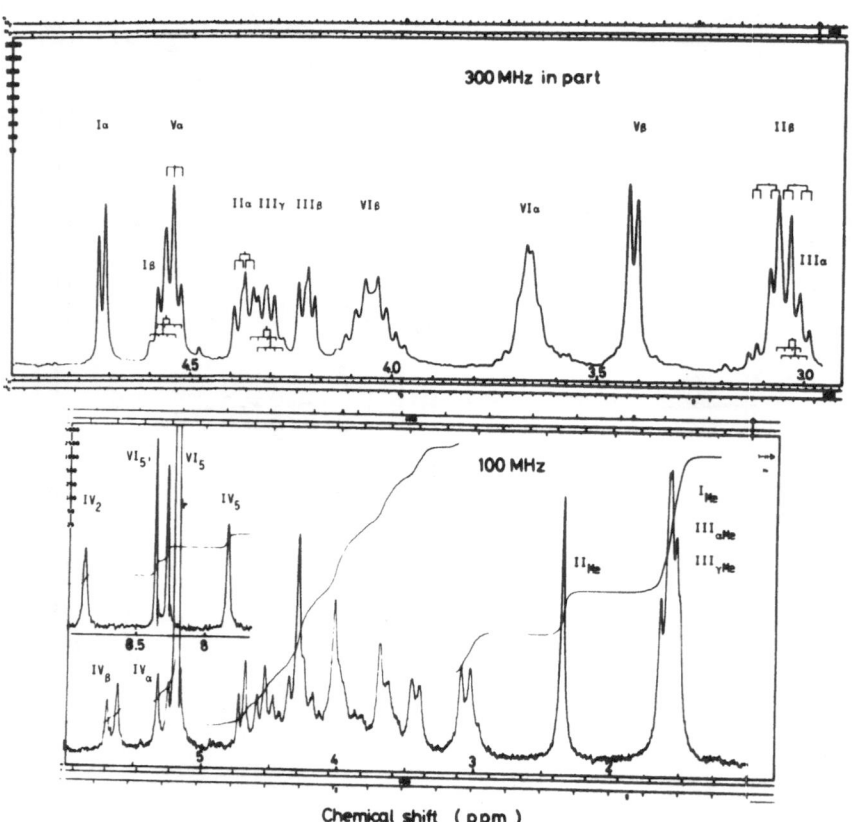

Fig. 4. ^{1}H-NMR spectra of deglyco-bleomycinic acid (P-6mo).

Table 2: Assignments of Resonances of Peptide Fragments[a]

	Tripeptide S δ (J, Hz)	Tetrapeptide S δ (J, Hz)	Pseudotripeptide δ (J, Hz)	Pseudotetrapeptide A δ (J, Hz)	P-6mo[b] δ (J, Hz)	
I	4.24 (d, 5)	4.64 (d, 5)			4.72 (d, 5)	αCH
	4.52 (d, q, 5, 6.5)	4.49 (d, q, 5, 6.5)			4.56 (m, 5, 6.5)	βCH
	1.63 (d, 6.5)	1.53 (d, 6.5)			1.58 (d, 6.5)	CH$_3$
II			4.88 (d, d, 6, 8)	4.87 (t, 6)	4.37 (d, d, 6, 8)	αCH
			3.30[c]	3.29[c]	3.06[c]	βCH$_2$
			2.52 (s)	2.48 (s)	2.37 (s)	5CH$_3$
III		3.08 (q, d, 7, 10)		2.86 (q, d, 6.5, 6)	3.03 (q, d, 6.5, 5)	αCH
		1.72 (d, 7)		1.51 (d, 6.5)	1.63 (d, 6.5)	αCH$_3$
		4.31 (d, d, 10, 2.5)		4.17 (t, 6)	4.21 (d, d, 5, 7)	βCH
		~3.8		4.39 (m, 6, 6.5)	4.31 (m, 6.5, 7)	γCH
		1.72 (d, 7)		1.57 (d, 6.5)	1.55 (d, 6.5)	γCH$_3$
IV			5.19 (d, 6)	5.26 (d, 7.8)	5.30 (d, 7.5)	αCH
			5.74 (d, 6)	5.73 (d, 7.8)	5.66 (d, 7.5)	βCH
			9.05 (br, ~1)	9.05 (br, ~1)	8.81 (d, ~1)	2H
			7.82 (br, ~1)	7.90 (br, ~1)	7.82 (d, ~1)	5H
V			4.55 (t, 6.5)	4.55 (t, 6.5)	4.54 (t, 7)	αCH
			3.93 (d, 6.5)	3.91 (d, 6.5)	3.42 (d, 7)	βCH$_2$
VI	8.43 (s)	8.47 (s)			8.32 (s)	5H
	8.54 (s)	8.58 (s)			8.39 (s)	5'H
Terminal amine	3.7 (m)	3.76 (m)			3.67 (m)	αCH$_2$
	~4 (m)	~4.1 (m)			4.05 (m)	βCH$_2$
	4.01 (t, 7)	4.02 (t, 6)				αCH$_2$
	2.61 (m)	2.62 (m)				βCH$_2$
	3.86 (t, 7)	3.86 (t, 7)				γCH$_2$
	3.40 (s)	3.39 (s)				S(CH$_3$)$_2$

[a] 100 MHz in D$_2$O, ppm from ext. TMS.
[b] Analysis was aided by spin decoupling experiments at 300 MHz.
[c] Chemical shifts are expressed at the center of AB portion of ABX pattern.

spectrum indicated the exact number of carbons present in the BLM molecule and gave information about their bonding characteristics.

A collection of ^{13}C-NMR spectra of well-characterized, related compounds has been analyzed in terms of their structural relationship with reference to known chemical shift data and established chemical shift rules, and each signal has been assigned to specific carbon atoms in the BLM molecule. Some ambiguities were resolved with the aid of selective proton decoupling experiments.

The ^{13}C-NMR spectra of BLM A_2'-c, of which the terminal amine is histamine, is shown in Fig. 5. The lower spectrum is the noise decoupled spectrum. There are 53 signals; only two sets of signals were coincident. There are 24 sp^2-carbons between δ110 and 180. Two signals around δ100 are assigned to the anomeric carbons. Between δ10 and 80 there are 29 sp^3-carbon signals. The off-resonance spectrum showed that there were four C–CH$_3$ signals at δ10–20 which appeared as quartets, eight methylene signals (marked with "t") which appeared as triplets, and 17 other signals corresponding to methine carbons which appeared as doublets. There are no tertiary carbons in the sp^3-region. In the sp^2-region, there were six doublet signals in the off-resonance spectrum which are marked with a "d". The others were singlets.

In the upper spectrum, an example of a selective proton decoupling experiment is shown for the assignment of the anomeric carbons. The selective low-power irradiation of the mannose anomeric proton at δ5.47 caused the anomeric carbon signal at δ99.8 to coalesce. In the same way, irradiation of the anomeric proton of gulose at δ5.77 gave the δ98.2 signal as a sharp singlet. Consequently, the anomeric carbon signals could be assigned definitively.

^{13}C-Chemical Shifts of the Sugar Moiety

Assignment of proton signals to the sugar moiety could not be achieved, even from the 300-MHz spectrum. However, assignment of the ^{13}C signals was achieved from the 25.2-MHz spectrum by the following analysis. In order to assign signals to the mannose moiety (Table 3), the spectra of methyl-α-D-mannopyranoside and its 3-O- and 2-O-carbamoyl derivatives were used as reference compounds. These three monosaccharide derivatives correspond to the mannose moieties of decarbamoyl-BLM, BLM, and iso-BLM, respectively. By reference to the carbamoyl substitution effects, the six carbon signals of the mannose moiety of BLM were assigned definitely.

To facilitate assignment of the carbon signals of the gulose moiety, the chemical shift of methyl-α-L-galactopyranoside, which was assigned by Perlin and others[10], was used as a standard. Methyl-α-L-gulopyranoside is different from the galactoside only in its stereochemistry at C-3; the effect of such an inversion on the ^{13}C-chemical shift was also studied in detail by Perlin. The interactions and expected shieldings according to his rules are shown in Table 4. As a result, the chemical shift of methyl-α-L-gulopyranoside can be assigned as shown in

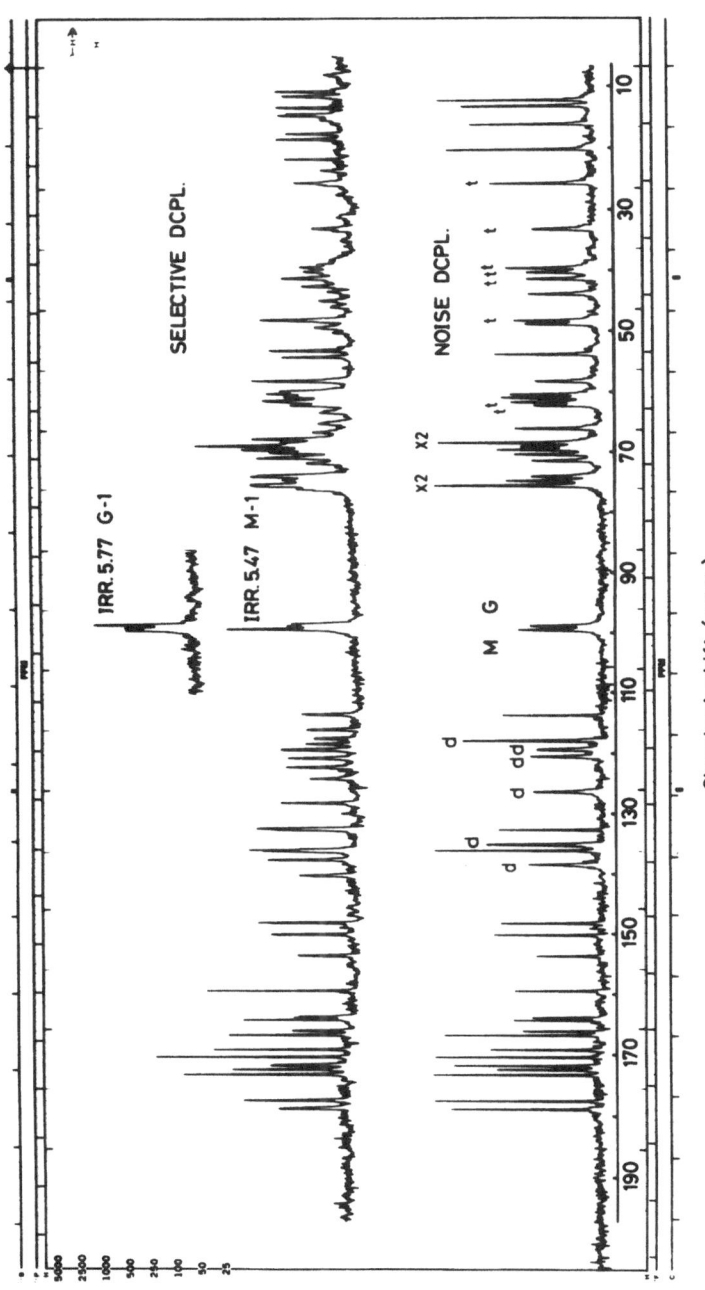

Chemical shift (ppm)

Fig. 5. ^{13}C-NMR spectra of bleomycin A_2'-c.

Table 3: ^{13}C-Chemical Shifts for Methyl Glycosides of Mannose and Its O-Carbamoyl Derivatives, and Mannose Moieties of Bleomycins

Compound	CO	Carbon						
		1	2	3	4	5	6	CH$_3$
Me-α-D-mannoside		101.6	71.4	70.8	67.6	73.4	61.8	55.7[a]
		(101.7)	(71.5)	(70.8)	(67.7)	(73.4)	(61.9)	(56.0)[b]
3-O-Carbamoyl derivative	159.0	101.4	69.1	75.2	65.4	73.4	61.7	55.6
Δ		+0.2	+2.3	−4.4	+2.2	0	+0.1	
2-O-Carbamoyl derivative	158.8	99.2	73.3	69.9	67.7	73.3	61.5	55.8
Δ		+2.4	−1.9	+0.9	−0.1	+0.1	+0.3	
De-Carbamoyl–BLM A$_2$'-c		98.9	71.4	70.2	67.6	74.3	61.9	
BLM A$_2$'-c	158.5	98.2	69.1	75.2	65.6	73.7	61.8	
Δ		+0.7	+2.3	−5.0	+2.0	+0.6	+0.1	
Iso-BLM A$_2$'-c	158.6	96.2	73.5	69.9	67.7	73.9	61.6	
Δ		+2.7	−2.1	+0.3	−0.1	+0.4	+0.3	

[a] δ-value: internal dioxane as δ67.4.
[b] Perlin et al. (10), δ-value: CS$_2$ as δ193.5.

Table 4:   Assignment of ^{13}C-Chemical Shifts of the Gulose Moiety of Bleomycin Estimated from Those of Galactose

Carbon	Me-a-L-galactoside[a]	Interaction[a]	Expected shielding[a]	Me-a-L-guloside	Gulose moiety of bleomycin
1	100.5	0// 0-3, -0// H-3	+, -	100.4	98.9
2	70.6	adj.3,	+	65.5 OH→OR$_2$	74.4
3	70.9	ax.0, 0// 0-1 0// H-5, -0 / 0-4 -H// 0-1	+, + +, - -	71.4	71.0
4	69.6	adj.3, -0 / 0-3	+, -	70.4	69.9
5	71.9	H// 0-3	+	67.3	68.6
6	62.5	0	0	62.0	61.2
CH$_3$	56.3			56.3	

[a]Perlin et al. [10].

Table 4. To estimate the chemical shifts of the gulose moiety of BLM, only the additional glycosidic effect at C-2 was required; this involved a downfield shift of about 8 ppm. Thus, the chemical shifts of the carbons in the gulose moiety could be assigned.

The assignment of peptide carbons was achieved by analysis of many spectra of peptide fragments and derivatives. Table 5 indicates the complete assignment of ^{13}C-NMR chemical shifts of BLM, its derivatives and fragments. This map also contains some information pertinent to the conformation of BLM. For example, the chemical shifts of Compounds I, III, VI, and VII, which comprise tetrapeptide S[11], are almost the same as those of the analogous carbon atoms in BLM and its derivatives. The chemical shifts of the carbons in the pseudodipeptide moieties of decarbamoyl BLM and iso-BLM are very similar, but slightly different from those of BLM. This suggests that in aqueous solution an interaction exists between the 3-O-carbamoyl group and pseudodipeptide moiety of bleomycin. In depyruvamide BLM three carbons ascribed to the V moiety are missing and the chamical shifts of the side chain carbons of II, as well as that of C-2 of the pyrimidine moiety, were altered significantly; in ring-closed BLM only six signals assigned to the pyrimidine-2 substituent carbons were shifted significantly, as shown in Fig. 6. Thus, the ^{13}C-chemical shift map is useful for structural determination of BLM-related antibiotics and unknown BLM derivatives.

The shortcoming of ^{13}C-NMR spectrometry is that a large amount of sample is required for measurement. Usually 40 mg of BLM or its derivatives are needed

Table 5: ^{13}C-Chemical Shift Map of Bleomycin

Group	Assignment	Depyruvamide BLM A₂'-c	Ring-closed BLM A₂'-c	Deamido BLM A₂'-c	Epi-BLM A₂'-c	Iso-BLM A₂'-c	Decarbamoyl-BLM A₂'-c	BLM A₂'-c	BLM A₂	BLM-acid	Deglyco-BLM-acid	Tetrapeptide S of BLM A₂	Tripeptide S of BLM A₂	Pseudotetrapeptide A	Pseudotripeptide	Amino acid
I	CO	172.5	172.5	172.5	172.5	172.5	172.5	172.5	172.5	172.5	172.5	172.3	168.6			175.0
	β-CH	67.9	68.0	67.9	68.0	67.9	67.9	67.9	68.0	67.9	67.9	67.9	66.9			66.7
	α-CH	59.9	59.9	59.9	59.9	59.9	59.9	59.9	59.9	60.0	59.9	60.0	59.6			61.1
	CH₃	19.9	20.0	19.9	20.0	19.9	19.9	19.9	20.0	19.9	19.9	19.9	19.8			19.3
II	S-CO	174.5	178.8	176.6	176.3	176.7	176.7	176.7	176.7	176.8	176.7			177.4	177.2	172.8
	R-CO	168.0	168.1	168.3	168.3	168.5	168.3	168.3	168.3	168.2	168.1			168.0	167.8	166.6
	2	160.6	165.1	165.5	166.1	166.3	166.1	165.2	165.2	165.7	165.9			161.1	160.4	157.1
	4	165.4	165.1	165.2	165.6	164.9	164.9	165.2	165.2	165.1	164.8			165.3	165.4	165.1
	6	152.0	152.2	153.0	151.3	154.0	153.9	152.9	152.9	152.8	153.7			153.2	153.2	146.0
	5	114.1	113.3	112.9	114.3	112.2	112.3	113.0	113.0	113.0	112.2			113.3	113.6	114.5
III	CH	52.5	56.7	54.5	53.7	53.6	53.4	53.3	53.3	53.2	53.1			51.1	50.8	49.5
	CH₂	37.1	44.8	40.7	42.0	41.1	41.1	41.1	41.0	41.0	40.9			38.7	38.5	36.9
	CH₃	11.9	11.9	11.9	11.9	11.8	11.8	11.8	11.9	11.9	11.8			11.9	11.9	12.3
	CO	178.0	178.1	178.1	178.0	178.1	178.1	178.1	178.1	178.1	178.0	177.0		182.6		182.7
	β-CH	75.2	75.1	75.2	75.3	75.2	75.2	75.2	75.2	75.2	75.0	72.9		75.6		73.6
	γ-CH	48.5	48.4	48.4	48.4	48.3	48.3	48.4	48.5	48.5	48.6	50.1		48.8		50.5
	α-CH	43.6	43.5	43.6	43.6	43.5	43.5	43.5	43.5	43.6	43.9	44.2		45.2		47.0
	α-CH₃	15.6	15.9	15.7	15.8	16.0	16.0	15.8	15.8	15.7	15.0	15.9		15.6		16.0
	γ-CH₃	13.0	12.8	12.9	12.8	12.8	12.9	12.9	12.9	13.0	13.8	11.9		13.4		11.7

Table 5: ¹³C-Chemical Shift Map of Bleomycin

Assignment	CO	2	4	5	β-CH	α-CH	CO	CH	CH₂	CO	2	2'	4	4'	5'	5	β-CH₂	α-CH₂
Depyruvamide BLM A₂'-c	169.2	136.9	133.9	118.8	68.6	57.4				171.1	163.6	163.1	149.4	147.6	125.7	119.8	40.0	32.8
Ring-closed BLM A₂'-c	169.9	137.7	136.3	118.4	68.6	57.9	174.6	58.1	49.9	170.7	163.0	162.6	149.5	147.5	125.3	119.5	40.0	32.9
Deamido BLM A₂'-c	169.4	137.2	134.2	118.7	68.5	57.7	173.3	60.5	47.3	171.0	163.4	163.0	149.4	147.5	125.7	119.7	40.0	32.8
Epi-BLM A₂'-c	169.4	137.8	135.4	117.9	68.0	57.3	171.5	61.5	48.1	171.0	163.4	163.0	149.4	147.6	125.6	119.7	40.0	32.9
Iso-BLM A₂'-c	169.7	137.7	135.6	118.4	67.9	57.4	172.1	60.4	48.3	170.9	163.3	162.9	149.4	147.5	125.5	119.6	40.0	32.8
Decarbamoyl-BLM A₂'-c	169.7	137.6	135.6	118.5	68.5	57.5	172.1	60.5	48.2	170.9	163.3	162.9	149.4	147.5	125.5	119.6	40.0	32.9
BLM A₂'-c	169.5	137.5	135.0	118.5	68.6	57.7	171.6	60.4	47.8	170.9	163.4	162.9	149.4	147.5	125.6	119.7	40.0	32.8
BLM A₂	169.4	137.4	134.7	118.6	68.6	57.7	171.6	60.5	47.8	171.0	163.7	163.0	149.4	147.5	125.7	119.8	40.0	32.9
BLM-acid	169.3	137.3	134.4	118.9	68.6	57.6	171.6	60.5	47.9	170.8	169.0	162.2	154.1	147.9	125.6	118.9	40.0	32.8
Deglyco-BLM-acid	169.5	135.9	134.9	117.6	66.5	58.4	171.6	60.5	47.8	170.9	168.9	162.2	154.1	148.0	125.6	118.9	39.9	32.8
Tetrapeptide S of BLM A₂										171.3	163.9	163.2	149.4	147.6	125.8	119.8	40.1	32.9
Tripeptide S of BLM A₂										170.9	163.6	162.9	149.3	147.5	125.7	119.8	40.3	32.7
Pseudotetrapeptide A	169.4	135.1	133.4	117.7	65.8	58.4	171.9	60.7	46.1									
Pseudotripeptide	174.7	134.4	133.2	117.2	66.9	59.5	171.8	60.7	46.1									
Amino acid	170.4	135.7	132.2	117.3	64.6	59.6	171.6	51.1	39.7	170.4	164.1	161.3	142.6	142.4	131.2	126.0	39.3	30.3

IV   V   VI

Table 5: ^{13}C-Chemical Shift Map of Bleomycin

Compound	VII (A2)				VII (A2'-c)					Gulose						Mannose						
Assignment	α-CH₂	γ-CH₂	CH₃	β-CH₂	2	4	5	α-CH₂	β-CH₂	1	2	3	4	5	6	CO	1	3	5	2	4	6
Depyruvamide BLM A₂'-c					133.8	131.5	117.1	39.2	25.3	98.5	74.4	71.0	69.9	68.3	61.3	158.6	98.9	75.2	73.2	69.1	65.5	61.8
Ring-closed BLM A₂'-c					135.7	134.9	117.6	40.0	27.0	98.6	74.4	70.8	70.0	68.0	61.2	158.5	98.6	75.1	74.7	69.2	65.6	61.8
Deamido BLM A₂'-c					134.0	131.6	117.1	39.2	25.3	98.4	74.4	70.9	69.9	68.2	61.2	158.5	98.7	75.2	73.5	69.1	65.6	61.8
Epi-BLM A₂'-c					134.1	131.8	117.1	39.2	25.5	98.4	74.7	70.3	70.2	68.0	61.5	158.1	97.2	75.1	71.9	69.0	65.4	61.7
Iso-BLM A₂'-c					134.7	132.7	117.2	39.5	25.9	98.5	74.2	70.1	69.9	67.9	61.0	158.6	96.2	69.9	73.9	73.5	67.7	61.6
Decarbamoyl-BLM A₂'-c					134.7	132.6	117.2	39.4	25.8	98.9	74.3	71.0	69.9	67.9	61.0		98.9	70.2	74.3	71.4	67.6	61.9
BLM A₂'-c					134.1	131.7	117.1	39.2	25.4	98.2	74.4	71.0	69.9	67.9	61.2	158.5	98.9	75.2	73.7	69.1	65.6	61.8
BLM A₂	41.8	38.6	25.7×2	24.6						98.3	74.4	71.0	70.0	68.0	61.2	158.5	98.8	75.2	73.6	69.2	65.6	61.8
BLM-acid										98.4	74.4	71.0	70.0	68.1	61.2	158.5	98.8	75.2	73.7	69.2	65.6	61.8
Deglyco-BLM-acid																						
Tetrapeptide S of BLM A₂	41.8	38.5	25.7×2	24.6																		
Tripeptide S of BLM A₂	41.8	38.5	25.7×2	24.6																		
Pseudotetrapeptide A																						
Pseudotripeptide																						
Amino acid																						

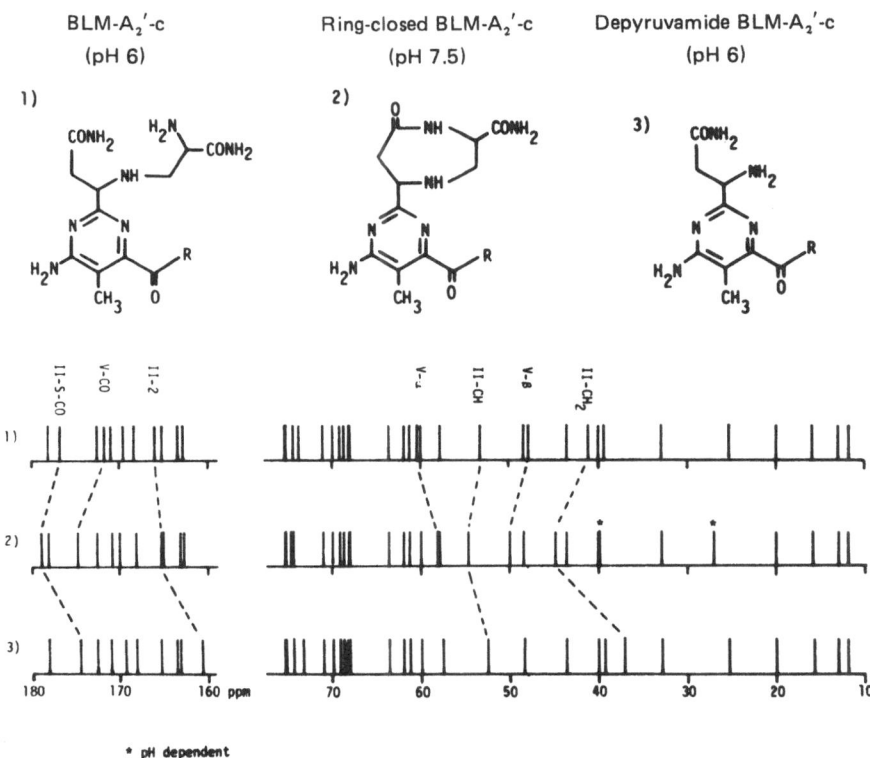

Fig. 6. ^{13}C-chemical shift changes in BLM A_2'-c, ring-closed BLM A_2'-c, and depyruvamide BLM A_2'-c.

to take a really good ^{13}C-NMR spectrum. The spectrum of BLM A_2 shown in Fig. 7 (lower tracing) was taken in a 5-mm sample tube containing 46.7 mg of the sample dissolved in 0.3 ml of D_2O. The number of transients was about 65,000, and it took about 18 hr to record the spectrum. The other spectrum of BLM A_2 in Fig. 7 (upper tracing) was taken in a 1.7-mm sample tube containing 2.93 mg of the sample dissolved in 28 μl of D_2O. The number of transients was about 172,000, and it took 38 hr to obtain the necessary data. The 53 signals, i.e., all of the signals of BLM A_2, can be distinguished in the upper spectrum. But if the structure of this sample were unknown, it would not be possible to conclude definitively that there were 53 signals in this spectrum.

In the future, our studies on BLM will be oriented toward biological and biochemical problems. Isolation of BLM or its biochemical modification products from biological reaction mixtures is hard work. Nevertheless, if a few milligrams of the metabolite can be isolated in pure form, and its ^{13}C-NMR spectrum taken, it will be possible to establish definitively what part of the bleomycin molecule has been modified by reference to our chemical shift map.

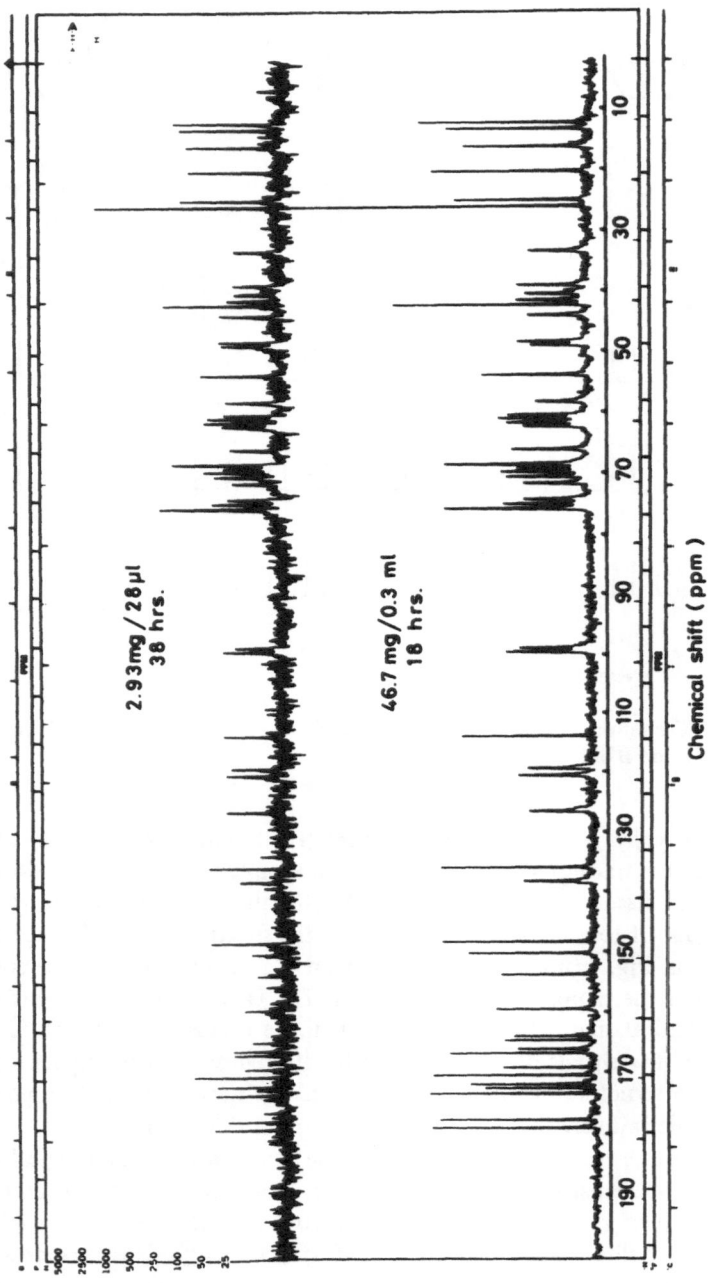

2.93 mg / 28 µl
38 hrs.

46.7 mg / 0.3 ml
18 hrs.

Chemical shift (ppm)

Fig. 7. ^{13}C-NMR spectra of bleomycin A$_2$.

References

1. H. Naganawa, Y. Muraoka, T. Takita, and H. Umezawa, *J. Antibiot. (Tokyo)*, **30**, 388 (1977).
2. S. Omoto, T. Takita, K. Maeda, H. Umezawa, and S. Umezawa, *J. Antibiot. (Tokyo)*, **25**, 752 (1972).
3. K. Bock and C. Pedersen, *J. Chem. Soc. Perkin Trans. II*, 293 (1974).
4. T. Takita, Y. Muraoka, K. Maeda, and H. Umezawa, *J. Antibiot. (Tokyo)*, **21**, 79 (1968).
5. Y. Muraoka, T. Takita, K. Maeda, and H. Umezawa, *J. Antibiot. (Tokyo)*, **23**, 252 (1970).
6. T. Takita, T. Yoshioka, Y. Muraoka, K. Maeda, and H. Umezawa, *J. Antibiot. (Tokyo)*, **24**, 795 (1971).
7. T. Takita, Y. Muraoka, A. Fujii, H. Itoh, K. Maeda, and H. Umezawa, *J. Antibiot. (Tokyo)*, **25**, 197 (1972).
8. H. Umezawa, *Biomedicine*, **18**, 459 (1973).
9. D. M. Chen, B. L. Hawkins, and J. D. Glickson, *Biochemistry*, **16**, 2731 (1977).
10. A. S. Perlin, B. Casu, and H. J. Koch, *Can. J. Chem.*, **48**, 2596 (1970).
11. H. Umezawa, this volume, p. 24.

^{1}H-NMR Study of the Metal-Binding Sites of Bleomycin

Norman J. Oppenheimer

Bleomycin has a high affinity for certain divalent metal ions, in particular, the 1:1 cupric complex is the natural form of the drug that is isolated[1,2] and the ferrous complex has been implicated as the possible active species causing bleomycin-catalyzed DNA strand scission[3,4]. Both cupic and ferrous ions are paramagnetic, hence proton nuclear magnetic resonance (NMR) is of limited value for the study of any of their complexes with bleomycin because of paramagnetic line broadening. Recently Dabrowiak et al.[5] have shown by ^{1}H-NMR that bleomycin interacts with zinc to form a tightly associated complex that is in slow exchange with unbound metal. In the present chapter, the greater resolution available at 360 MHz combined with the techniques of partially relaxed Fourier transform (PRFT) spectroscopy and homonuclear decoupling have been used to study the conformational changes associated with the formation of the zinc:bleomycin complex.

Experimental

NMR Parameters

Spectra were obtained at 360 MHz on a Bruker HXS-360 NMR spectrometer equipped with a Nicolet Technologies 1180 computer/Fourier transform system and a computer-controlled homonuclear decoupling accessory. Quadrature detection was employed and both 16K and 32K Fourier transforms were obtained with a spectral width of 3610 Hz. Partially relaxed spectra were obtained using the standard inversion/recovery pulse sequence[6] and adjusting the delay time (τ) for optimum nulling of the resonances.

The solutions of the unbuffered samples were twice lyophilized from 99.8% D_2O and then dissolved in 100% D_2O. Metal ions were added as their sulfate salts and the pD was adjusted with either D_2SO_4 or NaOD. The ferrous solutions were all handled under nitrogen to minimize air oxidation of the ferrous: bleomycin complex. The standard electrode correction has been employed: pD=meter reading + 0.4[7].

Results and Discussion

Assignments

A portion of the ¹H-NMR spectrum of metal-free bleomycin is shown in Fig. 1. The proton assignments for the resonances are based on homonuclear spin decoupling experiments and are consistent with those specific assignments that were made by Chen *et al.*[8]. The interpretations of the resonances in the spectrum of the zinc:bleomycin complex shown in Fig. 2 are more difficult.

This study is focused primarily on those resonances that yield coupling constant data important to defining the torsional geometries of the backbone of bleomycin. The 4-amino-3-hydroxy-2-methylvalerate and threonine resonances can be unambiguously assigned by homonuclear spin decoupling because of their unique spin coupling patterns. The hydroxyhistidine α- and β-proton resonances

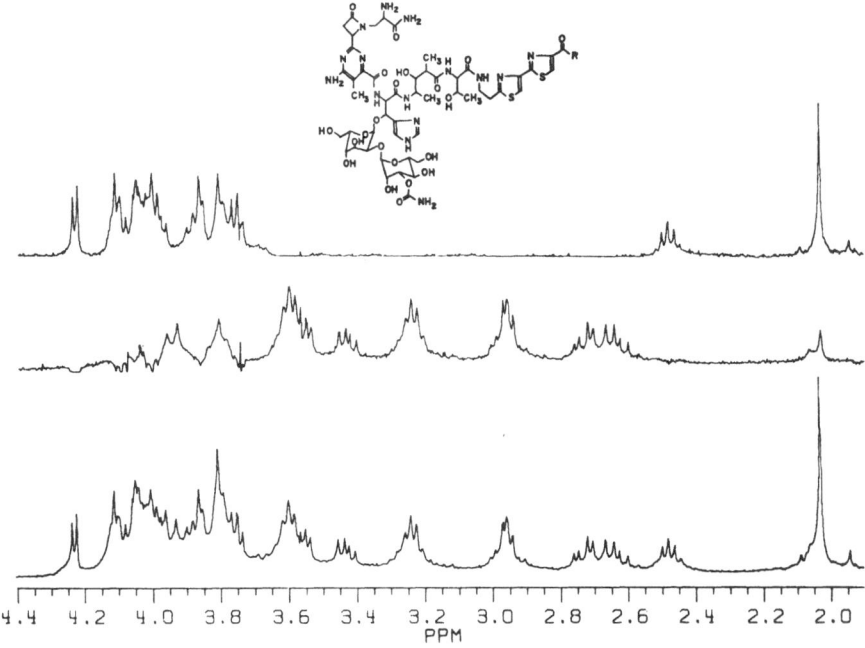

Fig. 1. A portion of the ¹H-NMR spectrum at 360 MHz of 5 m*M* bleomycinic acid (pD 5.4). The bottom spectrum is a normal Fourier transform spectrum (128 scans with 2.7 sec between acquisitions). The top and middle spectra show partially relaxed Fourier transform (PRFT) spectra with τ values of 240 and 510 msec, respectively. Thus in the top spectrum the methylene resonances are nulled, leaving only the methine resonances, and in the middle spectrum the methine resonances are nulled, allowing direct observation of the six sets of methylene resonances.

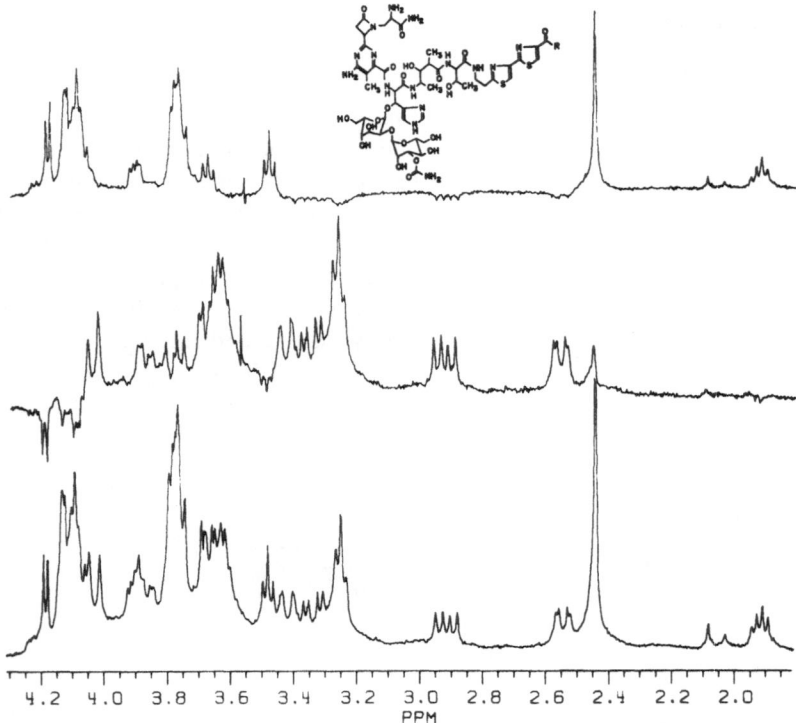

Fig. 2. A portion of the ¹H-NMR spectrum at 360 MHz of the 1:1 zinc:bleomy-cinic acid complex (pD 6.4). The bottom spectrum is a normal Fourier transform spectrum whereas the top and middle spectra are PRFTs showing the methine and methylene resonances, respectively. The parameters are the same as described in Fig. 1.

at 4.922 and 5.253 ppm were assigned by their chemical shift and more im-portantly their mutual spin splitting established by decoupling.

The methylene protons for bleomycinic acid and the zinc complex are readily observed in the partially relaxed Fourier transform (PRFT) spectrum shown in Fig. 3. Note that two of the methylene groups, the geminal protons designated 5 and 6, show a very large difference in chemical shift. The resonances at 3.335 and 2.913 ppm have been shown by spin decoupling to be vicinal to a methine at 4.585 ppm. These resonances (labeled 6) are attributed to the β-lactam moiety on the basis of the invariance of the coupling constants for the corresponding resonances in bleomycin and the zinc:bleomycin complex, a result consistent with a conformationally rigid structure such as the β-lactam. It therefore follows that the methylene proton resonances at 3.416 and 2.542 ppm (labeled 5) which are vicinally coupled to the methine resonance at 3.79 ppm arise from the β-aminoalanine moiety.

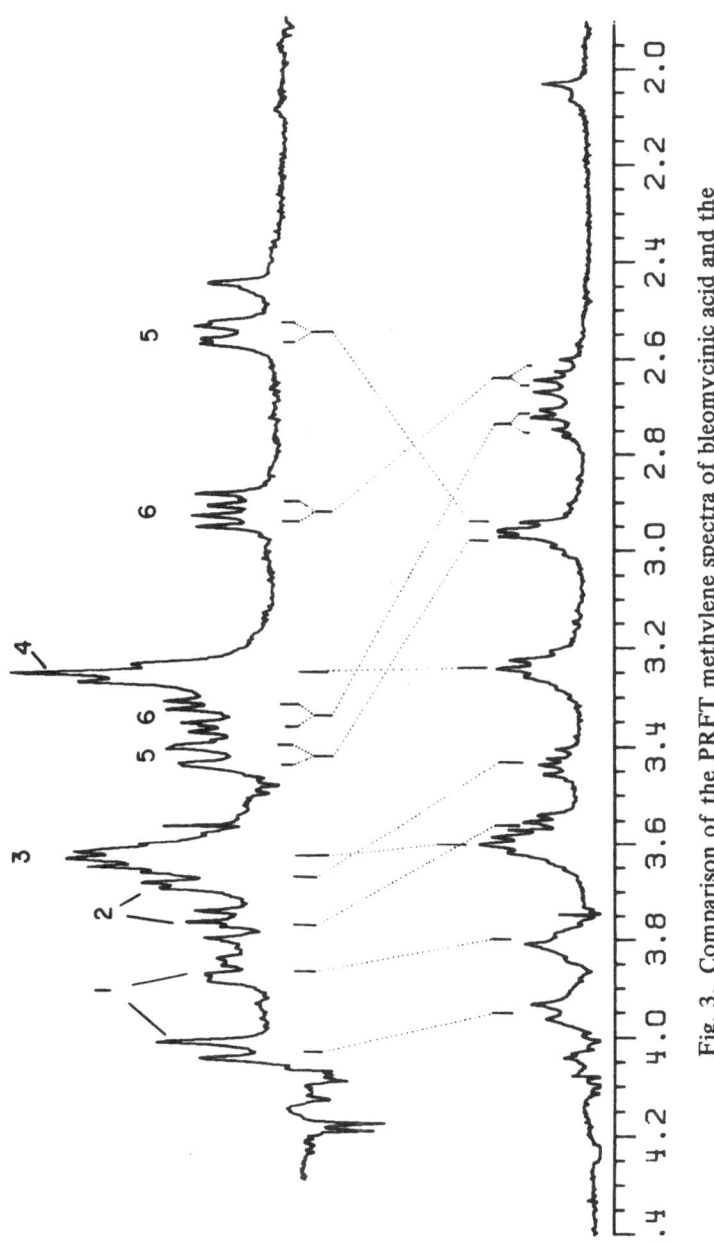

Fig. 3. Comparison of the PRFT methylene spectra of bleomycinic acid and the 1:1 zinc:bleomycin complex. Resonances 1 and 2 correspond to the 6 methylene protons of the sugar moieties and resonances 3 and 4 correspond to the two methylenes adjacent to the bithiazole ring. Resonances 5 are attributed to the methylene of the β-aminoalanine moiety and 6 to the methylene in the β-lactam.

Chemical Shift Changes Induced by Zinc Binding

Complexation with Zn^{2+} causes substantial changes in the chemical shifts among the ^1H resonances of bleomycin. The chemical shifts of a number of resonances in bleomycinic acid are compared in Table 1 to the corresponding resonances in the 1:1 complex with Zn^{2+}. Note that the largest change in chemical shifts occurs for the proton resonances of the following substituents: β-lactam ring, β-aminoalanine, pyrimidine, imidazole, valerate, and the 3 of the mannose. The chemical shifts induced in the remaining resonances are all less than 0.1 ppm. The sensitivity of the resonances to zinc binding is shown schematically in Fig. 4. These results are in agreement with the moieties implicated by Umezawa[9] as constituting the binding site.

The chemical shifts induced in the resonances of groups coordinated to zinc are caused by a combination of three separate effects: (i) the direct diamagnetic effects of the metal, (ii) the electrostatic effects of coordination to zinc resulting from the neutralization of charge (similar to protonation), and (iii) the conformational changes that fix certain spatial interactions among the substituents. Although a direct correlation between the observed chemical shifts and the separate contributions from each effect is not possible at this time, the resonances that are altered clearly reflect the general area of the metal-binding site.

Table 1:   Chemical Shifts[a]

	Bleomycin pD 5.4	Zinc:bleomycin pD 6.4
β-Lactam CH$_2$	2.723 2.638	3.335 2.913
CH	3.992	4.585
β-Aminoalanine CH$_2$	2.972 2.945	3.416 2.542
CH	4.016	3.79±0.01
Imidazole a	5.090	4.922
β	5.363	5.253
C$_2$H	8.018[b]	8.016
C$_4$H	7.397[b]	7.367
Valerate C$_2$H	2.522	1.908
C$_3$H	3.736	3.478
C$_4$H	3.898	3.674
C$_2$CH$_3$	1.141	1.014
CH$_3$	1.141	0.975
Pyrimidine CH$_3$	2.031	2.439
Gulose C1	5.286	5.351
Mannose C$_1$	5.025	4.944
C$_3$	4.709	N.D. < 4.2 ppm

[a]Values are reported relative to internal sodium 3-trimethylsilyl-(2,2,3,3-d_4)propionate and are accurate to 0.002 ppm.
[b]The chemical shifts of the imidazole protons are very sensitive to pD over the range 4.0–6.0.

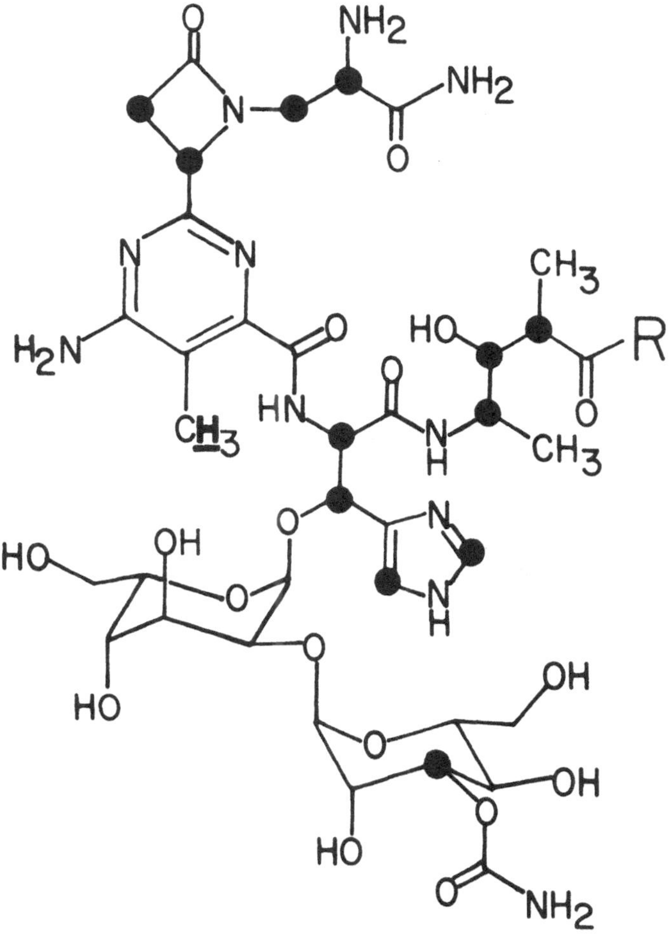

Fig. 4. Partial structure of bleomycin. (●) Proton resonances that show a greater than 0.1 ppm shift upon formation of the zinc:bleomycin complex. The resonances of the remainder of the molecule (R) show no significant shifts. Resonances of the sugar moieties show some shifts but those resonances have not been assigned and are therefore not designated.

Analysis of Coupling Constants

The bleomycin molecule is characterized by potentially very high degrees of torsional freedom around the backbone linkages. Thus measurement of the vicinal coupling constants for the appropriate resonances along the backbone will provide information about the dihedral angles between the coupled protons[10,11]. Furthermore, estimates of the population distribution of various rotational isomers around the bond can be made using the analysis of Blackburn, et al.[12].

The corresponding coupling constants for resonances in bleomycinic acid and the zinc:bleomycinic acid complex are compared in Table 2. Using estimates of the value for the trans coupling constant (10–12 Hz), the rotamer populations can be calculated and are also listed in Table 2. It is important to note that where a single vicinal coupling constant is available, only the relative populations of gauche and trans forms can be determined.

Complexation of Zn^{2+} to bleomycin causes large conformational changes in portions of the molecule as reflected by alterations in the vicinal coupling constants. An important point to note is that where a large change in coupling constants occurs, the result is a predominance of a single rotational isomer. That is, complexation with zinc confers a conformational rigidity on certain portions of the molecule whereas the remainder of the molecule is not affected very much.

There are two important sets of vicinal coupling constants that change upon complexation with zinc, the coupling between the α-proton and methylene protons of the β-aminoalanine moiety and the vicinal protons in the hydroxyhistidine moiety. Based on the analysis of the coupling constants, both these residues change from nearly equal populations of rotamers in the free bleomycinic acid to predominantly ($>$90%) gauche populations in the complex with zinc as shown in Figs. 5 and 6. In contrast, the resonances of the 4-amino-3-hydroxy-2-methylvaleric acid moiety show only small changes in coupling constants upon zinc binding, although they show large changes in chemical shift. Both chemical shift and coupling parameters for the remaining resonances of the side chain out to the bithiazole moiety are all unchanged, within experimental accuracy. It

Table 2: Conformational Analysis of Coupling Constants[a]

	Bleomycin pD 5.4		Bleomycin:Zn pD 6.4	
	^3J	Rotamer[b] population (%)	^3J	Rotamer[b] population (%)
β-Lactam	5.5	–	6.3	–
	8.9		8.4	
β-Aminoalanine	5.4	10 gg	2.0	$>$90 gg
	7.7		3.1	
Histidine	6.0	40 trans	2.7	$<$10 trans
Valerate J_{2-3}	6.0±0.3	40–45 trans	5.9±0.3	40–45 trans
J_{3-4}	6.0±0.3	40–45 trans	5.9±0.3	40–45 trans
Threonine	4.7	25 trans	4.9	30 trans
Glucose 1	3.6	–	4.3	–
Mannose 1	1.5±0.3	–	1–1.5	–

[a]Coupling constants are accurate to within ±0.2 Hz unless otherwise noted.
[b]Rotamer populations were calculated using the following parameters: J trans equals 10–12 Hz and J gauche equals 2–2.5 Hz. Thus changes in rotamer populations of less than 10% are not considered significant.

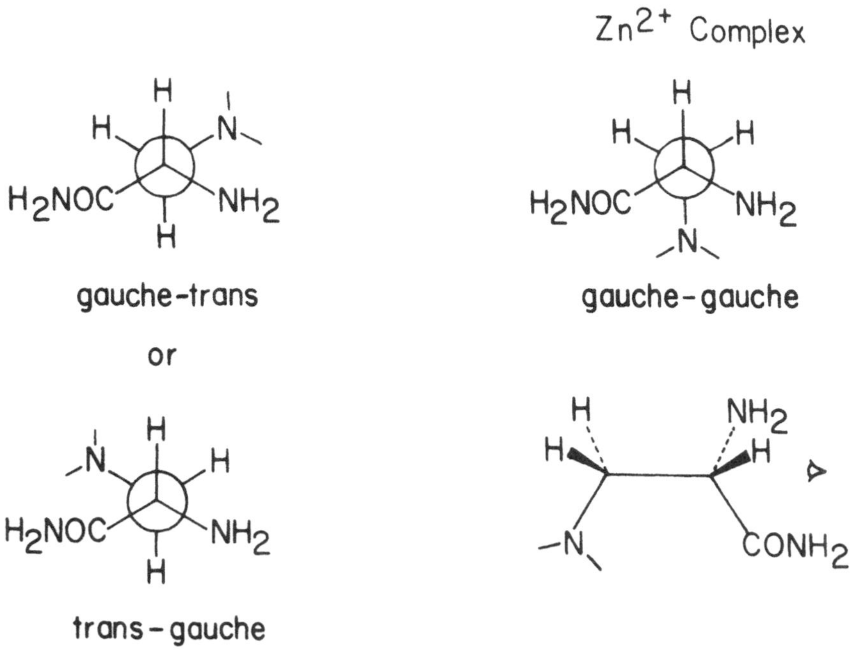

Fig. 5. The predominant rotamer populations for the β-aminoalanine moiety. In metal-free bleomycin both the gauche–trans and trans–gauche rotamers are almost equally populated and the gauche–gauche rotamer population is less than 10%. Upon formation of the zinc:bleomycin complex the gauche–gauche rotamer is favored exclusively, a rotamer that juxtaposes all three potential nitrogen ligands as shown.

should also be noted that although the data listed in Tables 1 and 2 are for bleomycinic acid, the results are identical for the mixture of bleomycin A_2 and B_2.

The preferred geometry of the β-aminoalanine moiety observed in the zinc: bleomycinic acid complex juxtaposes the ring nitrogen of the β-lactam with both the primary amine and the carboxamide group. This conformation represents an orientation that would be expected if some or all of these nitrogens were involved with complexation to the zinc. The hydroxyhistidine moiety shows a preference for the gauche conformation (see Fig. 6), consistent with maintaining a close proximity between the pyrimidine and imidazole moieties. The other gauche conformation, where the pyrimidine and imidazole moieties are trans, can probably be ruled out since it would not permit simultaneous binding of zinc to both these moieties.

Interactions with Cupric and Ferrous Ions

Addition of either $CuSO_4$ or $FeSO_4$ to solutions of bleomycin and adjustment of the pH to neutrality resulted in severe broadening of all the bleomycin

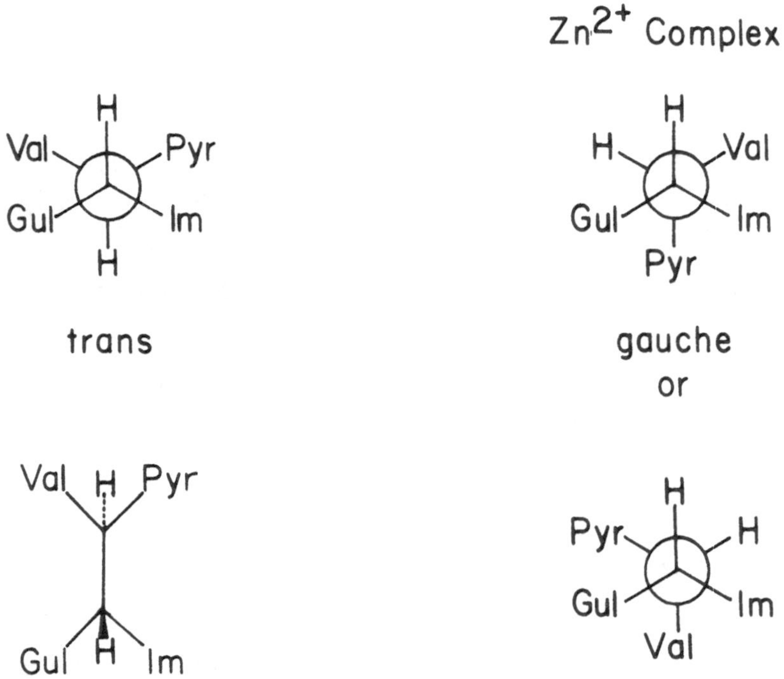

Zn^{2+} Complex

trans

gauche

or

Fig. 6. The predominant rotamer populations for the β-hydroxyhistidine moiety. Metal-free bleomycin favors a trans rotamer although considerable populations of gauche are present. Complexation with zinc results in an exclusively gauche rotamer. However, the analysis cannot distinguish between the two possible gauche rotamers shown, although only the top rotamer can allow both the pyrimidine and imidazole moieties to complex with zinc.

resonances as shown in Fig. 7. The result confirms the previous report by Dabrowiak *et al.*[5] that the copper:bleomycin complex gave detectable resonances only for the side chain residues. The copper in this complex is in slow exchange, since addition of less than stoichiometric amounts of copper results in the superposition of the sharp resonances of the uncomplexed bleomycin on the broad resonances of the copper:bleomycin complex. Copper was also observed to readily displace zinc from the zinc:bleomycin complex. In the time it took to add a stoichiometric amount of CuSO$_4$ to a solution of zinc:bleomycin, return the sample to the spectrometer, and acquire a spectrum (about 4 min), no resonances attributable to the zinc:bleomycin complex were detectable.

The complexation of bleomycin with ferrous ion required addition of FeSO$_4$ under a nitrogen atmosphere to prevent the rapid air oxidation of the ferrous: bleomycin complex. A portion of the spectrum of this complex is also shown in Fig. 7. Again the spectrum is characterized by broadened resonances, although the effect is not as severe as observed with copper. Substantial contact shifts of

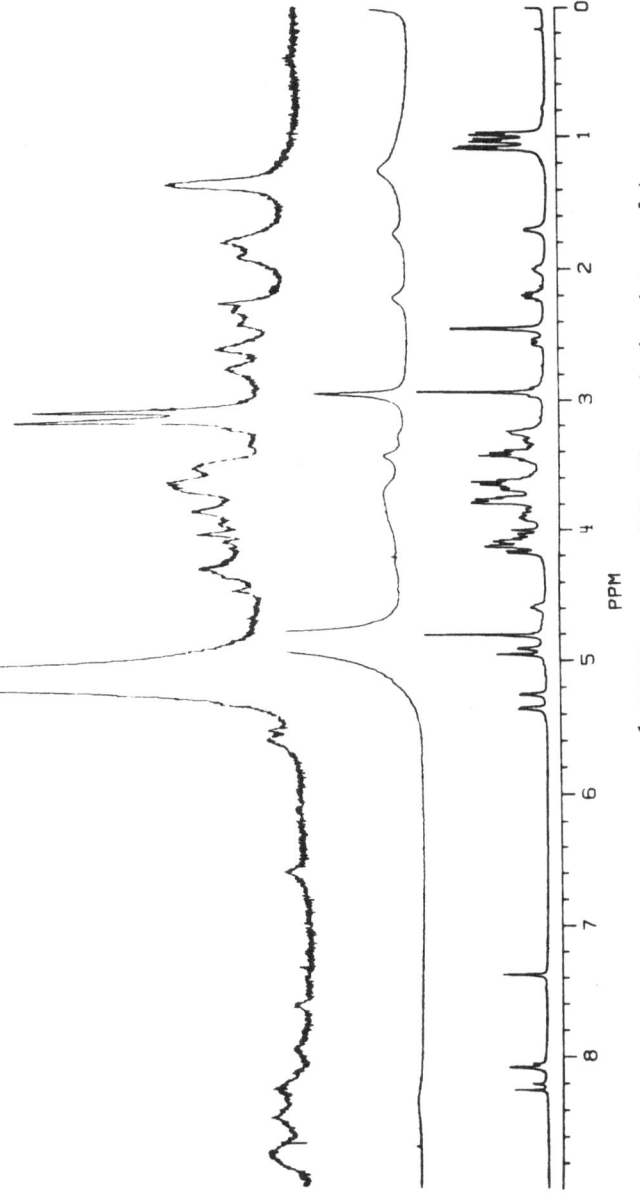

Fig. 7. Comparison of the ¹H-NMR spectra of the bleomycin (a mixture of A_2 and B_2 forms) complex with ferrous (top), cupric (middle), and zinc (bottom). Note that cupric ion causes the most paramagnetic line broadening whereas in the ferrous complex the resonances are relatively narrower and show contact shifts.

proton resonances are also evident. For example, resonances appear in the spectral region between 5.5 to 7 ppm where none are observed in either bleomycin or the zinc:bleomycin complex. The broadness of all the resonances, however, prevents their assignment at this time. The appearance of the doublet at 3.17 and 3.10 ppm, near the expected resonance frequency of the sulfonium methyls of bleomycin A_2, is most interesting. The two peaks may reflect magnetically distinct forms of the ferrous:bleomycin A_2 complex. It is not clear, however, whether this could result from two distinct metal-binding sites or to diastereotopic coordination of the bleomycin ligands to the ferrous ion. Alternatively, the second resonance could arise from the pyrimidine methyl group having been strongly deshielded because of its close proximity to the metal-binding site. Further experimentation will be needed to distinguish between these proposals.

Addition of Divalent Cations to Unbuffered Bleomycin

Previous investigators have reported a substantial drop in pH when a divalent cation is added to aqueous solutions of bleomycin. This aspect of divalent metal binding to bleomycin was studied by ^{1}H-NMR. The sulfates of either Zn^{2+}, Cu^{2+}, or Fe^{2+} were added to unbuffered solutions of bleomycin and the ^{1}H-NMR spectra were recorded. The results of these experiments are difficult to interpret.

The addition of the divalent metal ions results in substantial shifts of many proton resonances. Furthermore, the induced chemical shifts plateau at a 1:1 stoichiometry of divalent cation to bleomycin and a final pD of between 2.5 and 2.8. However, the three metals studied, Zn^{2+}, Cu^{2+}, and Fe^{2+} all show similar changes in chemical shifts and the observed changes are superimposible for the shift observed when the pH is lowered in the absence of any divalent metal ion (see Fig. 8). It should be pointed out that the pH of the divalent metal solutions themselves were all between 5.0 and 6.0 and thus cannot account for the low pH when added to solutions of bleomycin.

Even less understood is the absence of any line broadening of the resonances of bleomycin in the presence of nearly equimolar concentrations of the paramagnetic cupric or ferrous ions (see Fig. 8). If proton release, hence lowering of the pH, results from some complexation of the divalent metal ion with bleomycin, then why are no resonances broadened from the required interaction with the paramagnetic ions? Conversely, if the metals do not interact at all with bleomycin at low pH, then what is the origin of the released protons?

The ^{1}H-NMR spectra of these solutions did not show any change when incubated at 4°C for a week. In control experiments, the pH of the already formed tight metal:bleomycin complex was lowered to 2.5. This resulted in no observeable change in the ^{1}H-NMR spectrum for these solutions. Thus low pH appears to make interconversion between the complex and low pH "uncomplexed forms" kinetically unfavorable.

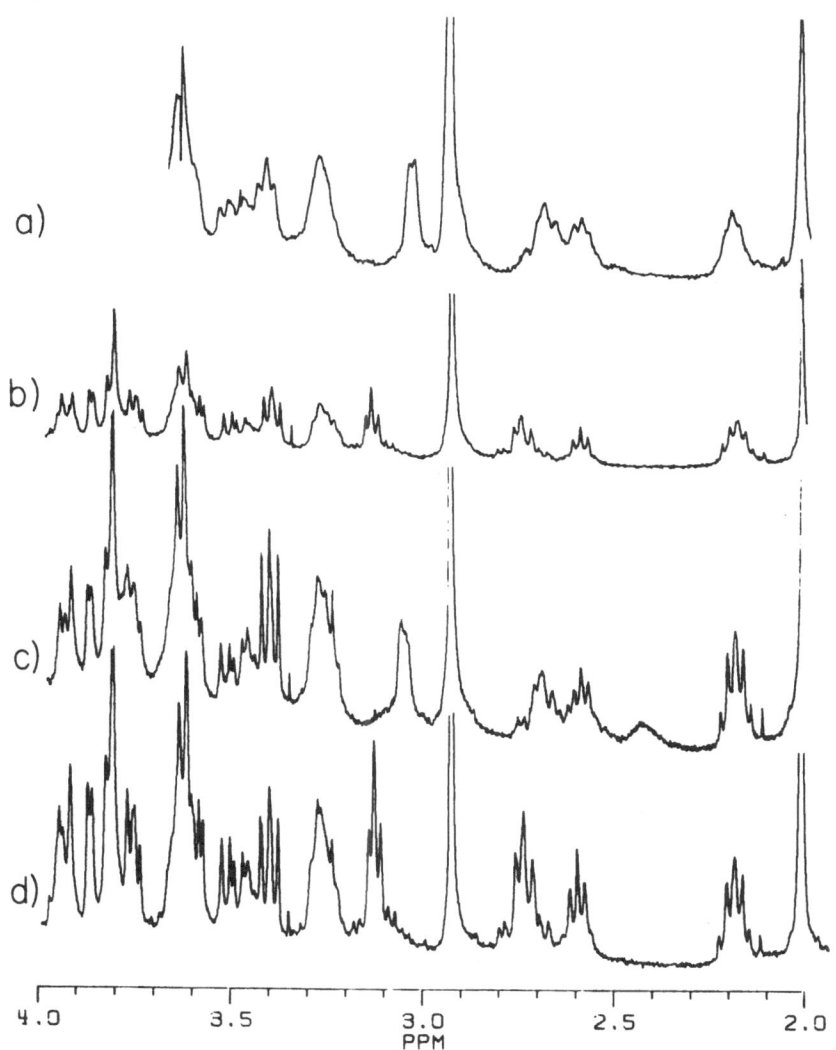

Fig. 8. Comparison of a portion of the ¹H-NMR of bleomycin showing the effect of adding divalent metal to an unbuffered solution of 5 mM bleomycin. The metal concentrations are 4 mM, thus the metal:bleomycin ratio is 0.8:1. All spectra are normalized except for spectrum (b) which is shown at half intensity. Spectrum (a) is for ferrous:bleomycin pD 4.2, (b) is for cupric bleomycin pD 3.8, (c) is for zinc:bleomycin pD 4.1, and (d) is for metal-free bleomycin pD 3.8.

Acknowledgments

I thank Mr. C. Garrett, Dr. D. V. Santi, and Dr. S. Hecht for providing samples and helpful discussions. This work was supported in part by NIH Grant GM–22982, UCSF NMR Research Resource Center Grant RR–00892, UCSF Computer Graphics Laboratory Grant RR–1081, and Stanford Magnetic Resonance Laboratory Research Resource Center Grants NSF–GP–23633 and NIH–RR–00711.

References

1. H. Umezawa, K. Maeda, T. Takeuchi, and Y. Okami, *J. Antibiot. (Tokyo)*, **19**, 200 (1966).
2. H. Umezawa, In *Antibiotics III* (J. W. Coccoran and F. E. Halm, Eds.), p. 21. Springer-Verlag, New York (1975).
3. E. A. Sausville, J. Peisach, and S. B. Horwitz, *Biochem. Biophys. Res. Commun.*, **73**, 814 (1976).
4. J. W. Lown and S.-K. Sim, *Biochem, Biophys. Res. Commun.*, **77**, 1150 (1977).
5. J. C. Dabrowiak, F. T. Greenaway, W. E. Longo, M. vanHusen, and S. T. Crooke, *Biochim. Biophys. Acta*, **517**, 517 (1978).
6. R. L. Vold, J. S. Waugh, M. P. Klein, and D. E. Phelps, *J. Chem. Phys.*, **48**, 3831 (1968).
7. P. K. Glasoe and F. A. Long, *J. Phys. Chem.*, **64**, 188 (1960).
8. D. M. Chen, B. L. Hawkins, and J. D. Glickson, *Biochemistry*, **16**, 2731 (1977).
9. H. Umezawa, *Fed. Proc.*, **33**, 2296 (1974).
10. M. Karplus, *J. Chem. Phys*, **30**, 11 (1959).
11. M. Karplus, *J. Am. Chem. Soc.*, **85**, 2870 (1963).
12. B. J. Blackburn, A. A. Grey, I. C. P. Smith, and F. E. Hruska, *Can. J. Chem.*, **48**, 2866 (1970).

The Metallobleomycins

J. C. Dabrowiak, F. T. Greenaway,
and F. S. Santillo

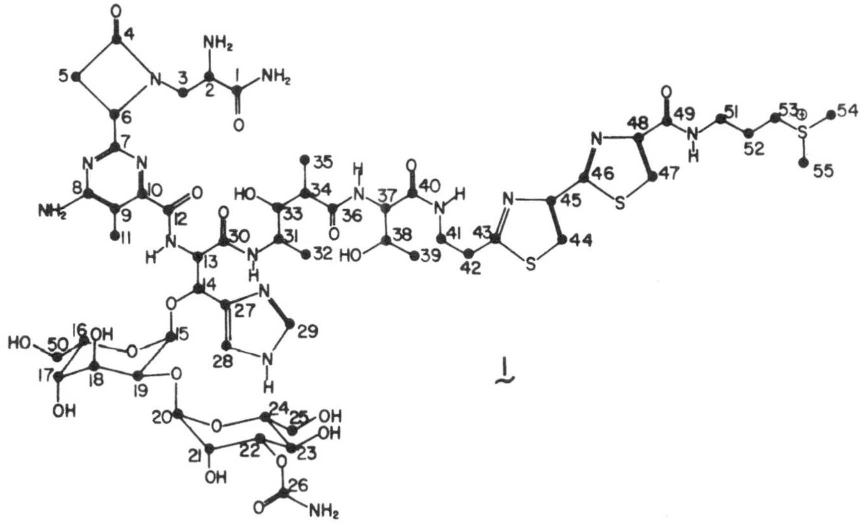

Bleomycin (BLM) is an antitumor antibiotic produced by a strain of *Streptomyces verticillus*[1]. Clinically employed bleomycin (Blenoxane) is a mixture of 11 active components. Of these, bleomycin A_2 (BLM A_2), **1**[2], is the most abundant, accounting for 55–70% (w/w) of the mixture. Bleomycin is of value in the treatment of squamous cell carcinomas, lymphomas, and testicular carcinomas[3,4]. The mechanism of action of bleomycin is not understood. Most of the early efforts as well as those in progress have focused on the interaction of bleomycin, most often bleomycin A_2, with DNA. The glycolpeptide is capable of 11 active components. Of these, bleomycin A_2 (BLM A_2), **1**[2], is the most abundant, accounting for 55–70% (w/w) of the mixture. Bleomycin is of value in the treatment of squamous cell carcinomas, lymphomas, and testicular carcimixture prevents DNA breakage. It is known that these metal ions complex to bleomycin and in so doing apparently inactivate the drug. The addition of Fe(II) salts to the reaction mixture, on the other hand, under conditions which lead to the formation of the Fe(II)–BLM complex, greatly enhances the DNA degrading ability of bleomycin[12,13].

Horwitz and co-workers[12] have shown that Fe(II)–BLM is air sensitive and that it is readily oxidized to Fe(III)–BLM in the atmosphere [Eq. (1)]. These workers speculated that a radical is formed in the oxidation process and that it is the radical that leads to DNA breakage.

$$\text{Fe(II)–BLM} \underset{\text{R-SH}}{\overset{\text{O}_2}{\rightleftharpoons}} \text{Fe(III)–BLM + radical} \tag{1}$$

Lown and Sim[13] have recently shown that DNA breakage by bleomycin can be prevented by free radical scavengers such as superoxide dismutase, catalase, and isopropanol. On the basis of these observations they speculated that the oxidation of the Fe(II)–BLM complex results in the formation of H_2O_2, O_2^- and $\cdot OH$ radicals. Sulfhydryl groups present in the cell [Eq. (1)] would offer a convenient means of reducing the trivalent iron–bleomycin complex to the divalent state thus making the system catalytic.

In experiments in which divalent iron has not been added to the reaction mixture, DNA breakage by bleomycin occurs but at a reduced rate[12,13]. The reason that it occurs at all has been attributed to the presence of trace amounts of iron contributed by DNA, bleomycin, or both. It should be pointed out that the degrading process by bleomycin can be quenched by the strong metal chelating agent EDTA[6,14,15]. The fact that the reaction exhibits an oxygen requirement is also supportive of the metal hypothesis[12,13].

Recently Umezawa and co-workers[16] have drawn attention to the possible role of the copper(II)–BLM complex in the DNA degrading process. It is known that both copper-free BLM and its metal complex inhibit DNA synthesis in cells[17] as well as prevent growth of animal and bacterial cells[1,17-20]. However, as was pointed out earlier, Cu(II)–BLM does not cause scission of DNA in vitro[11]. In the search for a cellular mechanism to explain the activity of Cu(II)–BLM, Umezawa and co-workers[16] identified a protein which is capable of reducing Cu(II) to Cu(I) in the bleomycin complex. They suggested that a sulfhydryl group of a protein carries out the reduction and that the resulting Cu(I) ion binds to a second protein in the cell [Eqs. (2) and (3)].

$$\text{Cu(II)–BLM} \xrightarrow{\text{protein-SH}} \text{Cu(I)–BLM} \tag{2}$$

$$\text{Cu(I)–BLM + protein} \longrightarrow \text{Cu(I)–protein + BLM} \tag{3}$$

The liberated bleomycin is then free to do DNA damage. No mention was made of radicals which may be produced in the reduction or of the possible role of a second metal ion [e.g., Fe(II)] in the degrading process.

In view of the importance of metal ions in the DNA degrading activity of bleomycin we initiated a study aimed at disclosing the physical and chemical properties of a group of metallobleomycins. The immediate objective was the elucidation of the structural aspects of the metal derivatives. Although the Zn(II) and Cu(II) complexes were the first and, as a result, the most extensively studied analogs, we have also uncovered a limited amount of structural information on the Co(II) and the biologically important Fe(II) and Fe(III) compounds.

Materials and Methods

Bleomycin A_2 and peptide M were supplied by Bristol Laboratories.

Most of the procedures have been described earlier in detail[21,22]. In general, the complexes were prepared by adding one equivalent of a particular metal perchlorate to a pH 8.0 phosphate buffer solution containing BLM A_2 or by adding NaOH to an aqueous solution containing a 1:1 ratio of the drug and the metal ion.

ESR spectra were determined on glycerol glasses of Cu(II)–BLM A_2 at 123 and 6°K. No spectral differences between the two temperatures were noted. The ESR spectrum of the Fe(III)–bleomycins was obtained at 123°K on a frozen water solution. The electrochemistry was carried out in a standard polarographic H cell using a previously described apparatus[23]. Both a pH 6.5 acetate buffer and a pH 8.0 phosphate buffer were used for the studies. No difference in the behavior of the drug in either buffer was noted. In order to cover the potential range +1.5 → -1.8 V, two electrodes were used. A carbon paste electrode was employed for the range +1.5 → -0.6 V while a Hg (DME) surface was used for the range +0.15 → -1.8 V. All potentials are referenced to a Ag/AgCl sat. NaCl system.

Results and Discussion

I. The Cu(II) and Zn(II) Bleomycins

Complexation of Cu(II) to BLM A_2 yields a blue ESR active compound[21]. At Cu(II)/BLM A_2 ratios of 1:1 or less only one species with $g_{\parallel} = 2.203$, $g_{\perp} = 2.058$, $|A_{\parallel}(Cu)| = 0.0186$ cm^{-1} can be detected (Fig. 1). At higher copper ratios the ESR spectrum showed the presence of the 1:1 complex as well as that of free $[Cu(II)(H_2O)_6]^{2+}$. Comparing the observed $g_{\parallel}$ and $A_{\parallel}$ values with those found for a series of square planar Cu(II) complexes having a variety of donor atoms[24] strongly suggests that Cu(II)–BLM A_2 possesses a square planar, Cu(II) N_4 or a Cu(II) N_3O, type of environment.

More specific information as to which atoms of the bleomycin molecule are utilized as donor atoms was obtained from potentiometric titration data[25]. Figure 2 shows the first derivative plots ($\Delta V/\Delta pH$ vs pH) of the titration curves of BLM A_2 and its Cu(II) and Zn(II) complexes. The inflections at pH 4.5 and 7.1 for BLM A_2 are each due to the release of one proton. On the basis of earlier published data[3] these inflections are associated with the deprotonation of the

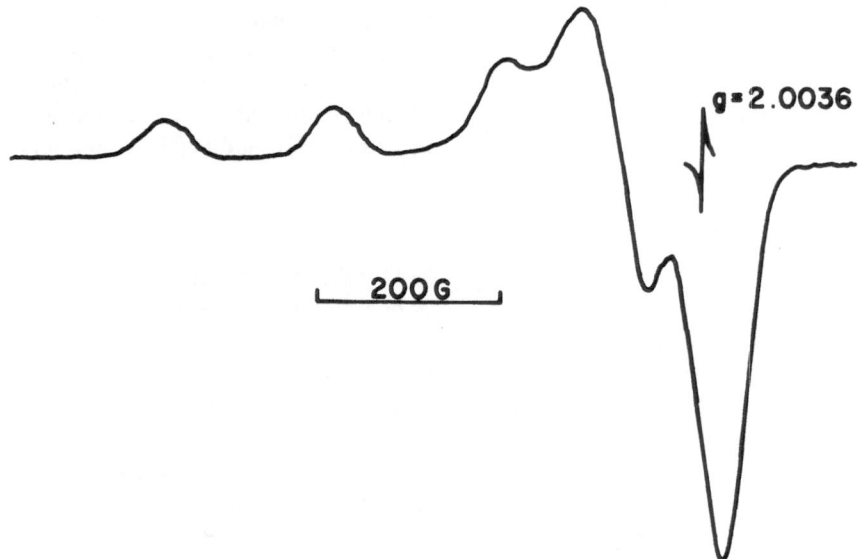

Fig. 1. The ESR spectrum of Cu(II)–BLM A$_2$ at 123°K.

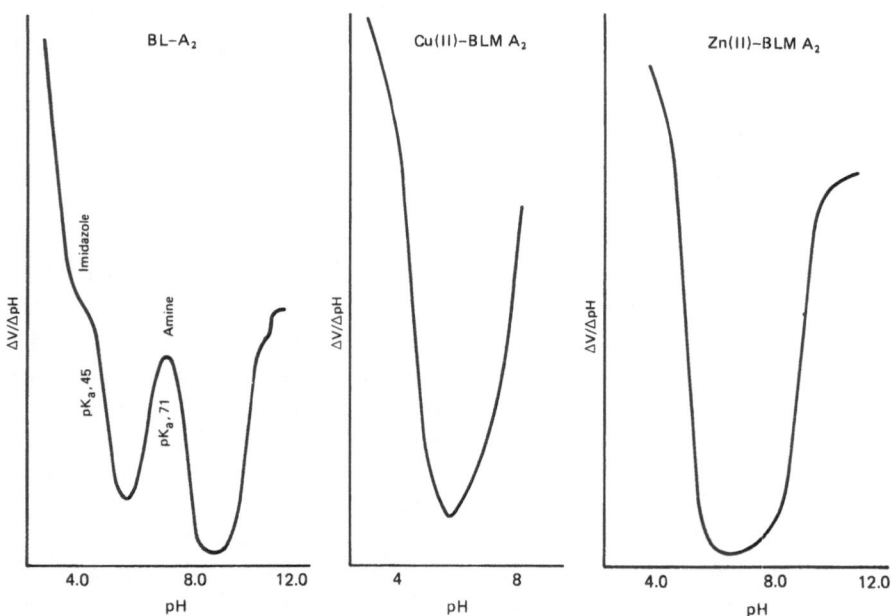

Fig. 2. First derivative plots ($\Delta V/\Delta$pH vs pH) of the titration curves of BLM A$_2$, Cu(II)–BLM A$_2$, and Zn(II)–BLM A$_2$.

protonated imidazole function [Eq. (4)] and the deprotonation of the protonated amino group [Eq. (5)], respectively.

$$\text{Im H}^+ \longrightarrow \text{Im} + \text{H}^+ \qquad \text{p}K_a, 4.5 \qquad (4)$$

$$-\text{NH}_3^+ \longrightarrow -\text{NH}_2 + \text{H}^+ \qquad \text{p}K_a, 7.1 \qquad (5)$$

The binding of either Zn(II) or Cu(II) to the drug results in the disappearance of
these two inflections from the potentiometric titration curve of the complex
(Fig. 2). Thus, these two functionalities must be metal-ligating groups.

Electronic absorption studies have also been useful in elucidating the metal
binding site of the antibiotic. The absorption spectrum of bleomycin A_2 is
shown in Fig. 3a. On the basis of previous work [26-28] the absorption at 287 nm
(ϵ_M = 14,000) contains the $\pi \to \pi^*$ electronic transition of the bithiazole residue
as well as the $n \to \pi^*$ electronic transition of the pyrimidine moiety of bleomycin. This band overlap was not acknowledged in the earlier published spectral
analysis [21]. The binding of either Cu(II) or Zn(II) causes the appearance of a
new band at 250 nm and a change in intensity as well as a slight change in the
position of the 287–nm band (287 nm, ϵ_M = 14,100 to 292 nm, ϵ_M = 17,400).
In the case of Cu(II) BLM A_2, a shoulder is found at 320 nm. These changes are
best observed in the difference spectra shown in Fig. 3b. Complexation studies
with peptide M, 2,[13,28] which does not contain the bithiazole residue, has
clearly shown that when Zn(II) or Cu(II) bind to the pyrimidine residue of the
peptide, an increase in the intensity as well as a decrease in the energy of the
$n \to \pi^*$ and the $\pi \to \pi^*$ electronic transitions of that chromophore result [25].
Since metal binding to bleomycin causes spectral changes which are similar to
those found for 2 (Fig. 3b), the pyrimidine residue of bleomycin is a metal-
ligating group. Thus, the 250–nm band in the spectra of the metallobleomycins
is the shifted $\pi \to \pi^*$ transition of the bound pyrimidine residue. The presence
of the shifted $n \to \pi^*$ transition is more difficult to detect. However, it appears
to be responsible for the apparent increase in intensity and slight shift in position of the 287–nm band.

The Cu(II) complex also exhibits a band in the visible region of the spectrum.
The intensity (ϵ_M = 120) and position (593 nm) of this band labels it as the
spin allowed $d-d$ electronic transitions of square planar Cu(II), i.e.,
$d_{xz}(d_{yz}) \to d_{x^2-y^2}$, $d_{xy} \to d_{x^2-y^2}$, $d_{z^2} \to d_{x^2-y^2}$ [29]. By analogy with other
square planar copper complexes [30], the copper ion in Cu(II)–BLM A_2 appears
to be in a square planar crystal field and very likely possesses a four nitrogen
environment. Thus the results of the visible absorption studies confirm the
analysis based on the earlier presented ESR data.

Investigations using proton NMR show that the imidazole and the pyrimidine
residues of the drug bind to zinc(II). Complexation of divalent zinc to bleomycin causes the C-2 and C-4 protons of the imidazole moiety to shift to lower
field (Fig. 4). Since the shift of the C-2 proton resonance is about three times

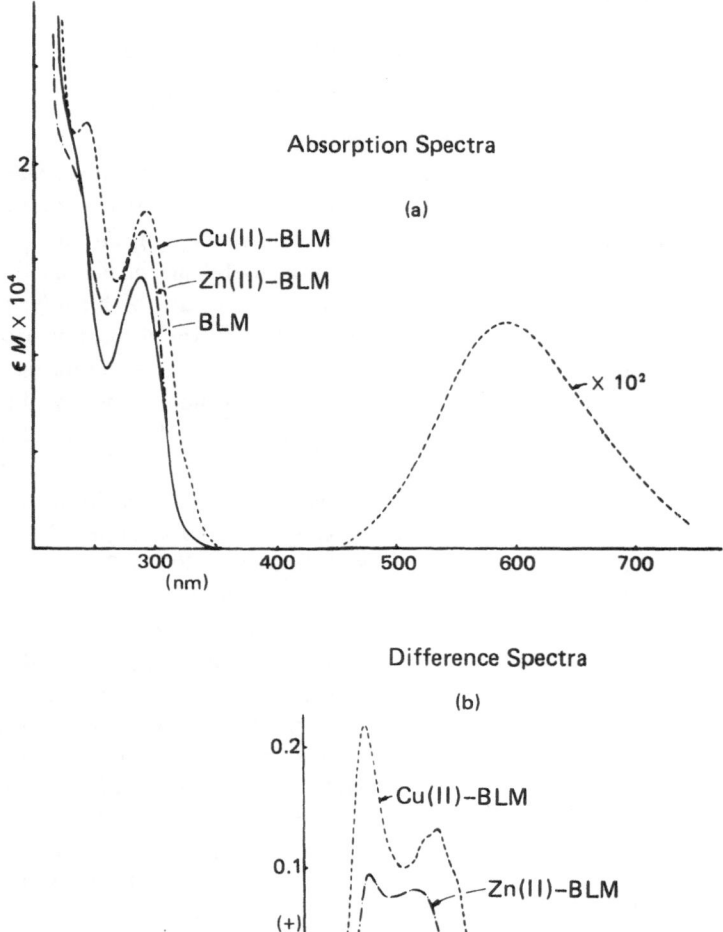

Fig. 3. (a) The absorption spectra of bleomycin A_2 (——), Cu(II)·bleomycin A_2 (----), and Zn(II)·bleomycin A_2 (-.-) in water. (b) The difference spectra of Cu(II)·bleomycin A_2 (----) and Zn(II)·bleomycin A_2 (-.-) versus bleomycin A_2.

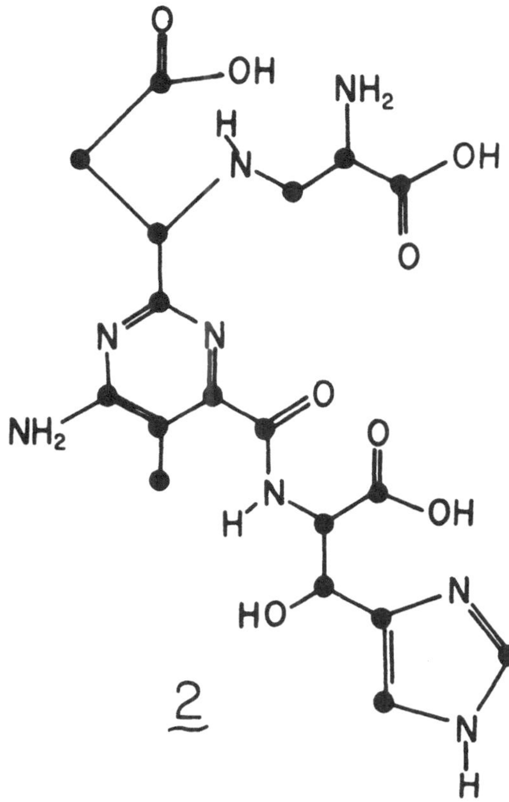

2

that observed for the C-4 resonance (C-2, δ7.86 → 8.11 vs C-4, δ7.32 → 7.40) the imidazole residue probably binds to the zinc ion via N(1) rather than N(3). The latter case should yield shifts of equal magnitudes for the two proton resonances. The small shifts observed for the C-5 and C-5' proton resonances upon metallation (Fig. 4) appear to be due to conformational and/or gross electronic changes which occur in the drug upon metal binding. The earlier presented visible absorption results and the ^{13}C-NMR studies, which will be subsequently discussed, clearly show that the bithiazole is not a metal-ligating group.

The binding of the pyrimidine residue causes a change in the position of the pyrimidine methyl resonance (C-11 of 1). This resonance shifts 0.41 ppm to lower field upon zinc binding (Fig. 4). Since the magnitude and sign of this shift are consistent with those observed for zinc binding to other nitrogen heterocycles[31], the pyrimidine residue is a metal-ligating group.

The establishment of the fourth metal binding group (after the amino group, the pyrimidine moiety, and the imidazole function) rests heavily on earlier

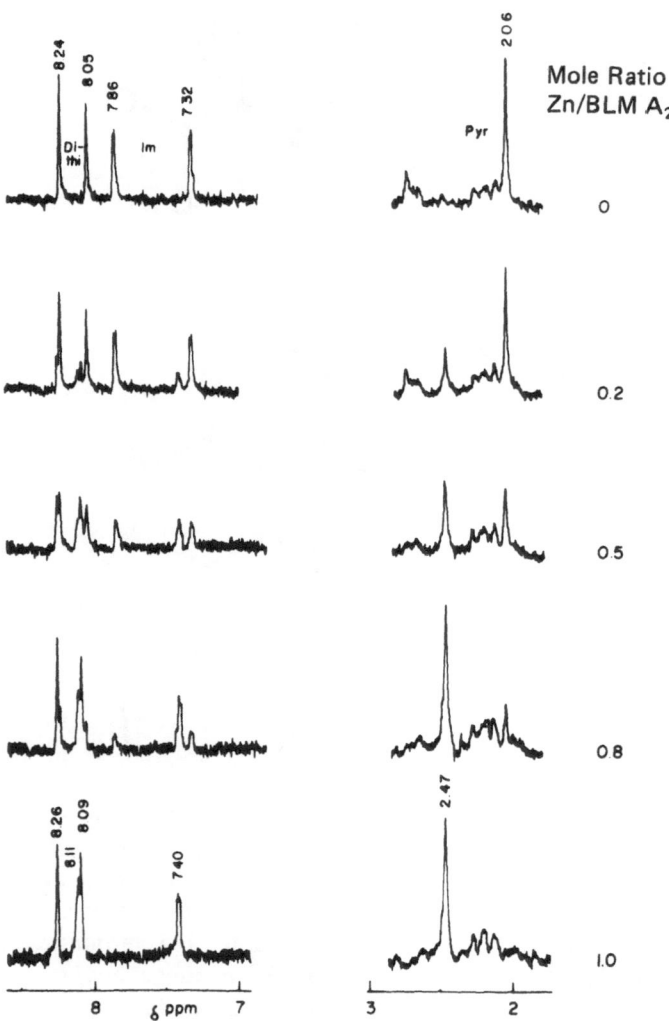

Fig. 4. The ^{1}H-NMR spectral changes which occur upon the formation of Zn(II)·BLM A$_2$.

chemical evidence presented by Umezawa[32] and on ^{13}C-NMR data collected in our laboratories[22]. Umezawa demonstrated that the migration of the carbamoyl function of the mannose moiety of BLM to give isobleomycin (migration from C-22 to carbon atom 21 of 1) does not occur for Cu(II)–BLM. On this basis he suggested that the carbamoyl function was a metal-ligating group. ^{13}C-NMR investigations on the drug and its metal complexes have supported this suggestion. Umezawa and co-workers[33] have also recently reported the complete carbon assignment of bleomycin A_2. The binding of Zn(II) to the antibiotic causes 42 of the 52 observable resonance lines to shift to new positions. Those atoms which sustain shifts are shown in Fig. 5a. Shifts in the mannose moiety as well as for the carbamoyl carbon atom (carbon atom 26) were observed. The binding of Cu(II) to bleomycin, on the other hand, eliminated nearly half of the ^{13}C-NMR resonances from the spectrum of the drug (Fig. 5b). While the sugar resonances are essentially unaffected by the presence of the paramagnetic ion, carbon atom 26, which if the carbamoyl nitrogen were bound to the copper ion, should be and is missing in the spectrum of Cu(II)–BLM A_2. Since neutral monodentate amide functions are poor metal donors[34,35], this group, if bound by the nitrogen atom, is very likely deprotonated in the metal complex. Reference to Fig. 2 shows that this deprotonation must occur below pH ~4.

II. The Iron Bleomycins

The importance of the iron–bleomycins in the mechanism of action of the pharmaceutical has been alluded to previously[12,13,16]. As a prerequisite to studying the ternary complex, iron–bleomycin–DNA, we have begun to examine the structural and chemical aspects of the binary system, iron–bleomycin. Our preliminary findings are outlined below.

The binding of Fe(II) to BLM A_2 yields a pinkish-orange air-sensitive complex. Titration data taken on Fe(II)–BLM A_2 indicates that like the Cu(II) and Zn(II) analogs, the imidazole function and the amino group of the α-amino carboxamide moiety bind to the metal ion. These pK_as (4.5 and 7.1) are missing from the titration curve of Fe(II)–BLM A_2. Moreover, electrochemical measurements (discussed in greater detail in the following section) show that the pyrimidine residue is also an iron-ligating group[36]. On the basis of these data the binding site for Fe(II) on the drug appears to be similar to that utilized by Cu(II) and Zn(II).

Neither the spin state nor the coordination number of the metal ion in Fe(II)–BLM A_2 is known. The divalent state of the metal is characterized by rich magnetic and stereochemical behavior and $S = 0$, 1, and 2 spin states[37] and coordination numbers of 4, 5, and 6 have been documented[29].

The Fe(II)–bleomycin complex is ESR quiet. This behavior is expected for the $S = 0$ state which is diamagnetic, but of the remaining spin possibilities, $S = 1$

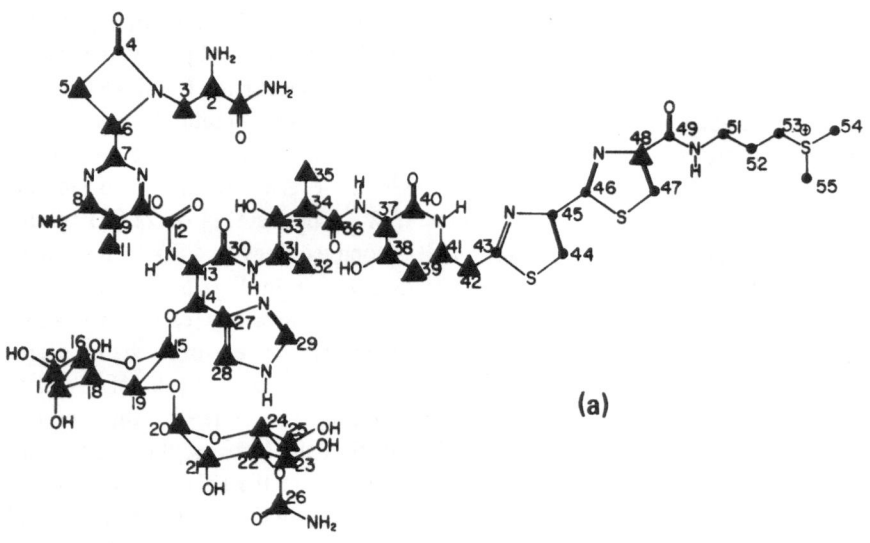

(a)

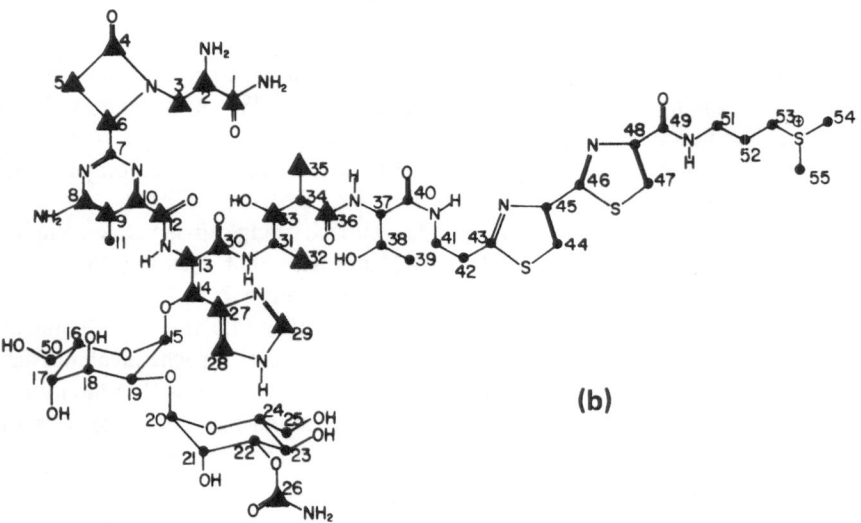

(b)

Fig. 5. The carbon resonances which shift when Zn(II) binds to bleomycin are indicated (▲) on the structure of bleomycin A₂ (a). The carbon resonances which are missing in the ¹³C-NMR spectrum of Cu(II)–BLM A₂ are indicated (▲) on the structure of the drug (b).

and $S = 2$, only the former is a resonable ESR candidate[38]. However, this state is very rare and, on that basis alone, it is not considered a likely possibility for Fe(II)–BLM A_2. The $S = 2$ state has a relatively short spin-lattice relaxation time and as such ESR spectra of high spin Fe(II) complexes are difficult to obtain.

The visible and near ultraviolet absorption spectrum of the complex exhibits two broad absorption bands both having molar extinction coefficients of less than ~600[13]. Although more confirmatory information is necessary, the position and intensity of the bands suggest that they are Laporte forbidden spin allowed transition of the d^6 metal ion. By analogy with other iron complexes, the most likely possibility is an $S = 0$, 6-coordinate Fe(II) environment for the bleomycin-bound iron ion[29].

Exposing Fe(II)–BLM A_2 to oxygen yields a number of ESR active species. ESR signatures for one high spin ($S = 5/2$) and two low spin ($S = 1/2$) iron(III)–bleomycin complexes can be observed (Fig. 6a). One of the $S = 1/2$ iron(III) complexes is unstable, and its ESR spectrum disappears at room temperature within a matter of minutes. The measured ESR g values for the unstable complex were $g_x = 2.41$, $g_y = 2.18$, and $g_z = 1.90$. The other low spin complex, $g_x = 2.25$, $g_y = 2.17$, and $g_z = 1.94$, appears to be stable indefinitely. No loss in signal for this complex was observed over a period of several days. The high spin complex, Fe(III)–BLM A_2 ($S = 5/2$), exhibits a single ESR transition at $g = 4.3$ indicating that the ion is bound to the drug and that it resides in a completely rhombic crystal field[39].

Under certain conditions the ESR spectrum of a nitroxide radical formed by reaction of a radical with N-tert-butyl-α-phenyl nitrone (PBN) can be observed in solutions containing Fe(III)–BLM formed by air oxidation of Fe(II)–BLM. The radical ESR spectrum (Fig. 6b) shows hyperfine coupling to one nitrogen and one β-hydrogen of the nitroxide. The ESR parameters are $g_{iso} = 2.0056$, $|A(H)|_{iso} = 16.0$ G, $|A(H)|_{iso} = 3.9$ G. Although the ESR parameters of nitroxide radicals are known to be solvent dependent, the observed parameters do not fall within the range expected for nitroxide radicals formed from either $\cdot$OH or O_2^- radicals[40].

The specific donor atoms utilized by bleomycin in binding Fe(III) are unknown. However, by analogy with Fe(III) porphyrins[41] the ($S = 1/2$) Fe(III)–bleomycin complexes must have a primarily nitrogen donor environment. Thus, on this basis, the site proposed for the Cu(II)– and Zn(II)–bleomycin complexes appears reasonable for the low spin Fe(III) analogs as well.

III. Electrochemistry of Bleomycin and Its Metalloderivatives

The proposed mechanism of action of the antibiotic underscores the importance of examining the electrochemical properties of the drug and its metalloderivatives. Initial investigations have shown that electrochemistry is not only useful for examining the redox characteristics of the system but it can be also used as

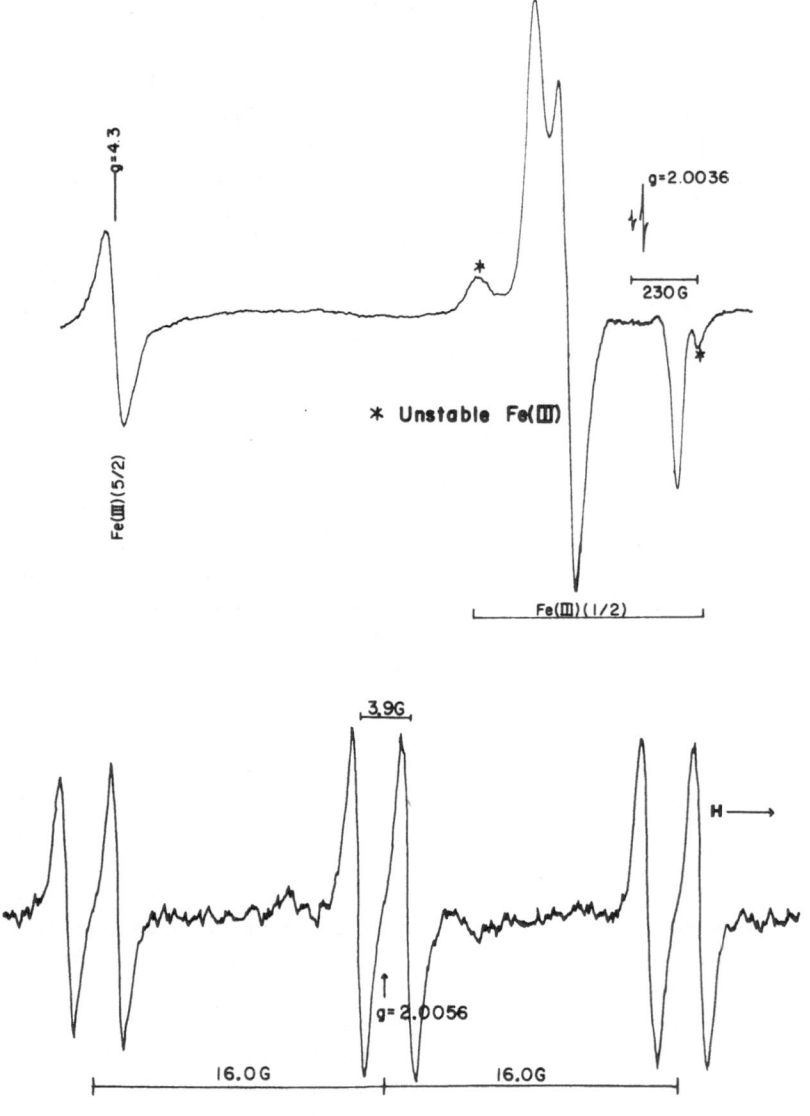

Fig. 6. (a) The ESR spectrum of oxygen-produced Fe(III)–BLM A$_2$. (b) The ESR of the radical produced in the oxidation of Fe(II)–BLM to Fe(III)–BLM by molecular oxygen in saturated solutions of PBN is shown (b). The oxidation was carried out in a pH 8.0 phosphate buffer solution.

a site specific probe of metallation and for studying other binding phenomena related to the antibiotic.

The filtered DC polarogram of bleomycin A$_2$ as its chloride salt is shown in

Fig. 7a. The drug shows a two electron reduction wave at –1.22 V but does not show an oxidation process in the accessible potential range. Using cyclic voltammetry (CV) the reduction process was found to be totally irreversible and the irreversibility was independent of the CV scan rate. Bleomycin A_2 chloride also exhibits a second large multielectron reduction wave at –1.48 V. Cyclic voltammetric experiments with a hanging mercury drop electrode (HDME) indicated that not only is this wave electrochemically irreversible but that the

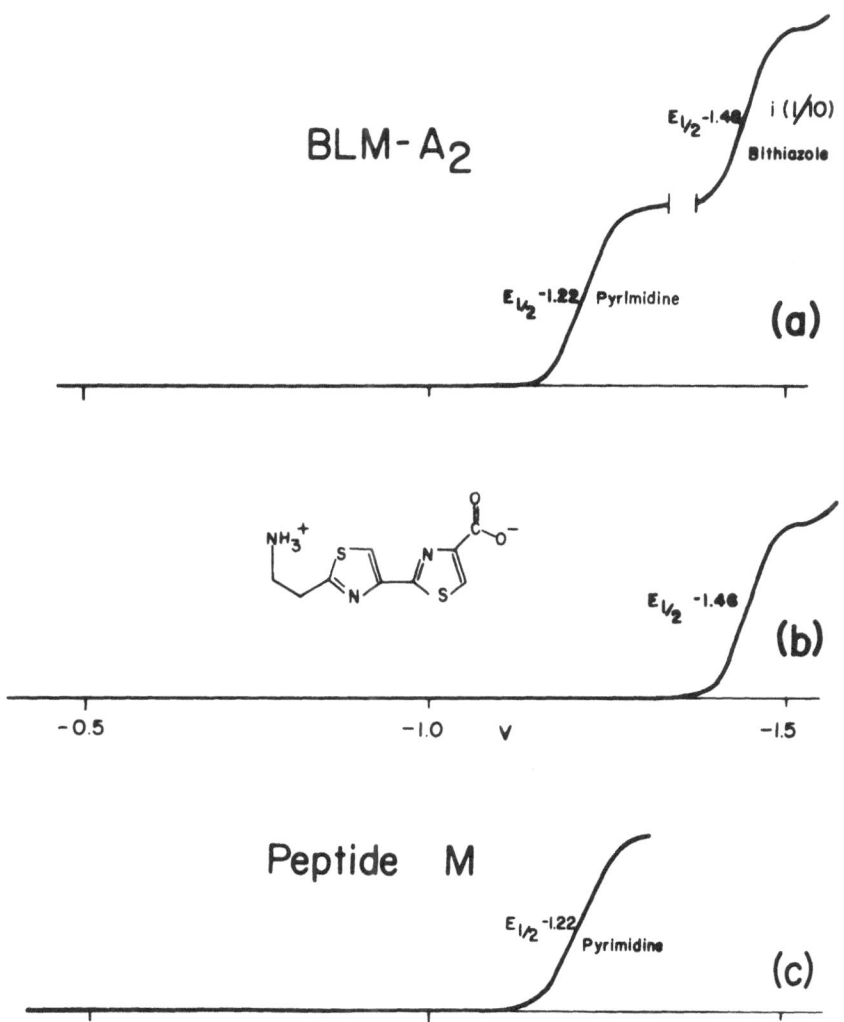

Fig. 7. The filtered DC polarograms of BLM A_2 Cl⁻ (a), bithiazole, 3, (b), and peptide M (c).

magnitude of the current associated with the wave is large, and appears to be related to the length of time that the Hg drop is in contact with the bleomycin solution. Thus the wave appears to be a catalytic wave[42].

The DC polarogram of the amino acid bithiazole, 3, is shown in Fig. 7b. The compound exhibits a single multielectron reduction process at –1.48 V. Since the position and the behavior of the wave toward an HDME electrode are similar to that observed for bleomycin, the multielectron wave at –1.46 V in the polarogram of the drug must be associated with a reduction process associated with the bithiazole moiety.

$$\underset{\widetilde{}}{3}$$

The assignment of the first reduction wave of bleomycin (–1.22 V) was accomplished via electrochemical studies on peptide M, 2. The peptide contains an open β-lactam ring system and the pyrimidine and imidazole moieties of the antibiotic. Both the bithiazole and the dimethylsulfonium residues of bleomycin are missing in peptide M. The DC polarogram of the cleavage fragment is shown in Fig. 7c. Identical to that of bleomycin, the fragment exhibits a two electron reduction wave at –1.22 V. Also noteworthy is that the polarogram (Fig. 7c) lacks the multielectron reduction process found for bleomycin A_2 at –1.46 V, thus confirming the assignment made earlier for the wave. The polarographic behavior of peptide M strongly suggests that the first two electron reduction wave of the antibiotic (Fig. 7a) is due to either the pyrimidine or the imidazole residue of the glycopeptide. Since imidazoles are generally difficult to reduce[36] but pyrimidines are known to exhibit rich electrochemical behavior[43], we have assigned the reduction wave at –1.22 V to the pyrimidine residue of the drug. Confirmation of this assignment has been obtained from the electrochemical–pH profile of the antibiotic. At acidic pHs the first reduction process of BLM A_2 splits into two waves. The DC polarogram at pH 3.4 for BLM A_2 is shown in Fig. 8 and a plot of the observed $E_{1/2}$ values of the pyrimidine wave as a function of the pH is shown in Fig. 9. The two waves found at low pH eventually coalesce to form a single wave at pH ~5. In the acid region the wave to less negative potentials is about three times as sensitive to pH effects as is the other wave. Since this behavior is very similar to that observed for a

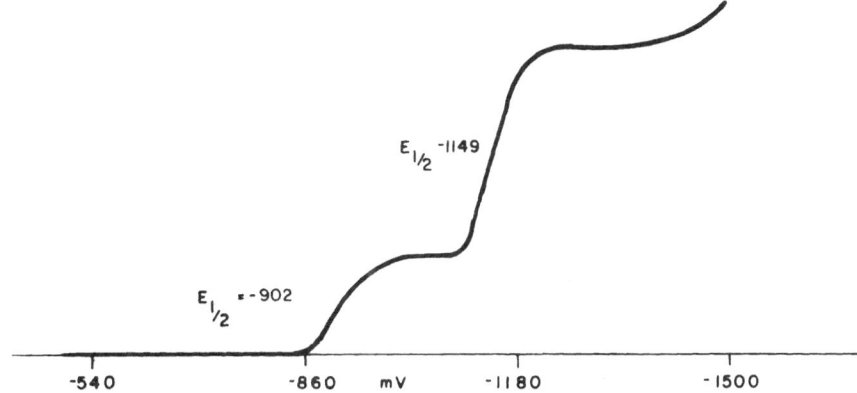

Fig. 8. The filtered DC polarogram of BLM A_2, Cl⁻ at pH = 3.4.

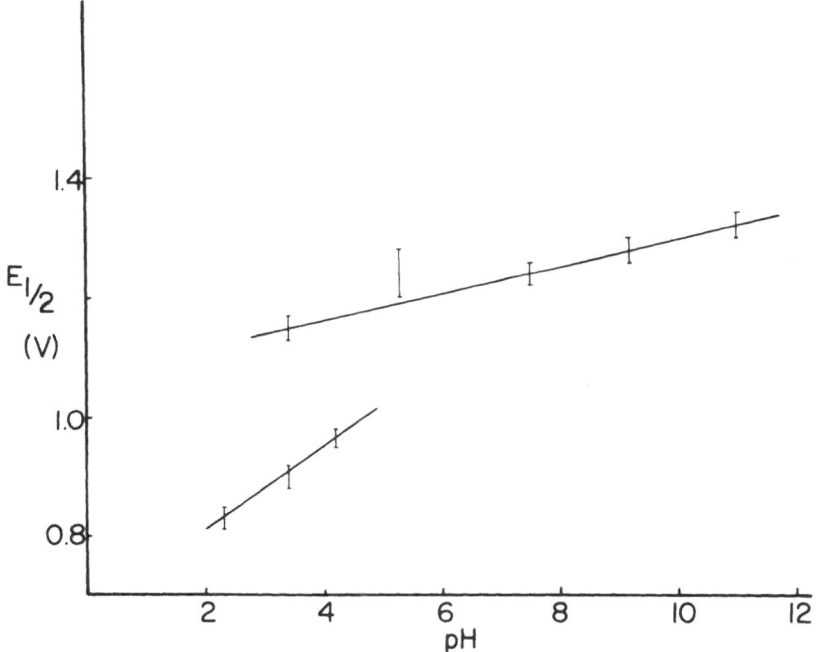

Fig. 9. A plot of the $E_{1/2}$ values of the pyrimidine wave of BLM A_2 Cl⁻ versus pH.

number of substituted pyrimidines[43, 44], the two electron reduction wave at –1.22 V in the polarogram of bleomycin must be due to the reduction of the pyrimidine moiety.

The binding of Fe(II), Co(II), and Zn(II) to the antibiotic causes the –1.22 V reduction wave to completely disappear from the polarogram of the drug. Identical behavior was observed for the Fe(III)–BLM A_2 complexes produced by the oxidation of Fe(II)–BLM A_2 with oxygen. The similarity in polarographic behavior of all of these metal complexes combined with the fact that the pyrimidine residue has been established as a metal-ligating group for Zn(II) (from NMR and visible absorption studies) shows that the pyrimidine residue binds to all of these ions. Unaffected by complexation is the bithiazole wave at –1.46 V.

The divalent Fe, Cu, and Zn bleomycins are passive to electrochemical oxidation. At first glance this behavior for Fe(II)–BLM A_2 appears to be incompatible with the documented[12,13,36] extreme air sensitivity of the complex. However, if the oxidation of the iron ion in Fe(II)–BLM A_2 proceeds via an inner sphere mechanism, i.e., direct binding of dioxygen to the coordinated iron ion, the polarographically active species is not the same species which is oxidized by the chemical oxidant. The lack of an oxidation wave for Fe(II)–BLM A_2 suggests that one of the iron coordinating groups is displaced by dioxygen (not necessarily one of the drug atoms) prior to metal oxidation.

Copper(II)–bleomycin displays polarographic behavior which is different than that observed for the above-mentioned metal complexes. The partial DC polarogram of Cu(II)–BLM A_2 is shown in Fig. 10. The complex exhibits a one electron metal centered reduction wave at –0.53 V, [Cu(II) ⟶ Cu(I)] as well as unshifted pyrimidine (–1.22 V) and bithiazole (–1.46 V) reduction waves. Since it is known that the pyrimidine residue is a Cu(II)-ligating group, the polarographic results appear to be at odds with the earlier presented spectroscopic analysis[21]. However the reduction of the metal ion occurs before the reduction of the pyrimidine moiety (–0.53 vs –1.22 V). Thus, at potentials more negative than $\sim$ –0.6 V the polarographic experiment is actually examining the reduction properties of Cu(I)–BLM A_2. These results indicate that while the pyrimidine binds to Cu(II) it does not bind to Cu(I).

Electrochemical efforts to study the Cu(II)–Cu(I) reduction couple before the pyrimidine–Cu(I) bond breaks have failed. CV experiments at rapid scan time have shown totally irreversible behavior (no reverse peak) for the Cu(II)–Cu(I) couple of Cu(II)–BLM A_2.

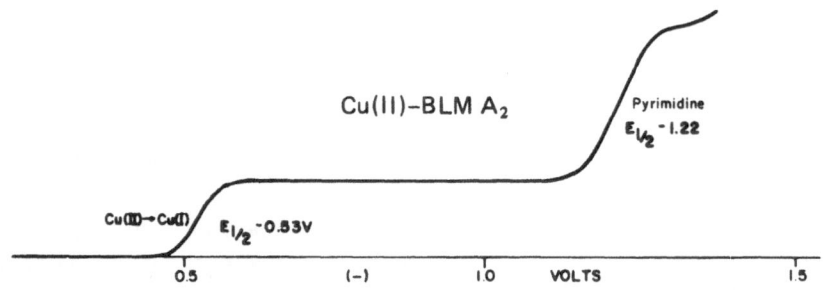

Fig. 10. The filtered DC polarogram of Cu(II)–BLM A_2.

IV. Metal Binding and the Mechanism of Action of the Drug

It has been proposed that bleomycin is part of a ternary complex formed between the drug, an iron ion, and DNA[12, 13]. This complex is capable of producing a radical and it is the radical which leads to DNA breakage within a tumor cell. The source of the radical is the oxygen-induced oxidation of the bleomycin-bound iron ion. Since divalent transition metal ions [Fe(II) included] bind to the β-lactam–pyrimidine–imidazole portion of the antibiotic while the bithiazole and the terminal R group of the drug are apparently used for DNA binding[45], the concept of DNA damage via a ternary complex, metal ion–drug–DNA, is intact. Bleomycin functional groups required for metal binding are not the same ones required for DNA binding.

Studies have shown that thymine is the most often released base in the bleomycin-promoted DNA-degrading process. Although specificity and a free radical process are not normally thought of as being compatible, the structural aspects of the ternary complex, metal ion–drug–DNA, and its effect on the specificity of the reaction are totally unknown. An inner sphere oxidation process followed by the directional release of a radical within a highly asymmetric environment could give rise to selectivity. It is hoped that through structural studies on the ternary complex a more detailed understanding of the functioning of the pharmaceutical will evolve.

Acknowledgments

We wish to thank Dr. S. T. Crooke of Bristol Laboratories for generously supplying samples of peptide M and Bleomycin A_2. We also wish to thank Mr. Robert Grulich for his help with the NMR studies on the drug. The bithiazole, 3, was generously supplied by Professor S. Hecht of the Department of Chemistry, Massachusetts Institute of Technology, Cambridge, Massachusetts.

References

1. H. Umezawa, K. Maeda, T. Takeuchi, and Y. Akami, *J. Antibiot. (Tokyo)*, Ser. A., **19**, 200 (1969).
2. A new structure for bleomycin and related antibiotics was proposed by Drs. Umezawa and Takita during the symposium.
3. H. Umezawa, *Biomedicine*, **18**, 459 (1973).
4. R. H. Blum, S. K. Carter, and K. A. Agre, *Cancer*, **31**, 903 (1973).
5. L. F. Povirk, W. Wübker, W. Kohnlein, and F. Hutchinson, *Nucl. Acid Res.*, **4**, 3573 (1977).
6. C. W. Haidle, *Mol. Pharmacol.*, **7**, 645 (1971).
7. C. W. Haidle, K. K. Weiss, and M. T. Kuo, *Mol. Pharmacol.*, **8**, 531 (1972).
8. M. T. Kuo and C. W. Haidle, *Biochim. Biophys. Acta*, **335**, 109 (1974).

9. W. E. G. Müller, Z. Yamazaki, H. J. Breter, and R. K. Zahn, *Eur. J. Biochem.*, **31**, 518 (1972).

10. A. Kono, Y. Matsushima, M. Kojima, and T. Maeda, *Chem. Pharm. Bull. (Tokyo)*, **25**, 1725 (1977).

11. K. Nagai, H. Yamaki, H. Suzuki, N. Tanaka, and H. Umezawa, *Biochim. Biophys. Acta*, **179**, 165 (1969).

12. E. A. Sausville, J. Peisach, and S. B. Horwitz, *Biochem. Biophys. Res. Commun.*, **73**, 814 (1976).

13. J. W. Lown and S. Sim, *Biochem. Biophys. Res. Commun.*, **77**, 1150 (1977).

14. I. Shirakawa, M. Azegami, S. Ishii, and H. Umezawa, *J. Antibiot. (Tokyo)*, **24**, 761 (1971).

15. M. Takeshita, A. P. Grollman, and S. B. Horwitz, *Virology*, **69**, 453 (1976).

16. K. Takahashi, O. Yoshioka, A. Matsuda, and H. Umezawa, *J. Antibiot. (Tokyo)*, **30**, 861 (1977).

17. H. Suzuki, K. Nagai, H. Yamaki, N. Tanaka, and H. Umezawa, *J. Antibiot. (Tokyo)*, **21**, 379 (1968).

18. T. Ichikawa, A. Matsuda, K. Miyamoto, M. Tsubosaki, T. Kaihara, K. Sakamoto, and H. Umezawa, *J. Antibiot. (Tokyo)*, **20**, 149 (1967).

19. M. Ishizuka, H. Takayama, T. Takeuchi, and H. Umezawa, *J. Antibiot. (Tokyo)*, **20**, 15 (1967).

20. H. Umezawa, M. Ishizuka, K. Kimura, J. Iwanaga, and T. Takeuchi, *J. Antibiot. (Tokyo)*, **21**, 592 (1968).

21. J. C. Dabrowiak, F. T. Greenaway, W. E. Longo, M. vanHusen, and S. T. Crooke, *Biochim. Biophys. Acta*, **517**, 517 (1978).

22. J. C. Dabrowiak, F. T. Greenaway, and R. Grulich, *Biochemistry*, **17**, 4090 (1978).

23. F. H. Fraser and D. J. Macero, *Chem. Instrum.*, **4**, 97 (1972).

24. J. Peisach and W. E. Blumburg, *Arch. Biochem. Biophys.*, **165**, 691 (1974).

25. J. C. Dabrowiak, W. E. Longo, and F. S. Santillo, unpublished results.

26. M. Konishi, K. Saito, K. Numata, T. Tsuno, K. Asama, H. Tsukiura, T. Naito, and H. Kawaguchi, *J. Antibiot. (Tokyo)*, **30**, 789 (1977).

27. G. Koyama, H. Nakamura, Y. Muraoka, T. Takita, K. Maeda, and H. Umezawa, *Tetrahedron Lett.*, **44**, 4635 (1968).

28. H. Asakura, M. Mori, and H. Umezawa, *J. Antibiot. (Tokyo)*, **28**, 537 (1975).

29. F. A. Cotton and G. Wilkinson, *Advanced Inorganic Chemistry*, 3rd ed. Interscience Publishers, New York, (1972).

30. E. J. Billo, *Inorg. Nucl. Chem. Lett.*, **10**, 613 (1974).

31. S. K. Gupta, T. S. Srivastava, *J. Inorg. Nucl. Chem.*, **32**, 1611 (1970).

32. H. Umezawa, *Fed. Proc.*, **33**, 2296 (1974).

33. H. Naganawa, Y. Muraoka, T. Takita, and H. Umezawa, *J. Antibiot. (Tokyo)*, **30**, 388 (1977).

34. T. F. Dorigatti and E. J. Billo, *J. Inorg. Nucl. Chem.*, **37**, 1515 (1975).

35. G. L. Eichhorn, *Inorganic Biochemistry*, Elsevier, New York (1973).

36. J. C. Dabrowiak and F. S. Santillo, unpublished results.

37. S. Koch, R. H. Holm, and R. B. Frankel, *J. Am. Chem. Soc.*, **97**, 6714 (1975).

38. B. R. McGarvey, *Trans. Metal Chem.*, **3**, 90 (1966).

39. W. E. Blumberg, In *Magnetic Resonance in Biological Systems* (A. Ehrenberg, B. E. Malmstrom, and T. Vanngard, Eds.), p. 119. Pergamon Press, London (1967).

40. E. G. Janzen, D. E. Nutter, Jr., E. R. Davis, B. J. Blackburn, J. L. Poyer, and P. B. McCay, *Can. J. Chem.*, **56**, 2237 (1978).

41. J. Peisach and W. E. Blumberg, *Adv. Chem. Ser.*, **100**, 271 (1971).

42. S. G. Mairanovskii, *J. Electroanal. Chem.*, **6**, 77 (1963).
43. B. Janik and P. J. Elving, *Chem. Rev.*, **68**, 295 (1968).
44. D. L. Smith and P. J. Elving, *J. Am. Chem. Soc.*, **84**, 2741 (1962).
45. M. Chien, A. P. Grollman, and S. B. Horwitz, *Biochemistry*, **16**, 364 (1977).

Metal Complex of Bleomycin and Its Implication for the Mechanism of Bleomycin Action

Tomohisa Takita

Bleomycin (BLM) is isolated from the fermentation broth of *Streptomyces verticillus* as a blue powder of its equimolar Cu(II) complex. This copper comes from inorganic cupric salt added to the fermentation medium; without addition of the copper, the fermentation yield of BLM is significantly reduced. The chelated copper can be removed from BLM by precipitation with hydrogen sulfide in methanol solution. Treatment of metal-free BLM with an excess of an inorganic cupric salt, such as cupric chloride, in neutral aqueous solution followed by chromatographic separation on CM-Sephadex regenerates the original equimolar Cu(II) complex in excellent yield, although there are many potential coordination sites in BLM.

In 1977, we reported the involvement of the α-amino group of the amine component of V [see Fig. 1 in Ref. [1]], the pyrimidine-ring nitrogen, the imidazole of IV, and the *O*-carbamoyl group in the copper complex [2]. Recently, Dabrowiak et al. [3] proposed a square-planar structure, which was constructed by arranging the four functional groups earlier assigned to us to the square-planar coordination sites using the initially proposed structure of BLM [4]. In this chapter the three-dimensional structure of the BLM–Cu(II) complex is presented based on our experimental results, which include the X-ray crystallographic analysis of P-3A Cu(II) complex, a biosynthetic intermediate structurally related to BLM. The implication of the metal complex in terms of the mechanism of BLM action is discussed.

The copper complex of BLM has not yet been crystallized. Therefore, the structure of the complex has been studied chemically and by spectroscopic methods. The copper complexes of all natural BLMs give the same electronic and CD spectra, which suggests that the terminal amine is not involved in metal binding. The UV spectrum of the Cu complex of BLM is different from that of the Cu complex of phleomycin [5,6] (Fig. 1). However, bleomycin and phleomycin give the same absorption in the visible region (Fig. 1). The difference UV absorption spectrum between copper-chelated and metal-free BLM is the same as that of phleomycin, as shown in Fig. 2. The spectra were superimposable, as

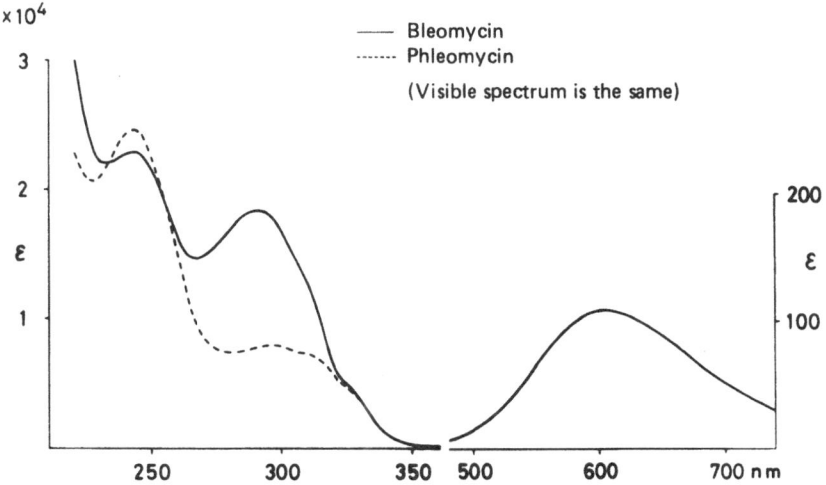

Fig. 1. Electronic absorption spectra of Cu complexes of bleomycin and phleomycin.

were the visible absorption spectra. This indicates that the bithiazole chromophore of BLM is not involved in the metal binding and that the Cu complexes of bleomycin and phleomycin have the same coordination geometry.

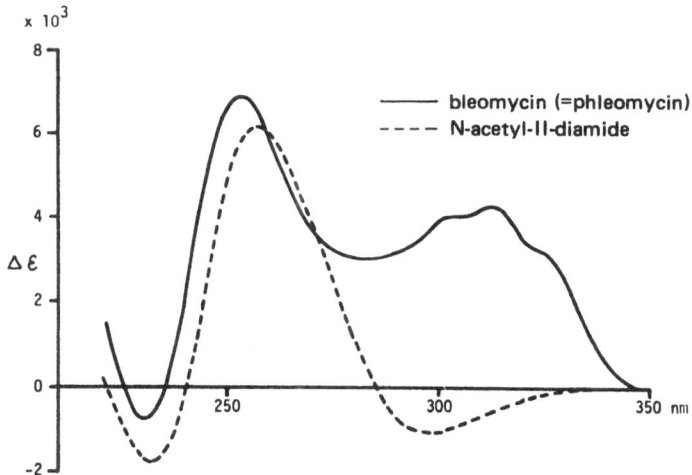

Fig. 2. Difference UV spectra of bleomycin between Cu-chelated and metal-free form, and of N-acetyl-II-diamide between 0.1 N HCℓ and H_2O.

The Coordination Sites of BLM–Cu(II) Complex as Assigned by Chemical and Spectroscopic Studies

(i) The free primary amino group of BLM [see Fig. 3[6]] was shown to be involved in the metal binding on the basis of the following two experimental results. The first of these was the disappearance of its dissociation constant. The pK_a value of this amino group in metal-free BLM A_2 is 7.4. However, in Cu-chelated A_2 there is no ionizable group between pK_a 4 and 9. The second observation is that this amino group can be acylated in metal-free BLM, but not in BLM–Cu complex, under (weakly basic) Schotten–Baumann reaction conditions.

(ii) Participation of the imidazole of compound IV in the metal binding was also suggested on the basis of two experimental results. The first was the disappearance of its dissociation constant at pK_a 4.7 upon addition of Cu(II) to metal-free BLM. The second was that the Pauly color reaction shown by metal-free BLM disappeared upon formation of the Cu chelate, as in the case of histidine. From the steric requirements imposed by the molecule as a whole, N^π but not N^τ of the imidazole could be shown to occupy the coordination site.

(iii) Participation of the pyrimidine in metal binding was suggested by formation of epi-BLM upon treatment of Cu-chelated BLM, but not metal-free BLM, under mildly alkaline conditions, as described below. An extremum at 253 nm in the difference UV spectrum between copper-chelated and metal-free BLM also

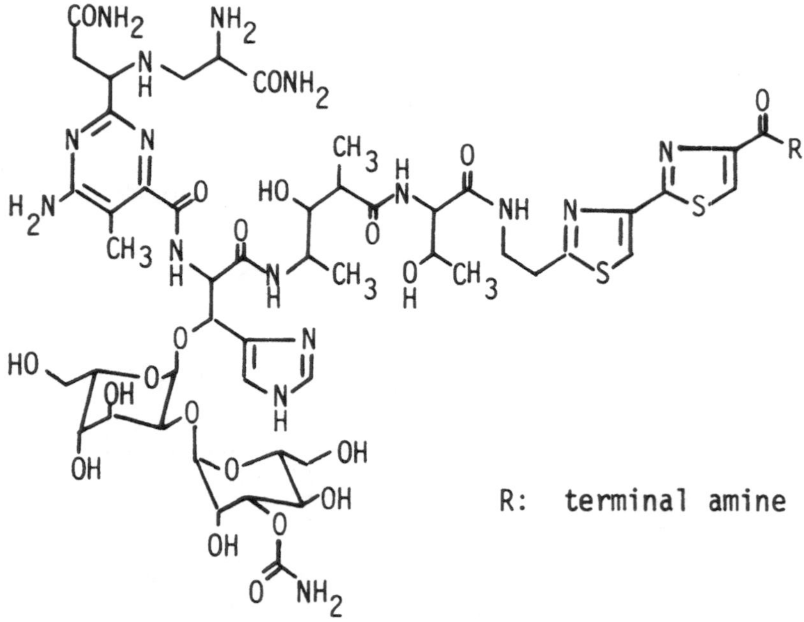

Fig. 3. Structure of bleomycin.

can be explained by formation of a coordination bond between the pyrimidine and copper (Fig. 2). This was shown by the difference UV spectrum of N-acetyl-II-diamide, a model compound for the pyrimidine chromophore. That is, an extremum at 256 nm in the UV spectrum of the model compound between the protonated and the intact pyrimidine chromophore corresponds to the extremum at 253 nm in the difference UV spectrum between the coordinated and the free pyrimidine chromophore of BLM. Parenthetically, it may be noted that another extremum at 314 nm with an inflection at each side is due to the charge transfer transition and a weak broad asymmetric absorption centered at about 600 nm (see Fig. 1) is due to the d–d transition.

(iv) The deprotonated amide nitrogen of IV was inferred to occupy the fourth coordination site. The existence of a deprotonated functional group as one of the coordination sites was suggested from the potentiometric titration and chromatographic behavior on CM-Sephadex. When CM-Sephadex chromatography is effected using pH 6.0 buffer, the retention time of the Cu complex of BLM is almost the same as that of its metal-free form. This means that at pH 6.0 metal-free BLM and its Cu complex have the same net positive charge. At this pH, metal-free BLM has one positive charge in addition to the terminal amine. Therefore, the Cu(II) complex of BLM must have a deprotonated functional group as one of the coordination sites. The amide nitrogen of IV is located in a favorable position for the coordination site to form 5- and 6-membered chelate rings. Therefore, the deprotonated amide nitrogen atom of IV was inferred to occupy one of the coordination sites.

(v) The fifth coordination site is presumably occupied by the oxygen atom of the carbamoyl group at the 3-O-position of the mannose. When metal-free BLM is kept in aqueous alcohol containing triethylamine at room temperature, the carbamoyl group at the 3-O-position of mannose migrates to the 2-O-position. This product was named iso-BLM[7]; an equilibrium exists between BLM and iso-BLM. The molar ratio of BLM to iso-BLM after equilibration was 2 to 3. On the other hand, when the Cu complex of BLM was maintained under the same conditions, epi-BLM, but not iso-BLM, was formed. Epi-BLM[2] differs from BLM only in the stereochemistry at the α-methine carbon of the pyrimidine substituent. The formation of epi-BLM can be explained by the increased acidity of the allylic methine proton induced by complex formation between N-1 of the pyrimidine and copper. The epimerization is an irreversible reaction, ostensibly because epi-BLM is thermodynamically more stable than BLM (Fig. 4).

When Cu(II) complexes of BLM, iso-BLM, and epi-BLM were eluted from a CM-Sephadex column with pH 4.5 buffer, they eluted in the order epi-BLM, BLM, and then iso-BLM. These retention times reflect the stabilities of the complexes. That is, iso-BLM is most easily protonated, so its retention time is the longest, while epi-BLM is protonated with the greatest difficulty, so its retention time is the shortest.

When iso-BLM Cu complex was maintained in basic, aqueous alcohol, it was transformed to epi-BLM via BLM (Fig. 5). This indicates that the O-carbamoyl

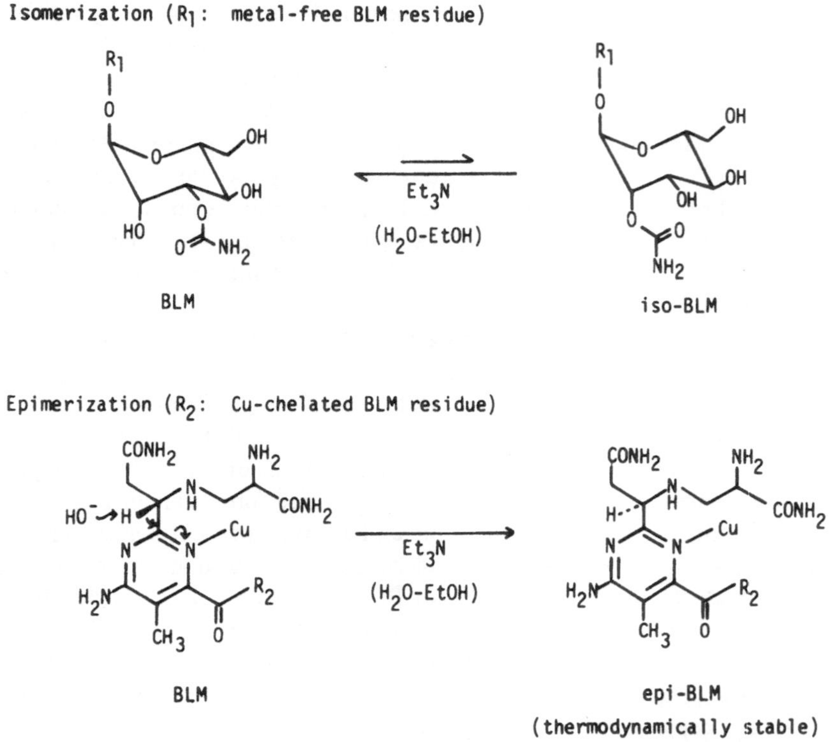
Isomerization (R_1: metal-free BLM residue)
Et$_3$N
(H$_2$O-EtOH)
BLM
iso-BLM
Epimerization (R_2: Cu-chelated BLM residue)
Et$_3$N
(H$_2$O-EtOH)
BLM
epi-BLM
(thermodynamically stable)

Fig. 4. Structural relationship between bleomycin, epibleomycin and isobleomycin.

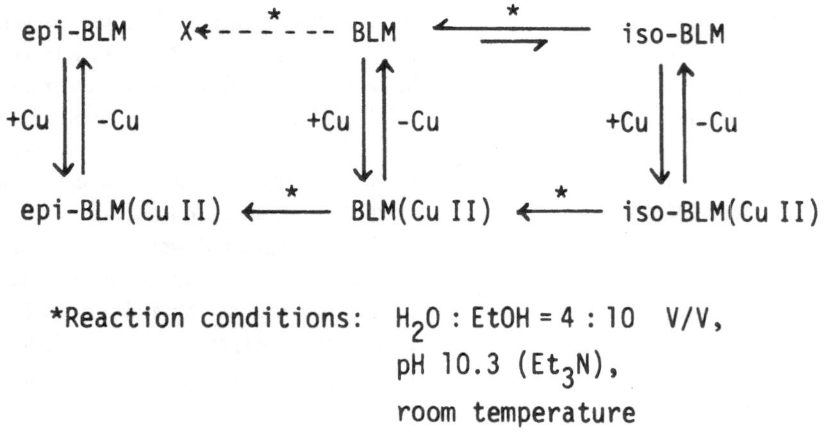
epi-BLM
X
BLM
iso-BLM
+Cu
-Cu
epi-BLM(Cu II)
BLM(Cu II)
iso-BLM(Cu II)
*Reaction conditions: H_2O : EtOH = 4 : 10 V/V,
pH 10.3 (Et_3N),
room temperature

Fig. 5. Conversion of bleomycin into isobleomycin and epibleomycin.

moiety of iso-BLM is free in its Cu complex, or does not take part in the coordination, while that of epi-BLM is fixed, i.e., takes part in the coordination. The carbamoyl group of BLM may be inferred to participate in the coordination, because the isomerization of the Cu complex of BLM cannot be observed in the initial stage of the reaction, although the rate of the isomerization of metal-free BLM is faster than that of the epimerization of the Cu complex of BLM.

The electronic absorption spectra of the copper complexes of BLM, iso-BLM, and epi-BLM were essentially the same, but the CD spectra were distinctly different from each other (Fig. 6). The difference in the CD spectra of BLM and epi-BLM is obviously explained by the difference in stereochemistry about a carbon present in a small chelate ring. The difference in the CD spectra of BLM and iso-BLM can be explained by strain of the chelate rings of BLM, caused by participation of the carbamoyl group in the coordination. If the carbamoyl group of BLM did not take part in the coordination, it would give the same CD spectrum as iso-BLM.

When the carbamoyl group of BLM–Cu complex participates in the coordination, the carbonyl oxygen, rather than the deprotonated amide nitrogen, probably occupies the coordination site. This seems likely because this carbamoyl group is remote from the other coordination sites and is like a monodentate amide, which usually coordinates via the carbonyl oxygen, whereas the coordination of the deprotonated amide nitrogen of IV described earlier is stabilized by formation of two 5- and 6-membered chelate rings.

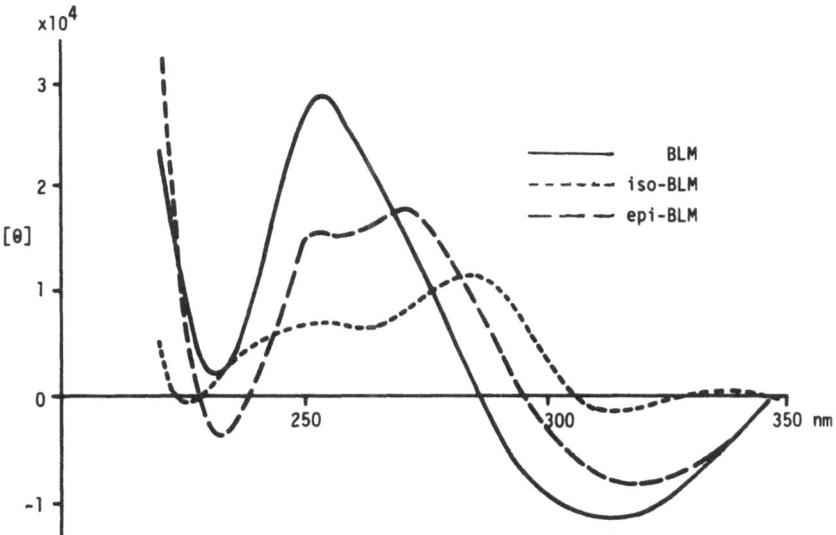

Fig. 6. CD spectra of Cu complexes of bleomycin, isobleomycin, and epibleomycin.

Three-Dimensional Structure of BLM–Cu(II) Complex

After the structural study of the copper complex of BLM by the chemical and spectrometric methods described above, the structure of P-3A copper complex, a biosynthetic intermediate structurally related to BLM, was clarified by X-ray crystallographic analysis[8]. The structure is shown schematically in Fig. 7. It has a square pyramidal coordination geometry with N^π and the deprotonated amide nitrogen of histidine, N-1 of the 4-aminopyrimidine ring, and the secondary amine as the square coordination sites and the primary amine as the apical coordination site. Four of the ligands of P-3A Cu(II) complex were also suggested to be ligated in BLM–Cu(II) complex, as described above. The secondary amine of BLM is also located in a favorable position to form two additional 5-membered chelate rings as P-3A Cu(II) complex. Thus BLM probably has the same coordination geometry as P-3A except for the carbamoyl group at the sixth coordination site[9] (Fig. 8). The bonding between the carbamoyl oxygen and copper appears to be very weak, because the ESR parameters of BLM and iso-BLM Cu(II) complexes were almost the same.

The Implication of the Metal Complex of Bleomycin for Its Mechanism of Action

Horwitz and co-workers[10, 11] recently reported that the conversion of SV40 DNA to acid-soluble products occurred at approximately equimolar levels of Fe(II), BLM, and DNA. Fe(III) did not substitute for Fe(II) in this reaction. Anaerobics inhibited the observed degradation of DNA by BLM and Fe(II). Optical spectral studies revealed that an oxygen-labile complex was formed between BLM and Fe(II).

We observed that metal-free BLM was decomposed in the presence of Fe(II) and oxygen, but not with Fe(III). This decomposition was also caused by a

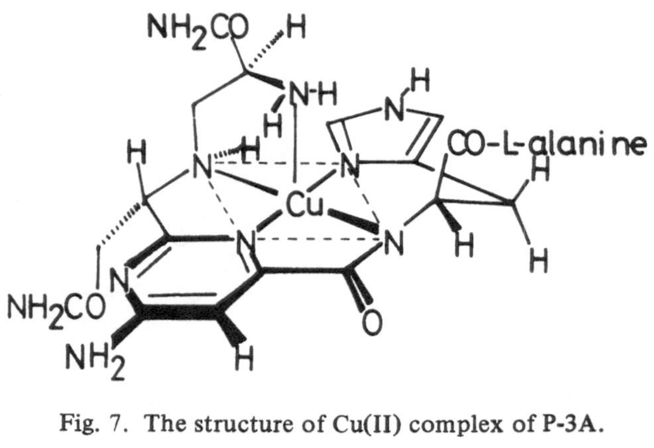

Fig. 7. The structure of Cu(II) complex of P-3A.

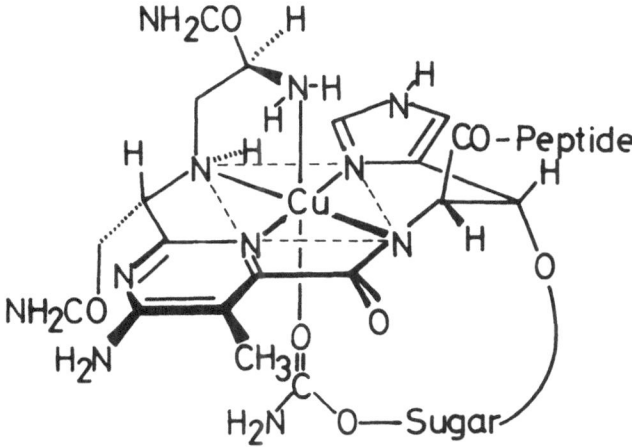

Fig. 8. The structure of the Cu(II) complex of bleomycin.

catalytic amount of Fe(II) when a reducing agent such as a sulfhydryl compound or ascorbic acid was present. However, BLM–Cu complex was not decomposed by Fe(II) and oxygen. We isolated the major product produced at the
initial stage of the decomposition of metal-free BLM. The ^{13}C-NMR spectrum
indicated that only the pseudodipeptide moiety had undergone reaction in the
presence of Fe(II) and oxygen and that the product lacked the pyrimidine substituent in the 2-position. The pyrimidine chromophore of this compound could
be decomposed completely by acid hydrolysis. We are studying the structure of
this reaction product. These results indicate that the active oxygen formed at
the sixth coordination site of BLM–Fe(II) complex reacts with BLM itself. It is
possible that the oxygen on BLM–Fe(II) complex is involved in the action of
BLM on DNA (Fig. 9).

If we assume that BLM–Fe(II) complex is involved in fragmentation of DNA
in cells, then the following two possibilities can be proposed. The first would be
that metal-free BLM in cells binds to DNA and thereafter Fe(II) is taken into
the metal-free BLM bound to DNA. The second would be that BLM–Cu complex[12] reaches the nucleus and binds to DNA and thereafter the copper in
the BLM–Cu complex bound to DNA is reductively removed[13] and replaced
by Fe(II).

BLM–[57]Co complex has been shown to bind to DNA of Ehrlich solid
tumor[14]; this complex neither inhibits the growth of the cells nor causes DNA
fragmentation, although bioactive metal-free BLM can be regenerated chemically from the bioinactive BLM–Co complex. It is anticipated that further study
of BLM–metal complexes will contribute to an understanding of the mechanisms
of cytotoxic and therapeutic action of BLM.

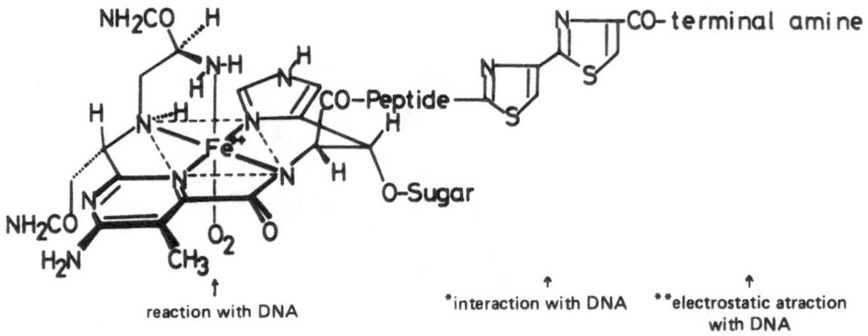

Fig. 9. The mechanism of bleomycin action. *, Interaction between bleomycin and DNA was suggested by Chien *et al.* (11); **, electrostatic attraction between terminal amine moiety of bleomycin and DNA was suggested by us[15].

References

1. T. Takita, this volume, p. 38.
2. Y. Muraoka, H. Kobayashi, A. Fujii, M. Kunishima, T. Fujii, Y. Nakayama, T. Takita, and H. Umezawa, *J. Antibiot. (Tokyo)*, **29**, 853 (1976).
3. J. C. Dabrowiak, F. T. Greenaway, W. E. Longo, M. vanHusen, and S. T. Crooke, *Biochim. Biophys. Acta*, **517**, 517 (1978).
4. T. Takita, T. Muraoka, T. Yoshioka, A. Fujii, K. Maeda, and H. Umezawa, *J. Antibiot. (Tokyo)*, **25**, 755 (1972).
5. T. Takita, Y. Muraoka, A. Fujii, H. Itoh, K. Maeda, and H. Umezawa, *J. Antibiot. (Tokyo)*, **25**, 197 (1972).
6. T. Takita, Y. Muraoka, T. Nakatani, A. Fujii, Y. Umezawa, H. Naganawa, and H. Umezawa, *J. Antibiot. (Tokyo)*, **31**, 801 (1978).
7. Y. Nakayama, M. Kunishima, S. Omoto, T. Takita, and H. Umezawa, *J. Antibiot. (Tokyo)*, **26**, 400 (1973).
8. Y. Iitaka, H. Nakamura, T. Nakatani, Y. Muraoka, A. Fujii, T. Takita, and H. Umezawa, *J. Antibiot. (Tokyo)*, **31**, 1070 (1978).
9. T. Takita, Y. Muraoka, T. Nakatani, A. Fujii, Y. Iitaka, and H. Umezawa, *J. Antibiot. (Tokyo)*, **31**, 1073 (1978).
10. E. A. Sausville, J. Peisach, and S. B. Horwitz, *Biochem. Biophys. Res. Commun.*, **73**, 814 (1976).
11. M. Chien, A. P. Grollman, and S. B. Horwitz, *Biochemistry*, **16**, 3641 (1977).
12. M. Kanao, S. Tomita, S. Ishida, A. Murakami, and H. Okada, *Chemotherapy (Tokyo)*, **21**, 1305 (1973).
13. K. Takahashi, O. Yoshioka, A. Matsuda, and H. Umezawa, *J. Antibiot. (Tokyo)*, **30**, 861 (1977).
14. A. Kono, *Chem. Pharm. Bull. (Tokyo)*, **25**, 2882 (1977).
15. H. Kasai, H. Naganawa, T. Takita, and H. Umezawa, *J. Antibiot. (Tokyo)*, **31**, 1316 (1978).

Electron Spin Resonance Studies of 1:1:1 Bleomycin–Cobalt(II)–Oxygen Adduct Complex

Yukio Sugiura

Bleomycin has been used extensively in the chemotherapy and radiodiagnosis of cancer[1]. One of the most important aspects of its chemistry is its ability to coordinate a variety of metals. Recently, bleomycin–metal complexes have attracted particular attention because of their interesting behavior: (i) the copper(II) complex is a metabolic form of bleomycin[2], (ii) γ-emitting metal complexes such as ^{57}Co, ^{99m}Tc, and ^{111}In have been investigated as a means of visualizing tumors[3], and (iii) an oxygen-labile iron(II) complex may play an important role in DNA degradation by bleomycin[4,5]. Sausville et al. reported that the conversion of SV40 DNA to acid-soluble products occurs at approximately equimolar levels of Fe(II), bleomycin, and DNA[4]. Takita et al. have proposed that an oxygen activated by the bleomycin-Fe(II) complex is involved in the action of bleomycin on DNA[5]. Additionally, it has been suggested that superoxide radical is one of the chemical mediators responsible for the enhancement of DNA chain breakage by bleomycin[6]. Herein, the 1:1 bleomycin A_2–Co(II) complex and its oxygen adduct complexes have been characterized by electron spin resonance (ESR) spectroscopy; the present results should serve to facilitate an understanding of the degradation of DNA by the bleomycin–Fe(II)–O_2 complex.

Experimental

Purified bleomycin A_2 was kindly supplied by Nippon Kayaku Co. Ltd, and a standard Co(II) solution was prepared from reagent grade material and standardized by EDTA complexation. Deionized water was used throughout the experiments. X-Band ESR spectra were obtained with a JES-FE-3X spectrometer operating with 100-kHz magnetic field modulation. ESR operating frequencies were measured with a Takeda–Riken microwave frequency counter. The g values were determined taking Li-TCNQ ($g = 2.0026$) as a standard, and the magnetic fields were calibrated by the splitting of Mn(II) in MgO($\Delta H_{3-4} = 86.9$ G).

165

Results and Discussion

It is known that the bleomycin–Co(II) complex changes easily to the corresponding stable Co(III) complex by air oxidation[7]. Under anaerobic conditions, however, the 1:1 bleomycin A_2–Co(II) complex clearly shows an ESR spectrum that indicates a nearly axial symmetry about the Co(II) ion in frozen aqueous solution ($t = 77°K$) (see Fig. 1A). The one axial ligand donor is a nitrogen atom, as demonstrated by the ^{14}N ($I = 1$) three-line superhyperfine pattern

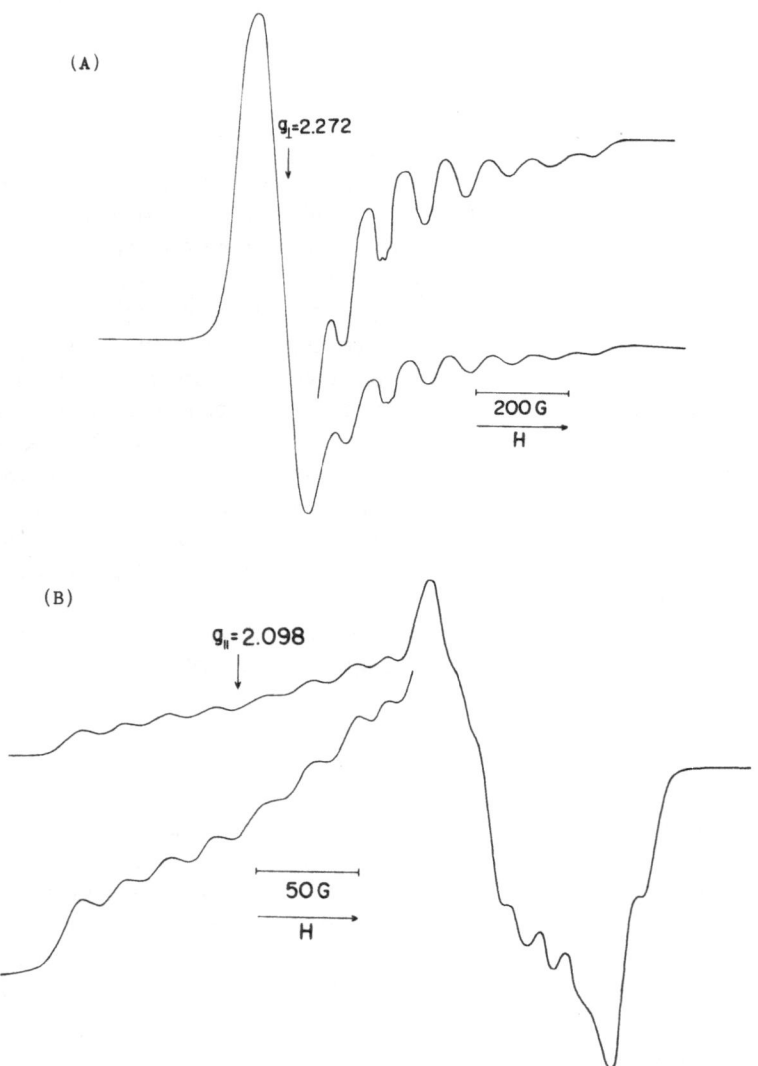

Fig. 1. ESR spectra of bleomycin–Co(II) (A) and bleomycin–Co(II)–O_2 (B) complexes at 77°K.

split, superimposed on the eight-line ^{59}Co ($I = 7/2$) parallel hyperfine pattern with splittings, $A_\parallel^{Co}$. The X-band ESR feature of the 1:1 bleomycin–Co(II) complex is very similar to those of the 1:1 stoichiometric complexes of low-spin Co(II)–Schiff bases with nitrogenous bases[8], deoxy-Co(II)-myoglobin[9], and deoxy-Co(II)-peroxidase[10]. The observed relationship of $g_\perp > g_\parallel \cong 2.0$, the presence of superhyperfine splitting from one axial ^{14}N atom and the apparent absence of superhyperfine splitting from the in-plane ^{14}N atoms indicate that the bleomycin–Co(II) complex is a low-spin Co(II) ($3d^7$, $S = 1/2$) complex in a pentacoordinated square-pyramidal configuration and that the unpaired electron is in the $3d_z2$ orbital[8]. The $A_\perp^{Co}$ value is estimated to be less than 12.5 G, as judged from the peak-to-peak line width of the $g_\perp$ extremum. The A_{iso}^{Co} of the bleomycin–Co(II) complex is estimated to be approximately 40 G from the equation of $A_{iso} = (2A_\perp + A_\parallel)/3$.

If oxygen is admitted to a sample tube containing a solution of the 1:1 bleomycin–Co(II) complex, the ESR spectrum shown in Fig. 1B is obtained upon cooling to 77°K. The ESR characteristics of the bleomycin–Co(II)–O_2 complex resemble closely those of various monooxygenated low-spin Co(II) complexes (see Table 1). The A_{iso}^{Co} value is estimated to be 15.0 G for the bleomycin–Co(II)–O_2 complex, from $A_\perp^{Co}$ and $A_\parallel^{Co}$ values. The effective g values, the relationship of $g_\parallel > g_\perp \cong 2.00$, and the relatively small A_{iso}^{Co} value of the bleomycin–Co(II)–O_2 complex suggest a considerable delocalization of the unpaired electron from the Co(II) ion. Hoffman et al. estimated that an A_{iso}^{Co} value of 10–14 G corresponds to about 90% transfer of the spin density from Co(II) to oxygen in their ESR study of monooxygenated Co(II)–Schiff bases complexes[8]. Therefore, the paramagnetic center of the bleomycin–Co(II)–O_2 complex may also be formally described as Co(III)–O_2^-. No ESR absorptions of the bleomycin–Co(II)–O_2 complex are observed at 293°K,

Table 1: ESR Paramaters of Some Cobalt–Oxygen Complexes

Compound	$g_\perp$	$g_\parallel$	$A_\perp^{Co}$ (G)	$A_\parallel^{Co}$ (G)	$A_\parallel^{N}$ (G)	Reference
Co(bleomycin)	2.272	2.025		92.5	13	This work
Co(bleomycin)(O_2)	2.007	2.098	12.4	20.2		
Co(acacen)py	2.44	2.011		97.53	15.7	(8)
Co(acacen)py(O_2)	1.999	2.082	10.73	19.64		
Co(acacen)CN-py	2.39	2.013		101.17	16.24	(8)
Co(acacen)CN-py(O_2)	1.997	2.080	10.30	20.11		
Co(3-methoxysalen)py$_2$	2.33	2.038		78.0	12.5	(11)
Co(3-methoxysalen)py(O_2)	1.997	2.079	10.25	17.3		
Co(Ts-phthalocyanine)	2.27	2.068		107		(12)
Co(Ts-phthalocyanine)(O_2)	2.004	2.075	7.9	14.9		
Deoxy-Co-myoglobin	2.32	2.03	6.2	75	17	(9)
Oxy-Co-myoglobin	2.007	2.08	9	16		
Deoxy-Co-peroxidase	2.34	2.03	10.5	78	18.3	(10)
Oxy-Co-peroxidase	2.01	2.10	13.6	23.2		

consistent with the fact that the ESR absorption of O_2^- is not conventionally observable at room temperature.

The ESR signal of the bleomycin–Co(II) complex disappeared completely upon reaction with nitric oxide. Similar phenomena have been observed in the case of nitric oxide–Co(II)–hemoproteins[13]. The loss of the ESR absorption can be interpreted as a spin pairing or charge-transfer between the metal and the liganded nitric oxide. On the other hand, the bleomycin–Fe(II)–O_2 complex is ESR inactive. The bleomycin–Fe(II) complex reacts with nitric oxide to form a paramagnetic complex, which exhibits an intense ESR absorption with $g_x =$ 2.041, $g_z = 2.008$, and $g_y = 1.976$. Nitric oxide is thus a useful spin-label probe for clarifying the nature of the oxygen-binding site of the bleomycin–Fe(II) complex. Detailed analyses through isotope substitutions with ^{15}NO and ^{57}Fe reveal that the bleomycin–Fe(II) complex forms an oxygen adduct complex similar to that of the bleomycin–Co(II) complex[14].

In conclusion, the present ESR results indicate strongly that the 1:1 bleomycin–Co(II) complex has a square-pyramidal configuration with an axial nitrogen donor, and that oxygen is incorporated into the vacant sixth position of the bleomycin–Co(II) complex (see Fig. 2). The potentiometric result also suggests that the 1:1 bleomycin–Co(II) complex has a coordination arrangement similar to that of the 1:1 bleomycin–Cu(II) complex[15]. For the O_2 adduct of the bleomycin–Co(II) complex, a qualitative molecular orbital scheme as shown in Fig. 3 is proposed. In this case, the Co(II) ion has an unpaired electron in d_{z^2} with high enough energy to spin pair with an odd electron in the oxygen antibonding orbital.

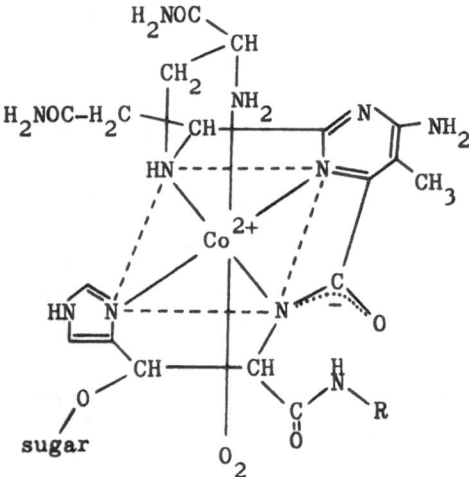

Fig. 2. Bleomycin–Co(II)–O_2 complex.

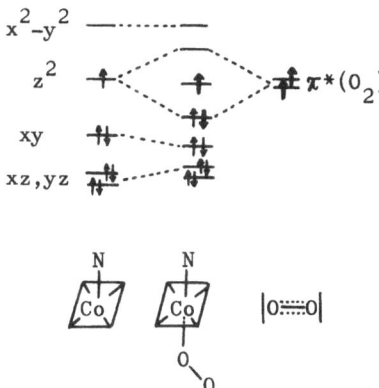

Fig. 3. Molecular orbital model for the coordination of O_2 to bleomycin–Co(II) complex.

Acknowledgment

Thanks are due to Prof. K. Ishizu for helpful discussions.

References

1. H. Umezawa, *Fed. Proc.*, **33**, 2296 (1974).
2. M. Kanao, S. Tomita, S. Ishihara, A. Murakami, and H. Okada, *Chemotherapy (Tokyo)*, **21**, 1305 (1973).
3. J. J. Rasker, M. A. P. C. Poll, H. Beekhuis, M. G. Woldring, and H. O. Nieweg, *J. Nucl. Med.*, **16**, 1058 (1975).
4. E. A. Sausville, J. Peisach, and S. B. Horwitz, *Biochem. Biophys. Res. Commun.*, **73**, 814 (1976).
5. T. Takita, Y. Muraoka, T. Nakatani, A. Fujii, Y. Iitaka, and H. Umezawa, *J. Antibiot. (Tokyo)*, **31**, 1073 (1978).
6. R. Ishida and T. Takahashi, *Biochem. Biophys. Res. Commun.*, **66**, 1432 (1975).
7. A. D. Nunn, *Int. J. Nucl. Med. Biol.*, **4**, 216 (1977).
8. B. M. Hoffman, D. L. Diemente, and F. Basolo, *J. Am. Chem. Soc.*, **92**, 61 (1970).
9. T. Yonetani, H. Yamamoto, and T. Iizuka, *J. Biol. Chem.*, **249**, 2169 (1974).
10. M-Y. R. Wang, B. M. Hoffman, and P. F. Hollenberg, *J. Biol. Chem.*, **252**, 6268 (1977).
11. D. Diemente, B. M. Hoffman, and F. Basolo, *J. Chem. Soc. Chem. Commun.*, 467 (1970).
12. E. W. Abel, *J. Chem. Soc. Chem. Commun.*, 449 (1971).
13. T. Yonetani, H. Yamamoto, J. E. Erman, J. S. Leigh, and G. H. Reed, *J. Biol. Chem.*, **247**, 2447 (1972).
14. Y. Sugiura and K. Ishiza, *J. Inorg. Biochem.*, in press.
15. Y. Sugiura, K, Ishiza, and K. Miyoshi, *J. Antibiot. (Tokyo)*, **32**, 453 (1979).

A Role for Iron in the Degradation
of DNA by Bleomycin

SUSAN B. HORWITZ, EDWARD A. SAUSVILLE,
AND JACK PEISACH

Bleomycin is a glycopeptide antibiotic with antineoplastic activity produced by cultures of *Streptomyces verticillus*. Although it was isolated by Umezawa and his collaborators[1,2] as a copper chelate, it is used clinically as a metal-free preparation. Studies with bleomycin have suggested that its cytotoxic properties are related to effects of the compound on the physical integrity of DNA in cells. In cultures of bacterial or eukaryotic cells, a major effect of bleomycin is the introduction of strand breaks into DNA[3]. Breakage of DNA may be demonstrated after incubation of bleomycin with purified DNA. If this *in vitro* degradation reaction is to proceed with greatest efficiency, it requires the presence of a reducing agent, although a small amount of DNA breakage may be observed in the absence of reducing agent[4,5].

Many reducing agents have been found to stimulate the breakage of DNA *in vitro* by bleomycin. These include mercaptoethanol, ascorbate, NaBH$_4$, NADPH, and H$_2$O$_2$[6]. When efficient DNA breakage occurs in the presence of 2-mercaptoethanol, distinctive products are generated. These include oligonucleotides of average size 10–13 residues[7], free bases[8], and an aldehyde product whose derivative with 2-thiobarbituric acid resembles that produced by malondialdehyde[9].

It is of interest to examine the mechanism by which various reducing agents stimulate the degradation of DNA *in vitro* by bleomycin. This chapter summarizes the evidence that all reducing agents act by allowing continued reduction of contaminating Fe in reaction mixtures to which Fe has not been added. The reduced metal ion may then be available to combine with bleomycin and participate in DNA degradation by an oxidative mechanism. In this way, bleomycin may be visualized both as a chelating agent and a DNA-binding molecule. We propose that both of these properties of bleomycin are involved in the degradation of DNA *in vitro*. This view of the action of bleomycin takes into account its ability to bind various metal ions[1,2,10] and to bind to DNA[11].

Evidence for the Involvement of Fe in the
Degradation of DNA by Bleomycin

Many substances have been found to stimulate or inhibit the degradation of
DNA by bleomycin, but there are conflicting reports in the literature concerning the activities of these materials. Nagai *et al*.[12] found that Co(II), Zn(II), and
Cu(II) inhibited the action of bleomycin in the presence of 2-mercaptoethanol in
experiments which measured a decrease in the melting temperature (T_m) of
DNA. Where breakage of DNA has been followed directly, this result was confirmed[13]. It is generally accepted that high concentrations of EDTA can inhibit
the degradation of DNA[14-16].

Ishida and Takahashi[17] noted that Fe(II) and Fe(III) stimulated the degradation of DNA by bleomycin in the presence of 2-mercaptoethanol, in contrast to
the report of Müller *et al*.[18] that neither Fe(III) nor EDTA in the presence of
dithiothreitol altered the degradation of DNA by bleomycin. These findings
prompted us to investigate whether Fe(II) could be the species through which
2-mercaptoethanol and other reducing agents acted to stimulate the degradation
of DNA by bleomycin. If this view was correct, then Fe(II) added in the absence
of a thiol should efficiently promote the reaction, whereas Fe(III) should not
facilitate DNA degradation. In the presence of a thiol, however, both Fe(II) and
Fe(III) should stimulate the reaction, since Fe(II) could be produced by the
reduction of Fe(III) by the thiol compound. This hypothesis was tested and it
was observed that Fe(II) does indeed act with bleomycin in the absence of thiol
to degrade DNA to acid-soluble products whereas Fe(III) has no effect[19]. In
the absence of drug, Fe(II) does not produce highly efficient damage to DNA,
although limited breakage of DNA does occur in the presence of Fe(II) alone.
Neither Cu(II), Zn(II), nor Co(II) could replace Fe(II) in eliciting the degradation of DNA with bleomycin in the presence or absence of thiols, although each
of these three metals inhibits the action of Fe(II) with bleomycin.

In an effort to determine whether Fe(II) could be the active species in stimulating DNA degradation in reaction mixtures which do not receive exogenous
metal ion, experiments with chelating agents were conducted. Deferoxamine, a
specific and powerful iron chelator,[20] is considerably more potent than EDTA
in inhibiting the efficient degradation of DNA to acid-soluble products by bleomycin and 2-mercaptoethanol. DNA degradation observed with bleomycin and
ascorbate, dithiothreitol, H_2O_2, and NADPH is in all cases sensitive to inhibition
by EDTA, and deferoxamine is for the most part more efficient in eliciting this
inhibition. Even the limited degradation of DNA which is observed in the absence
of reducing agent or added metal is sensitive to inhibition by chelators, with
deferoxamine again being demonstrably more potent than EDTA in eliciting
this effect. Where iron concentrations are measured directly by atomic adsorption spectroscopy, significant quantities of iron are found in our reaction mixtures[21].

These observations demonstrate that Fe(II) can specifically act with bleomycin to degrade DNA. Fe is present in significant quantities in reaction mixtures of the type commonly used to study degradation of DNA by bleomycin. In the absence of added metal ion, these reactions are sensitive to chelators in a way that is consistent with a role for contaminating Fe. The fact that chelators could inhibit the reaction indicates that it is necessary for bleomycin to combine with a metal ion, presumably Fe(II), at some point in the course of the reaction. If this were true, then a complex between Fe(II) and bleomycin should be demonstrable. Under anaerobic conditions, Fe(II) and bleomycin form a 1:1 complex with an optical absorption maximum near 475 nm (Fig. 1A and B). Upon the addition of oxygen, an intense absorption in the near ultraviolet appears. The addition of dithionite regenerates the original spectrum that was obtained under anaerobic conditions.

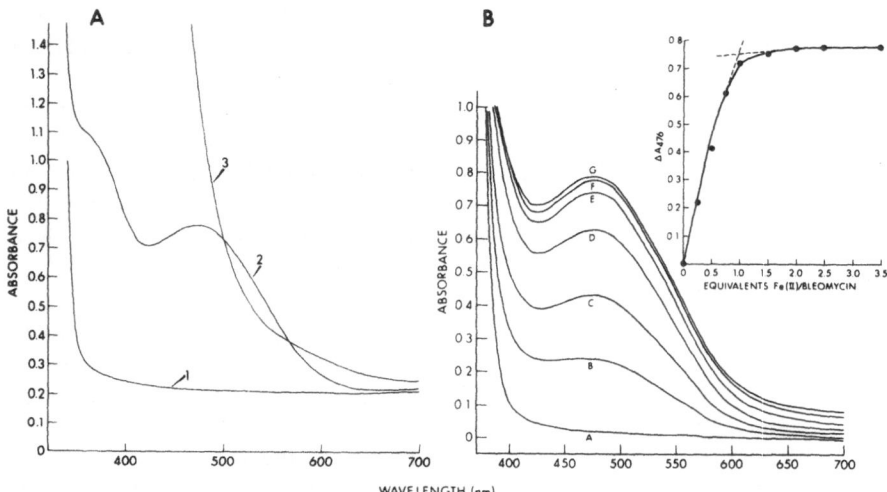

Fig. 1. (A) Anaerobic spectra of Fe(II)–bleomycin in Hepes buffer. Three milliliters of 2 mM bleomycin in 0.05 M Hepes buffer, pH 7.0, contained in the main compartment of a Thunberg cuvette was made anaerobic with purified argon (Curve 1). On addition of 60 μl of 0.1 M Fe(II) from the sidearm, Curve 2 was observed. On opening the cuvette to air, Curve 3 was recorded. The reference sample was H_2O. Fe(II) in Hepes buffer at the concentrations employed does not absorb significantly in this spectral region. (B) Anaerobic titration of bleomycin with Fe(II) in Hepes buffer. In the main compartment of a Thunberg cuvette a 3-ml solution consisting of 2 mM bleomycin in 0.05 M Hepes, pH 7, was made anaerobic with purified argon. From the sidearm of the cuvette, 5–8 mg of sodium dithionite was added. To this solution, an anaerobic solution of Fe(II) was introduced through a rubber septum stopper to the following final concentrations (M): A, O; B, 5 × 10^{-4}; C, 1 × 10^{-3}; D, 1.5 × 10^{-3}; E, 2 × 10^{-3}; F, 3 × 10^{-3}; G, 4 × 10^{-3}, 5 × 10^{-3}, and 7 × 10^{-3}. The inset plots the observed A_{476} as a function of the ratio of Fe(II) to bleomycin.

In 1975, Onishi *et al.*[6] had shown that in the absence of added metal, the degradation of DNA by bleomycin, studied in the presence of various organic reducing agents, was sensitive to O_2. These results suggested that O_2 or one of its reduction products is necessary for DNA degradation by the drug. Our spectroscopic results demonstrating complex formation between Fe(II) and bleomycin prompted us to examine whether breakage of DNA by bleomycin and Fe(II) is also sensitive to O_2. Our experiments indicate that efficient DNA degradation to acid-soluble products by Fe(II)–bleomycin does require the presence of O_2[22]. Thus we relate the spectroscopic changes observed upon the addition of O_2 to Fe(II)–bleomycin to a process which is also important in DNA degradation by the drug and the metal ion.

Using a difference spectroscopic technique, we have examined the effect of added Zn(II) and Fe(II) on the optical absorption of bleomycin in the ultraviolet (Fig. 2). We find that Fe(II) perturbs the optical spectrum of bleomycin in a way similar to that produced by Zn(II). These results suggest that Zn(II) and Fe(II) have a common binding site in the bleomycin molecule and that the inhibition by Zn(II) on the effect of Fe(II) on DNA breakage by bleomycin is due to competition for this metal-binding site. Dabrowiak *et al.*[10] have presented evidence that Zn(II) binds to bleomycin through a site not involving the bithiazole portion of the molecule. Further evidence that the bithiazole residues do not bind to Fe(II) derives from the observation that Fe(II)–phleomycin (phleomycin differs from bleomycin only in the bithiazole residue) has a virtually identical visible spectrum and sensitivity to oxygen as does Fe(II)–bleomycin (Sausville and Horwitz, unpublished observations).

Properties of the Degradation of DNA by Fe(II)–Bleomycin

When bleomycin is added to an aerobic solution containing DNA, and then Fe(II) is added in the absence of any thiols or other reducing agents, degradation of DNA occurs. At bleomycin/DNA nucleotide ratios of less than 1, breaks appear in the DNA to an extent far greater than may be attributed to any action of Fe(II) alone. At a bleomycin/DNA nucleotide ratio of 1–2, acid-soluble DNA degradation products are observed. DNA degradation as measured by this assay is complete in less than 1 min at 37 or $0°C$. Increasing the ionic strength of the buffers utilized in our studies significantly decreases DNA degradation[22]. This latter result is in accord with the prior observation that increased ionic strength decreases the binding of bleomycin to DNA[11], but also may be due to competition between the buffer anions and bleomycin for Fe(II). DNA degradation by bleomycin and Fe(II) is relatively insensitive to changes in temperature from 2 to $60°C$, although some decline in apparent activity at the higher temperature is noted.

The degradation of DNA by Fe(II) and bleomycin, as measured by the generation of acid-soluble products, is remarkably dependent on the nature of the

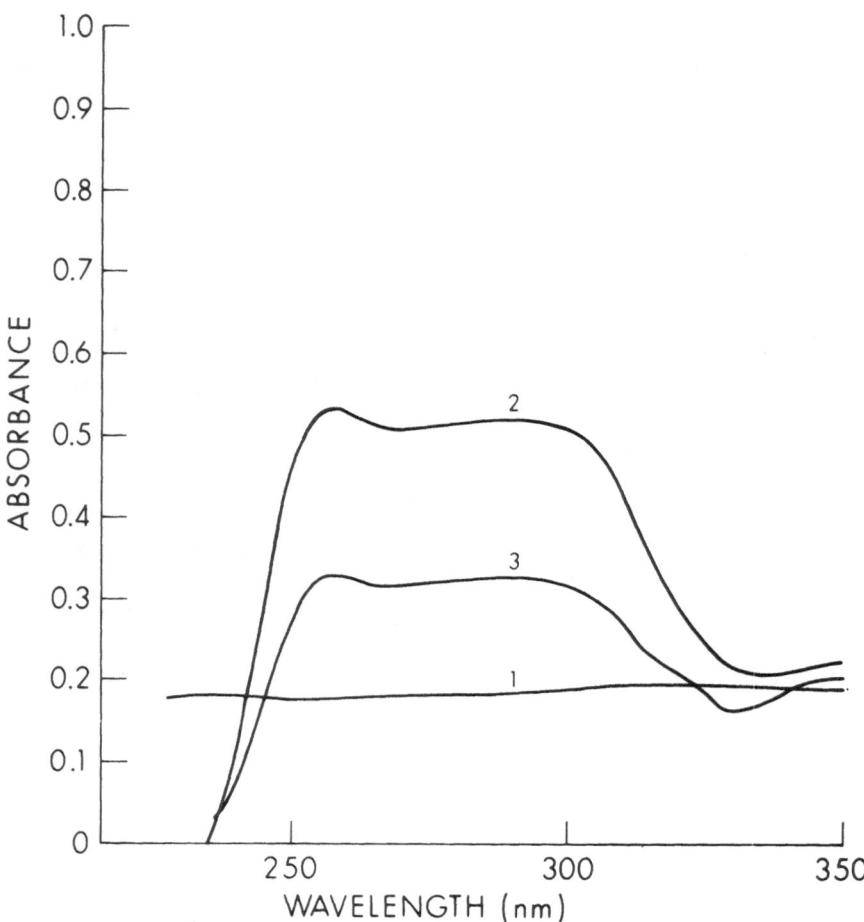

Fig. 2. Difference spectra between Fe(II)–bleomycin, Zn(II)–bleomycin, and bleomycin. In Curve 1, the spectrum of 10^{-4} M Fe(II) or 10^{-4} M Zn(II) in 0.05 M Hepes buffer, pH 7.2, versus Hepes buffer is recorded. In Curve 2, the optical difference spectrum between a solution containing 10^{-4} M Fe(II) and 10^{-4} M bleomycin versus 10^{-4} M bleomycin was studied. All spectra with Fe(II) were conducted in Thunberg cuvettes under an argon atmosphere. In Curve 3, the optical difference spectrum between a solution containing 10^{-4} M Zn(II) and 10^{-4} M bleomycin versus 10^{-4} M bleomycin was recorded. Both reference and sample solutions were prepared in 0.05 M Hepes buffer, pH 7.2. The salts used were Fe $(NH_4)_2(SO_4)_2 \cdot 6H_2O$ and $ZnCl_2$.

buffer used in the reaction, with a less evident dependence on pH (Fig. 3A). While possible explanations for this effect are considered elsewhere[22], it is clear that activity becomes significant between pH 4 and 6. It is observed that Fe(II)–bleomycin complex formation is sensitive to pH, with an apparent pK of 5.2 (Fig. 3B). Thus the pH dependence of the degradation of DNA by Fe(II)–bleomycin correlates closely with the pH sensitivity of complex formation between Fe(II) and bleomycin.

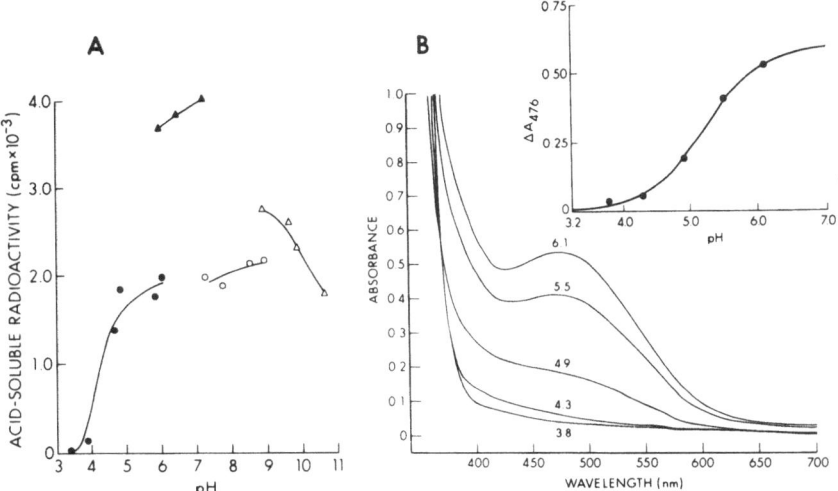

Fig. 3. (A) Effect of pH and buffer composition on the degradation of DNA by Fe(II) and bleomycin. Reaction mixtures contained 4.0 μM adenovirus [^{3}H] DNA (3800 cpm), 50 μg/ml bleomycin, and 1×10^{-4} M Fe(II) in a buffer which was 0.05 M in either acetate (●), phosphate (▲), Tris–HCl (○), or glycine–NaOH (△). After equilibration at 37°C, the Fe(II) was added and incubation continued for 15 min. Acid-soluble radioactivity was determined. (B) pH dependence of Fe(II)–bleomycin complex formation. In the main compartment of a Thunberg cuvette a solution (3 ml) consisted of 2 mM bleomycin in 0.05 M acetate at the indicated pH plus 5 – 8 mg dithionite under an argon atmosphere. Fe(II) was added from the sidearm to a concentration of 2 mM. The inset plots A_{476} as a function of pH. Absorbance attributed to Fe(II)–bleomycin at differing pH was fitted by a least-squares analysis for the determination of pK (26).

Since oxygen is required for the highly efficient degradation of DNA by bleomycin and Fe(II), it was desirable to determine if the presence of bleomycin altered the reaction of Fe(II) with oxygen. Kurimura et al.[23] presented data which indicated that chelating agents increase the rate of Fe(II) oxidation. When Fe(II) is added to phosphate buffer equilibrated with air, vigorous consumption of oxygen is observed. When bleomycin is added, the rate of oxygen uptake is increased and, in addition, more oxygen is consumed (Fig. 4A). Oxygen consumption both in the presence and absence of bleomycin occurs with a time course[22] that is commensurate with the oxygen dependence of DNA degradation by bleomycin and Fe(II). The presence of DNA does not alter the increase of oxygen consumption brought about by bleomycin addition (Fig. 4B). Equilibrium is approached with respect to O_2 consumption within 2–4 min after addition of Fe(II). The ΔP_{O_2} observed at completion of the initial decline in P_{O_2} (arrows in Fig. 4A and B) indicate the oxygen tensions used for calculation of apparent stoichiometry of O_2 consumed in response to addition of Fe(II).

Experiments were done in which Fe(II) was added in differing amounts to solutions with or without bleomycin. In the absence of DNA (Fig. 4C), the stoichiometry of O_2 consumption is relatively constant at 0.5 mole of O_2 con-

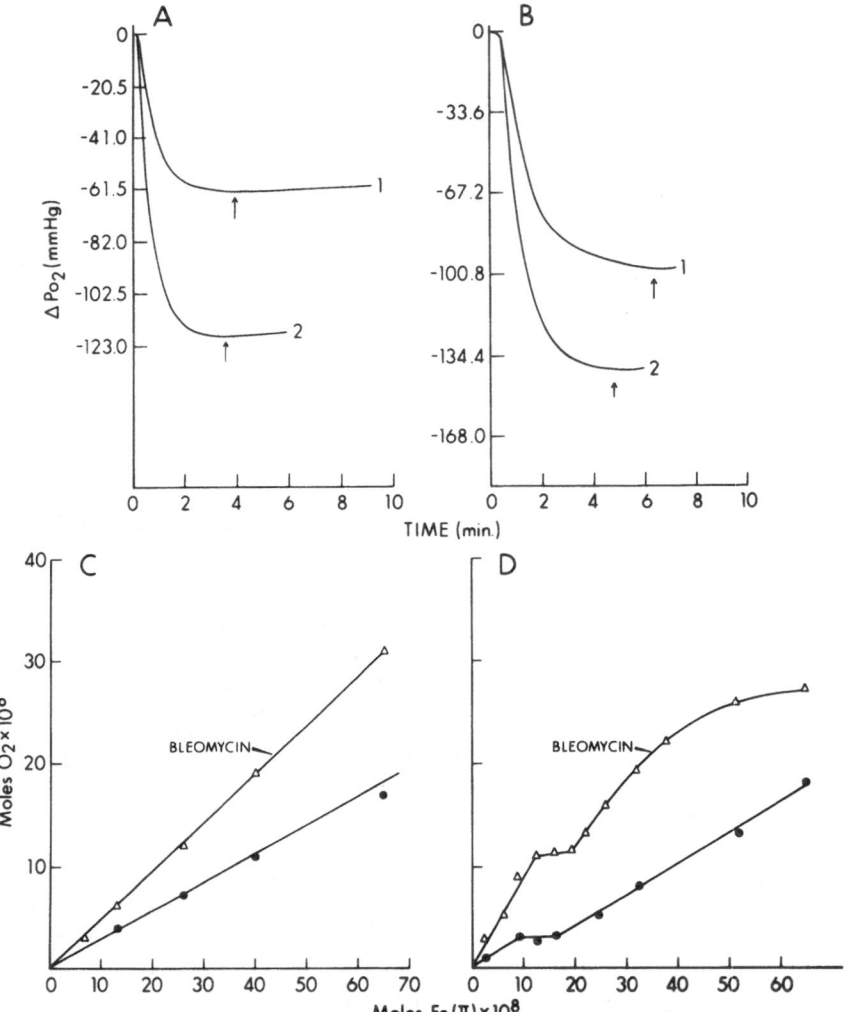

Fig. 4. Consumption of oxygen by Fe(II) and Fe(II)–bleomycin. (A) Reaction mixtures contained in a final volume of 1.3 ml: 0.05 M phosphate buffer, pH 7.0, 5 × 10^{-4} M Fe(II) and O bleomycin (Curve 1) or 6.5 × 10^{-4} M bleomycin (Curve 2). Reactions were initiated with addition of Fe(II). (B) Reaction mixtures contained in a final volume of 1.3 ml: 2.7 × 10^{-3} M calf thymus DNA, 0.019 M phosphate, pH 7.0, 5 × 10^{-4} M Fe(II) and O bleomycin (Curve 1) or 2 × 10^{-4} M bleomycin (Curve 2). Reactions were initiated by addition of Fe(II). (C) Reactions were conducted at 0.05 M phosphate, pH 7.0, in a volume of 1.3 ml in an oxygen electrode cell at 30°C. Fe(II) was added in 0.013 ml at $t=0$ min in the indicated amounts. O_2 consumption at near equilibrium was recorded in the presence (△) or absence (●) of 6.5 × 10^{-4} M bleomycin. (D) Reactions were conducted as in (C), and contained 0.019 M phosphate, pH 7.0, 2.7 × 10^{-3} M DNA, 2 × 10^{-4} M bleomycin, and the indicated amounts of Fe(II), (△); alternatively, reaction mixtures contained 2.9 × 10^{-3} M DNA, 0.019 M phosphate, pH 7.0, and the indicated amounts of Fe(II) (●).

sumed per mole of Fe(II) added in the presence of bleomycin whereas it is about 0.2 mole O_2 consumed per mole of Fe(II) in the absence of bleomycin. In the presence of DNA and bleomycin and with less than 1×10^{-7} moles of added Fe(II), approximately one mole of O_2 is consumed per mole of Fe(II) added (Fig. 4D). When greater than 1×10^{-7} mole of Fe(II) is added, the apparent stoichiometry of O_2 uptake changes to about 0.5 mole of O_2 consumed per mole of Fe(II) at greater than 6×10^{-7} moles of Fe(II). One may conclude from these experiments that, except at low concentrations of bleomycin in the presence of DNA, there is always less O_2 consumed than Fe(II) added. Thus, although bleomycin clearly alters the pattern of oxygen utilization in the oxidation of Fe(II) in the presence of DNA, models in which one oxygen molecule is consumed per mole of Fe(II) and bleomycin are tenable only when applied to reaction mixtures containing low concentrations of Fe(II).

Products of the Degradation of DNA by Fe(II)–Bleomycin

Release of Bases

A characteristic reaction of bleomycin in systems employing organic reducing agents in the study of DNA degradation is the release of free base residues[8]. To examine this phenomenon in our system, high-pressure liquid chromatography was used. Pyrimidines are released to a greater extent than purines, and approximately one base in four is released in the reaction which occurs at 1 bleomycin:1 DNA nucleotide:1 Fe(II) (Table 1). Takeshita et al.[24] have also studied the release of bases in the presence of organic reducing agents but in the absence of

Table 1: Release of Bases During Reaction of DNA with Bleomycin and Fe(II)[a]

Base	Percentage release[b]
Thymine	13.7 ± 1.9
Cytosine	7.8 ± 0.8
Adenine	4.2 ± 0.6
Guanine	2.8 ± 1.2

[a]Reaction mixtures contained in a final volume of 2 ml: $2.4 \times 10^{-4}\,M$ calf thymus DNA, $2.3 \times 10^{-4}\,M$ bleomycin, 0.019 M phosphate buffer, pH 7.0, and $2.3 \times 10^{-4}\,M$ Fe(II). Prior to addition of Fe(II), components were equilibrated with 100% O_2 at 22°C for 20 min. After addition of Fe(II) incubation was continued for 15 min at 37°C. The reaction mixtures were analyzed for free bases by high-pressure liquid chromatography as previously described[22].
[b]100 × (moles of base detected)/(moles of base in DNA).

added metal ion. They also find that pyrimidines are released to a greater extent than purines. Thus, the products of DNA degradation with bleomycin in the absence of added iron are the same as in the presence of added iron. Such data support the idea that Fe(II) is the active species in systems containing organic reducing agents which act with bleomycin to degrade DNA.

Formation of Oligonucleotides

DEAE cellulose chromatography was used to examine the oligonucleotides found in limit digests of ^{32}P-labeled DNA incubated with bleomycin and Fe(II) (Fig. 5). In reactions where 80% of the label had been converted to an acid-soluble form, the median oligonucleotide is of the order of 7 – 10 nucleotide phosphate residues in size, although an appreciable fraction ($\sim$ 30%) migrates between 10 and 20 residues in size and is eluted with 2 M NaCl. This latter fraction is completely refractory to the further action of bleomycin and Fe(II). In this reaction mixture > 98% of the ^{32}P label could be adsorbed to charcoal, and therefore inorganic phosphate is not produced in appreciable quantities during the reaction. Mononucleotides and dinucleotides are also not found as principal products. Since the DNA used in this experiment contained labeled thymine residues, base release could be demonstrated. The ^{3}H-radiolabeled material in fractions 10–16 comi-

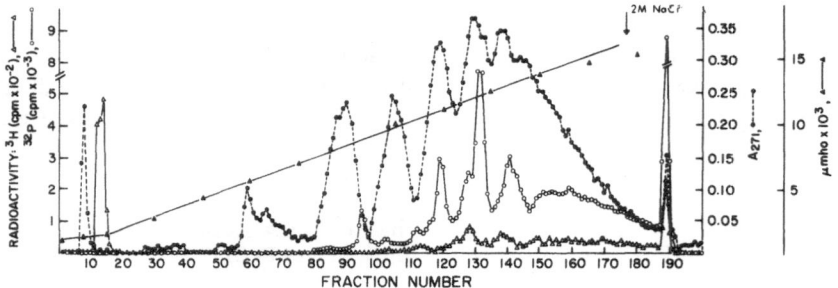

Fig. 5. DEAE cellulose chromatography of the limit product of Fe(II)–bleomycin degradation of DNA. The reaction mixture contained in a volume of 1.0 ml: 5 mM phosphate buffer, pH 7.0, 21 μM adenovirus-2 [^{3}H] thymine ^{32}P DNA, 333 μg/ml bleomycin, and 2 $\times$ 10^{-4} M Fe(II). The solution was saturated with O$_2$ prior to the start of reaction; incubation was for 1 hr, at 37°C, with readdition of bleomycin and Fe(II) at 0.5 hr (to a final concentration of 550 μg/ml and 4.2 $\times$ 10^{-4} M, respectively). At the end of the reaction, the solution was made 0.01 M in EDTA, and to it was added 8.8 mg of calf thymus DNA which had been extensively digested with pancreatic DNAase. After dilution to an appropriate buffer concentration (0.02 M) and addition of urea to 7 M, the sample was applied to a 1.8 $\times$ 27-cm column of DEAE cellulose equilibrated with 0.02 M Tris–HCl buffer, pH 7.6, containing 7 M urea. After washing with 60 ml of the latter buffer, nucleotides were eluted with a 0 – 0.3 M NaCl gradient (2 liters), and fractions of 11 ml were collected at 1 ml/min. Nucleotides not eluted with the gradient were removed by 2 M NaCl in 0.02 M Tris–HCl, pH 7.6, 7 M urea. A_{271} (●), μmho (▲), ^{32}P (○), and ^{3}H (△) in aliquots of column fractions as shown.

grates with authentic thymine in our high-pressure liquid chromatography system.

An attempt was made to study the nature of the oligonucleotides produced in the reaction. Studies with nuclease digestions and phosphatase treatment[22] revealed that these oligonucleotides are highly atypical in their sensitivity to nucleases and do not represent homogeneous structures with respect to end groups such as are obtained following nucleolytic degradation of DNA.

The results of these studies correlate with previous efforts to determine the size of nucleotides resulting from the degradation of DNA by bleomycin and 2-mercaptoethanol in the absence of added metal ion. Such studies also revealed that phosphate and mononucleotides were not detected as products of the reaction, and the average size of the limit oligonucleotides was 10–13 residues as measured by analytical ultracentrifugation[7,8].

Formation of Aldehydes

Kuo and Haidle[9] detected a product of the degradation of DNA by bleomycin and 2-mercaptoethanol which optically resembled malondialdehyde when reacted with 2-thiobarbituric acid. The reaction of bleomycin and Fe(II) with DNA generates a species which reacts with 2-thiobarbituric acid, and the occurrence of this species is dependent on the presence of DNA, bleomycin, and Fe(II). Fe(III) cannot substitute for Fe(II). What is notable is that the generation of this species at 37°C occurs rapidly and a greater amount of this species is formed when the reaction is conducted at elevated oxygen tensions[22].

The generation of the aldehyde allows an independent approach to the stoichiometry of Fe(II) with respect to bleomycin in the degradation of DNA. Reactions were conducted in the presence of an excess of DNA in relation to bleomycin, with varying amounts of added Fe(II). At three concentrations of bleomycin, the production of the aldehyde reaches a plateau at concentrations of Fe(II) which are equivalent to the bleomycin present in the reaction mixture (Fig. 6A). Thus, the amount of aldehyde product formed is approximately stoichiometric with the Fe(II) added.

The chemical nature of the chromophore, produced on heating the products of DNA degradation by bleomycin and Fe(II) with 2-thiobarbituric acid, has been explored (Fig. 6B–E). When authentic malondialdehyde is reacted with 2-thiobarbituric acid, a chromophore is produced with $\gamma_{max} = 532$ nm and an inflection at about 498 nm (Fig. 6B, curve 2). In contrast, reaction mixtures containing excess DNA which had been treated with bleomycin and Fe(II) also contain a species with a chromophore at 532 nm, and a shoulder near 500 nm, but a second absorption maximum at 454 nm is seen as well (Fig. 6B, curve 1). The reaction of DNA [not treated with bleomycin or Fe(II)] with 2-thiobarbituric acid produces a chromophore with $\lambda_{max} = 498$ and 454 nm (Fig. 6C, curve 1), while bleomycin reacted with 2-thiobarbituric acid does not yield a chromophore in the visible region (Fig. 6C, curve 2). The experiments in Fig. 6D demonstrate that if reaction mixtures containing DNA which had been treated with bleomycin and Fe(II) are adsorbed to DEAE cellulose (but not to CM-Sephadex) prior to reaction with 2-thiobarbituric acid, the species in reaction mixtures which yields appreciable absorbances at 454 and 498 nm is completely removed, while the species giving rise to the chromophore at 532 nm is recovered in > 80% yield. Authentic malondialdehyde is not removed from solution by DEAE cellu-

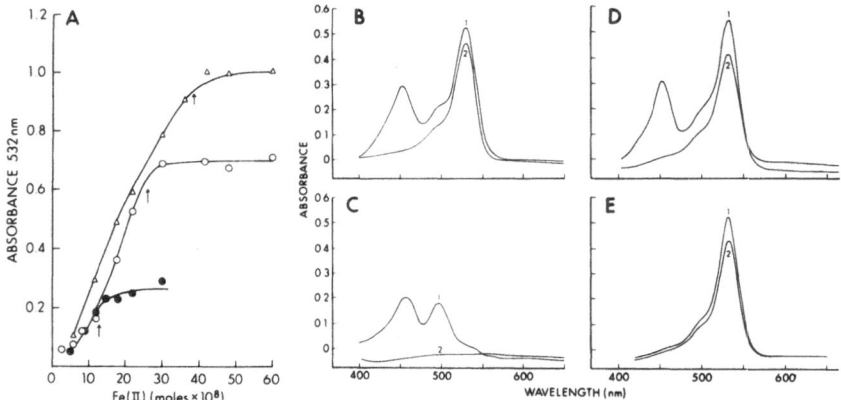

Fig. 6. Generation of a malondialdehyde-like product after reaction of bleomycin and Fe(II) with DNA. (A) The malondialdehyde-like product was assayed with 2-thiobarbituric acid (22). After heating for 20 min at 95°C, the characteristic chromophore (λ_{max} = 532 nm) was measured. Reactions contained in a volume of 1.3 ml: 0.019 M phosphate buffer, pH 7.0, 2.9 × 10^{-3} M calf thymus DNA, Fe(II) in the indicated amount and 155 μg/ml (= 0.13 μmole, ●), 310 μg/ml (= 0.26 μmole, ○), or 465 μg/ml (= 0.39 μmole, △) of bleomycin. Reactions were initiated by addition of Fe(II) at 37°C. Iron was added rapidly in a volume of 0.13 ml, and the tubes were shaken for 15 – 30 sec to effect mixing. Incubation was continued for 15 min. An aliquot was removed from each reaction mixture for determination of malondialdehyde-like material. The arrows indicate the amount of Fe which is approximately equimolar to the bleomycin present in reaction mixtures. (B) Reaction mixtures contained in a final volume of 1.3 ml: 2.9 × 10^{-3} M calf thymus DNA, 0.019 M phosphate buffer, pH 7.0, 1 × 10^{-4} M bleomycin, and 1.7 × 10^{-4} M Fe(II). After incubation for 15 min, 37°C, 0.5 ml was diluted with 0.5 ml H_2O. Then, 0.5 ml of the diluted solution was reacted with 2-thiobarbituric acid (Curve 1). Authentic malondialdehyde (2.9 μM) was reacted with 2-thiobarbuturic acid (Curve 2). (C) DNA (0.5 ml of 7.5 × 10^{-3} M) was mixed with 1.5 ml of 0.6% 2-thiobarbituric acid (Curve 1). Curve 2 presents the spectrum of 0.5 ml of 2 × 10^{-3} M bleomycin treated analogously. (D) The reaction of (B) was diluted into an equal volume of 80% v/v in H_2O of packed CM Sephadex (Curve 1) or DEAE cellulose (Curve 2); the solution was shaken for 20 min and the phases were separated. A portion (0.5 ml) of the supernatant was reacted with 2-thiobarbituric acid. (E) Malondialdehyde (23 μM) was added (0.5 ml) to 0.4 ml of 80% v/v of packed CM-Sephadex (Curve 1) or DEAE cellulose (Curve 2), and treated subsequently as described in (D). Visible spectra of reactions with 2-thiobarbituric acid are presented. The reference solution for (B), (C), (D), and (E) consisted of 0.5 ml of 0.6% 2-thiobarbituric acid to which was added 0.5 ml of H_2O followed by heating for 20 min at 95°C.

lose (Fig. 6E). These experiments are consistent with the possibility that the species which reacts with 2-thiobarbituric acid to yield the chromophore at 454 nm derives from oligonucleotides, as it resembles in its spectroscopic absorption properties that species produced from DNA when reacted with 2-thiobarbituric acid. In contrast, the species with λ_{max} = 532 nm greatly resembles that produced after reaction of malondialdehyde with 2-thiobarbituric acid. It should be noted that adsorption to DEAE cellulose preceded the heating step in the determination of malondialdehyde-like substances. Thus, the species which reacts with 2-thiobarbituric acid is apparently released during the reaction, and does not require the action of heat to generate it from damaged DNA. Despite the great similarity of the aldehyde product obtained in the degradation of DNA by bleomycin and Fe(II) to authentic malondialdehyde, it is not yet proven conclusively that this is the species obtained.

A Model for the Degradation of DNA by Bleomycin and Fe(II)

Our experiments are consistent with the idea that Fe(II) is a necessary cofactor of bleomycin in the degradation of DNA, even in reaction mixtures to which the metal ion has not been added exogenously. One way in which this might occur is described in Fig. 7. According to this view bleomycin binds to DNA through a mechanism not requiring metal ion. When bound to DNA, a portion of the bleomycin molecule is available to bind Fe(II). Oxidation of metal ion in the ternary complex may then occur, producing a variety of reduced species derived from molecular oxygen which may be deleterious to the physical integrity of DNA. Where Fe(II) has not been added to the reaction mixture, endogenous Fe(III) produced in the oxidation described above may be reduced by added reducing agents and be made available for subsequent DNA damaging events.

Additional evidence in support of this model derives from the observation that bleomycin can bind to DNA in the absence of added metal ion. During the interaction with DNA, proton magnetic resonances ascribed to the bithiazole moiety are broadened, whereas those of the β-hydroxyhistidinyl moiety are not affected[11]. Binding to DNA of tripeptide S, a major cleavage product of bleomycin which contains the bithiazole residue but no other aromatic residue, may be demonstrated by a fluorometric technique, although this compound is not active in the degradation of DNA[25]. The binding of tripeptide S occurs with an affinity constant closely resembling that of bleomycin[11]. These observations indicate that it is through the bithiazole moiety or a residue close to it that binding to DNA occurs. In contrast, our own spectroscopic studies and those of Dabrowiak[10] indicate that metal ions including Fe(II), Cu(II), and Zn(II) bind in a manner not involving the bithiazole-containing residue. From these studies, the possibility that the drug molecule possesses a DNA-binding site that is distinct from its major metal-binding site is suggested. The essence of our proposal is that both DNA binding and metal binding, specifically the binding of Fe(II), are closely related in the degradation of DNA by this antibiotic, although these functions may reside in differing portions of the molecule.

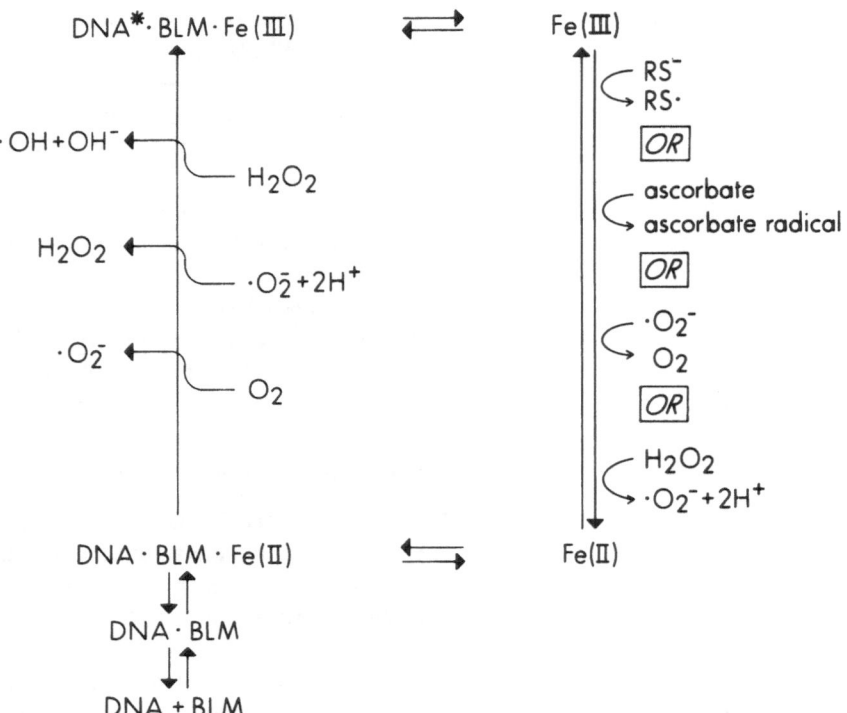

Fig. 7. Model proposed for the degradation of DNA by bleomycin. The antibiotic (BLM) binds to DNA. Fe(II) combines with BLM either before or after it binds to DNA. Oxidation of DNA–BLM–Fe(II) ultimately results in damaged DNA (DNA*). Fe(III) can be reduced to Fe(II) in the presence of a number of reducing agents and thereby take part in many DNA-breaking events. The reduction of Fe(III) shown here does not presuppose a specific site of binding of the metal ion during its reduction to Fe(II). Also, the schemes for oxygen reduction showing the formation of individual species are only suggested. Any one of the products might attack DNA. However, a concerted mechanism involving a quaternary complex of DNA, bleomycin, Fe(II), and O_2, or one of its reduction products, cannot be ruled out at this time.

The importance of the observations summarized here with respect to bleomycin are that this drug may be representative of a class of molecules which act to effect DNA degradation *in vitro* by directing or allowing the oxidation of a metal ion to occur vicinal to a site where damage to DNA may readily result. Further studies are necessary to determine whether aspects of this mechanism may be related to the cytotoxic effects of this drug in cells.

Acknowledgements
This research was supported by United States Public Health Service Grant No. CA 15784. SBH is a recipient of an Irma T. Hirschl Career Scientist Award.

References

1. H. Umezawa, K. Maeda, T. Takeuchi, and Y. Okami, *J. Antibiot. (Tokyo)*, **19**, 200 (1966).
2. H. Umezawa, Y. Suhara, T. Takita and K. Maeda, *J. Antibiot. (Tokyo)*, **19**, 210 (1966).
3. H. Suzuki, K. Nagai, H. Yamaki, N. Tanaka, and H. Umezawa, *J. Antibiot. (Tokyo)*, **22**, 446 (1969).
4. C. W. Haidle, *Mol. Pharmacol.*, **7**, 645 (1971).
5. H. Umezawa, H. Asakura, K. Oda, S. Hori, and M. Hori *J. Antibiot. (Tokyo)*, **26**, 521 (1973).
6. J. Onishi, H. Iwata, and Y. Takagi, *J. Biochem. (Toyko)*. **77**, 745 (1975).
7. M. T. Kuo, C. W. Haidle, and L. D. Inners, *Biophys. J.*, **13**, 1296 (1973).
8. C. W. Haidle, K. K. Weiss, and M. T. Kuo, *Mol. Pharmacol.*, **8**. 531 (1972).
9. M. T. Kuo and C. W. Haidle, *Biochim. Biophys. Acta*, **335**, 109 (1974).
10. J. Dabrowiak, F. T. Greenaway, W. E. Longo, M. Van Husen, and S. T. Crooke, *Biochim. Biophys. Acta*, **517**, 517 (1977).
11. M. Chien, A. P. Grollman, and S. B. Horwitz, *Biochemistry*, **16**, 3641 (1977).
12. K. Nagai, H. Yamaki, H. Suzuki, N. Tanaka, and H. Umezawa, *Biochim. Biophys. Acta*, **179**, 165 (1969).
13. I. Shirakawa, M. Azegami, S. Ishii, and H. Umezawa, *J. Antibiot. (Tokyo)*, **24**, 761 (1971).
14. H. Suzuki, K. Nagai, E. Akutsu, H. Yamaki, N. Tanaka, and H. Umezawa, *J. Antibiot. (Tokyo)*, **23**, 473 (1970).
15. M. Takeshita, A. P. Grollman, and S. B. Horwitz, *Virology*, **69**, 453 (1976).
16. J. C. Bearden, Jr., R. S. Lloyd, and C. W. Haidle, *Biochem. Biophys. Res. Commun.*, **75**, 442 (1977).
17. R. Ishida and T. Takahashi, *Biochem. Biophys. Res. Commun.*, **66**, 1432 (1975).
18. W. E. G. Müller, Z. Yamazaki, H. J. Breter, and R. K. Zahn, *Eur. J. Biochem.*, **31**, 518 (1972).
19. E. A. Sausville, J. Peisach, and S. B. Horwitz, *Biochem. Biophys. Res. Commun.*, **73**, 814 (1976).
20. T. Emery, *Adv. Enzymol.*, **35**, 135 (1971).
21. E. A. Sausville, J. Peisach, and S. B. Horwitz, *Biochemistry*, **17**, 2740 (1978).
22. E. A. Sausville, R. Stein, J. Peisach, and S. B. Horwitz, *Biochemistry*, **17**, 2746 (1978).
23. Y. Kurimura, R. Ochiai, and N. Matsuura, *Bull. Chem. Soc. Japan*, **41**, 2234 (1968).
24. M. Takeshita, S. B. Horwitz, and A. P. Grollman, 1977 Annual Meeting American Society for Microbiology, **Abstracts A-70**, p. 12.
25. M. Takeshita, S. B. Horwitz, and A. P. Grollman, *Ann. N.Y. Acad. Sci.*, **284**, 367 (1977).
26. J. Peisach and G. J. Mannering, *Mol. Pharmacol.*, **11**, 818 (1975).

Contribution of the Superoxide Anion-Hydroxyl Radical Pathway to the Cleavage of DNA by Bleomycin

J. WILLIAM LOWN

The bleomycins (BLM) are a group of glycopeptide antibiotics which exhibit clinically useful antineoplastic activity and are isolated from *Streptomyces verticillus*[1,2]. They are effective against squamous cell carcinoma, Hodgkin's disease, and malignant lymphoma and are particularly valuable since they show no immunosuppression[2]. The mode of action has attracted a great deal of attention but the molecular mechanism of action is by no means clear. BLM shows interference with nucleic acid metabolism as evidenced by (i) inhibition of DNA synthesis[3], (ii) interference with mitosis[2], and (iii) DNA breakage[4]. Recently BLM has been shown to release nucleosomes from chromatin[5]. In addition there is interference with enzyme action, e.g., inhibition of ligase action[6], but stimulation of nuclease activity[7]. BLM also causes liberation of DNA from DNA-membrane complexes[8]. Thus considerable biochemical evidence suggests that the principal cell target is DNA which is degraded by BLM both *in vivo* and *in vitro*[2,4,9-11]. DNA cleavage is enhanced in the presence of reducing agents such as 2-mercaptoethanol, dithiothreitol, or ascorbate and by the addition of Fe(II) and requires oxygen[2,11-13]. The DNA cleavage is inhibited by Cu, Zn, Co, Mg, and EDTA[14]. These observations led to the suggestion of a direct role of trace metal ions, particularly of Fe, in the mode of action of the antibiotic[12,13] which would be in accord with a general role suggested for Fe in tissue disorders and cancer control[15]. Clinical grade BLM was shown to contain 0.02 mole% Fe by atomic absorption[16].

Many of the reducing agents which had been employed hitherto with BLM nick DNA themselves. These include dithiothreitol[17], 2-mercaptoethanol[17], cysteine[18], ascorbate[19], the xanthine/xanthine oxidase system[19], and hydrogen peroxide in the presence of trace metal ions[19]. The ethidium fluorescence assay in conjunction with PM2–CCC–DNA[20] has proved useful for detecting DNA strand scission produced by several antitumor antibiotics[11,21-25]. The method is sensitive [resolution 1 nick may be detected in a DNA of 6×10^7 daltons[21-25]] and can readily be adapted to measure reaction rates and to establish chemical mechanisms by adding selective inhibitors.

We demonstrated using this assay that 10 mM 2-mercaptoethanol in deionized water cleaves 32% PM2–CCC–DNA in 60 min at 37°C. Dithiothreitol (1 m*M*) behaves in a similar manner (Fig. 1). It was shown by enzymatic inhibition that the reaction proceeds via O_2^{-}, and by spin trapping techniques it was shown that OH˙ radicals are produced (Fig. 2). Since these reducing agents nick DNA by a mechanism similar to that proposed for the antibiotic they are inappropriate for examining a possible free radical component of BLM action.

However NADH was found to be suitable since it causes only very slow break-age of DNA at very high concentrations[17]. Employing the ethidium fluorescence assay with PM2–CCC–DNA it was shown that a freshly prepared $5 \times 10^{-7} M$ solution of BLM in deionized water nicks 70% of the DNA in 15 min at pH 9.0 (see Fig. 3) but is quite slow at pH 7 and loses its nicking properties on pro-longed exposure to air[16]. That this is due to oxidation of the traces of Fe(II) is indicated by addition of 0.02 mole% of $FeSO_4$ to the inactivated BLM which is

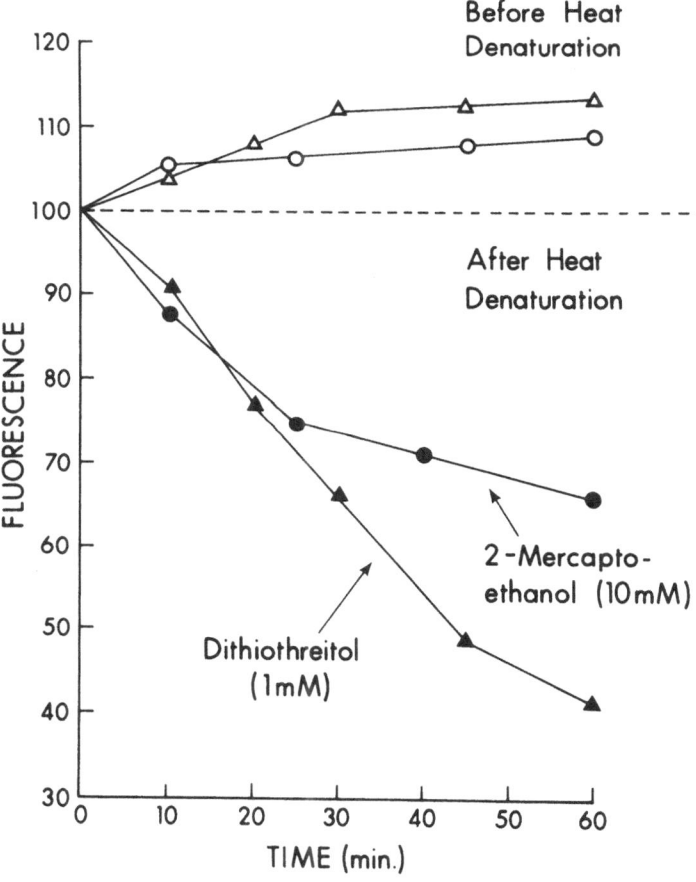

Fig. 1. Nicking of DNA by thiols.

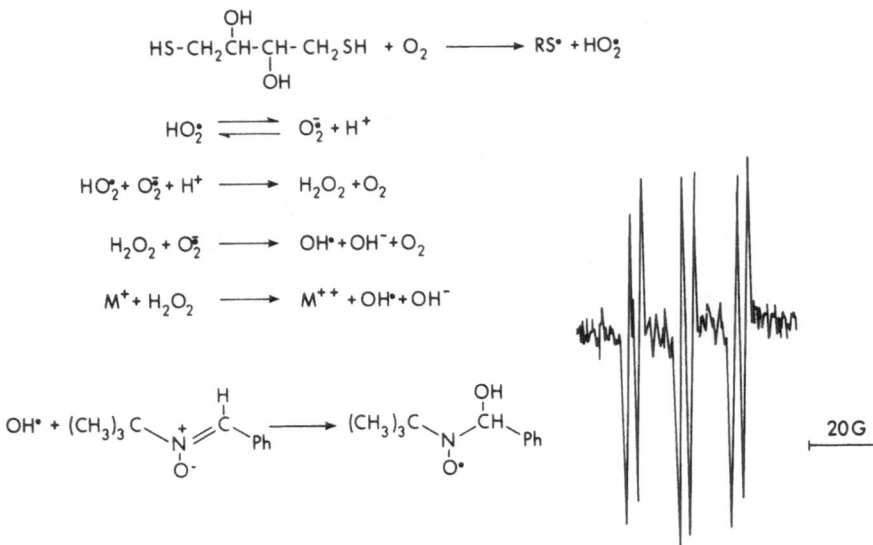

Fig. 2. Spin trapping of hydroxyl radical from oxidation of dithiothreitol.

then capable of cleaving 80% PM2–CCC–DNA in 20 min. Alternatively, continuous re-reduction of the trace of Fe may be effected by NADH. Addition of 2×10^{-4} M NADH to 5×10^{-5} M BLM produces 75% nicking of PM2–DNA in 20 min at 37°C and is suppressed by catalase or superoxide dismutase or more effectively by the two enzymes together.

In addition the nicking is suppressed by general free radical scavengers, e.g., isopropyl alcohol, sodium benzoate, or mannitol. The behavior of BLM with DNA is closely paralleled by 6×10^{-6} M $FeSO_4$ which nicks 80% DNA in 10 min. Again the DNA may be protected against cleavage by superoxide dismutase and catalase and by free radical scavengers. For a given concentration of $FeSO_4$ the rate and extent of nicking is increased by adding an equivalent of BLM (see Fig. 4). This suggests that the Fe is sequestered by the antibiotic and bound to the DNA where the cleavage process is therefore more efficient. The results obtained are analogous to those obtained with the anthracyclines[22]. For example, reduced daunorubicin nicks DNA more efficiently than an equimolar concentration of the aglycone. The difference is attributed to the known stronger binding of the parent antibiotic to the DNA.

It has been reported that BLM releases thymine residues from DNA preferentially[26,27]. The evidence for any base selective binding of BLM to DNA is conflicting, perhaps because it is difficult to assess the binding by conventional means owing to the concurrent strand breakage.

Psoralen, a furocoumarin, forms interstrand cross-links with DNA with 360 nm irradiation by reaction specifically at thymine residues[28,29]. Olivomycin

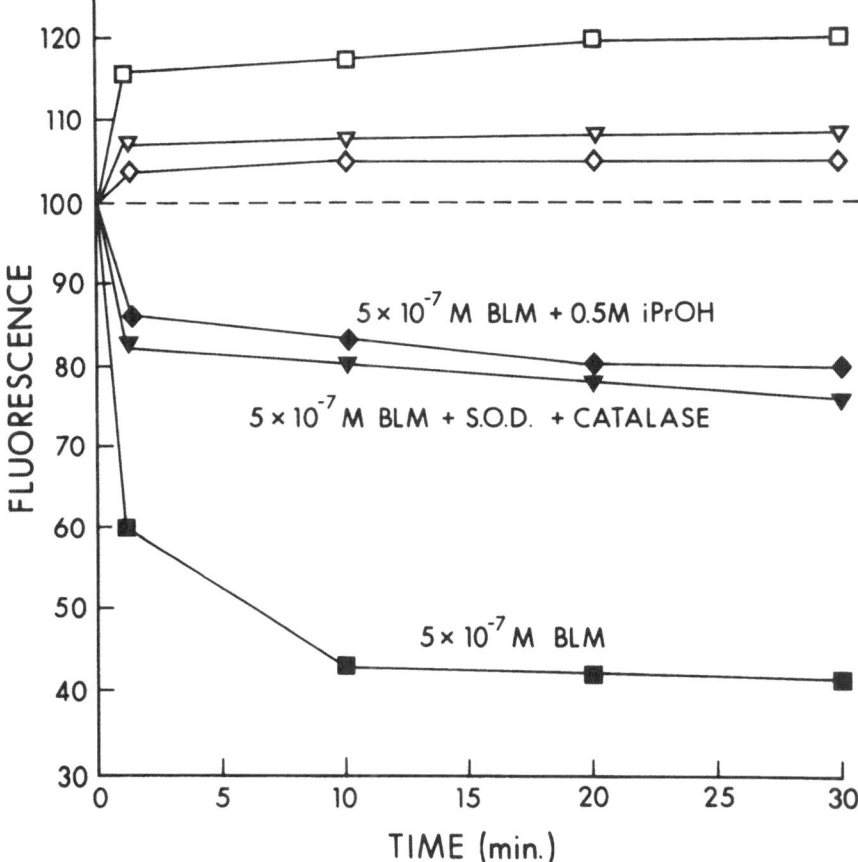

Fig. 3. Data of Ref. (11) showing nicking of DNA by BLM and its selective inhibition.

and chromomycin A_3, known G,C specific binding antibiotics, have no effect on this process[29]. However, if one protects the DNA against scission with mannitol then BLM at 10^{-5} M strongly inhibits this process like netropsin, a known A,T specific binding antibiotic (Fig. 5). This result may indicate a preference for binding of BLM to A,T-rich sites perhaps accounting for the observed preferential release of thymine.

The intercalation of some of the aromatic moieties of bleomycin into duplex DNA is conceivable. To test for this the relaxation of supercoiled covalently closed circular PM2–DNA by calf thymus topoisomerase[22,30] was examined in the presence of daunorubicin and of BLM, with protection against scission in the latter case by sodium benzoate. Whereas the anthracycline binds and

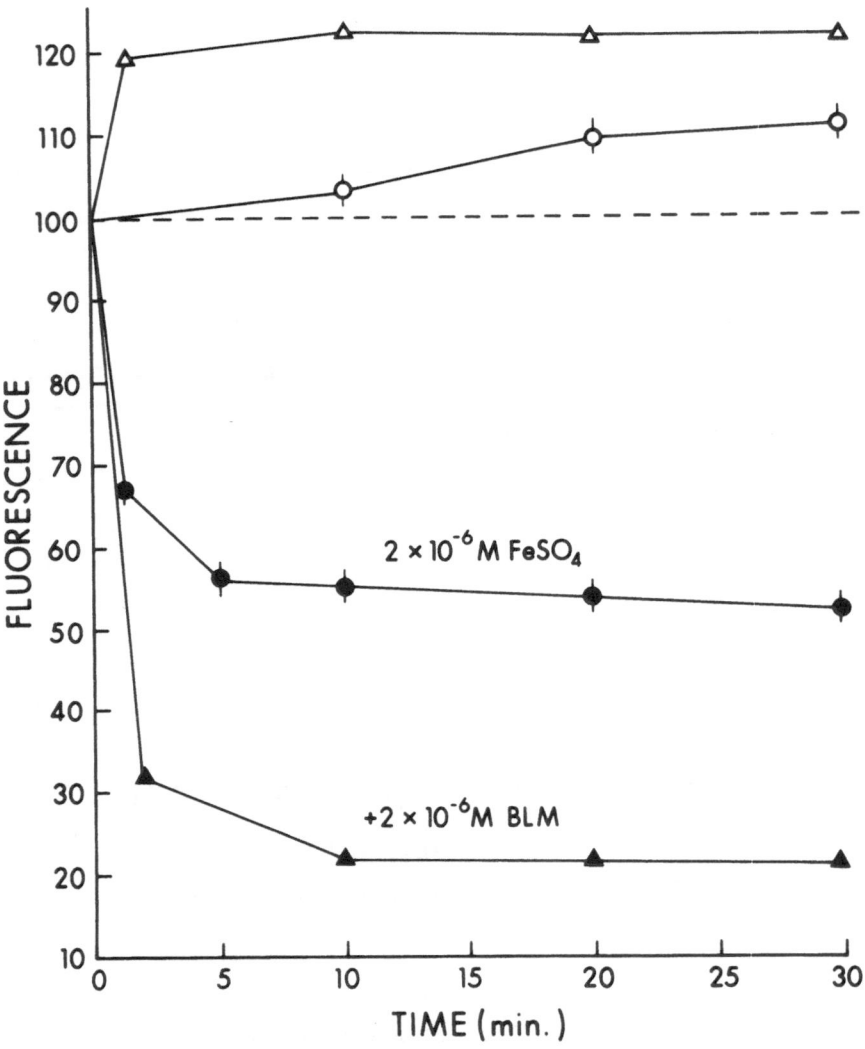

Fig. 4. Data of Ref. (11) showing effects of sequestering of Fe(II) on DNA nicking by BLM.

relaxes DNA readily[22], BLM has no effect on the superhelical density of PM2-DNA and so gives no evidence of intercalation. BLM in this case evidently does not interfere with the ligase action of the topoisomerase which is presumably coupled to the endonuclease activity.

BLM is isolated from the microorganisms as the copper complex (1) so the possible role of copper in the degradation of DNA was investigated. A 1.6 ×

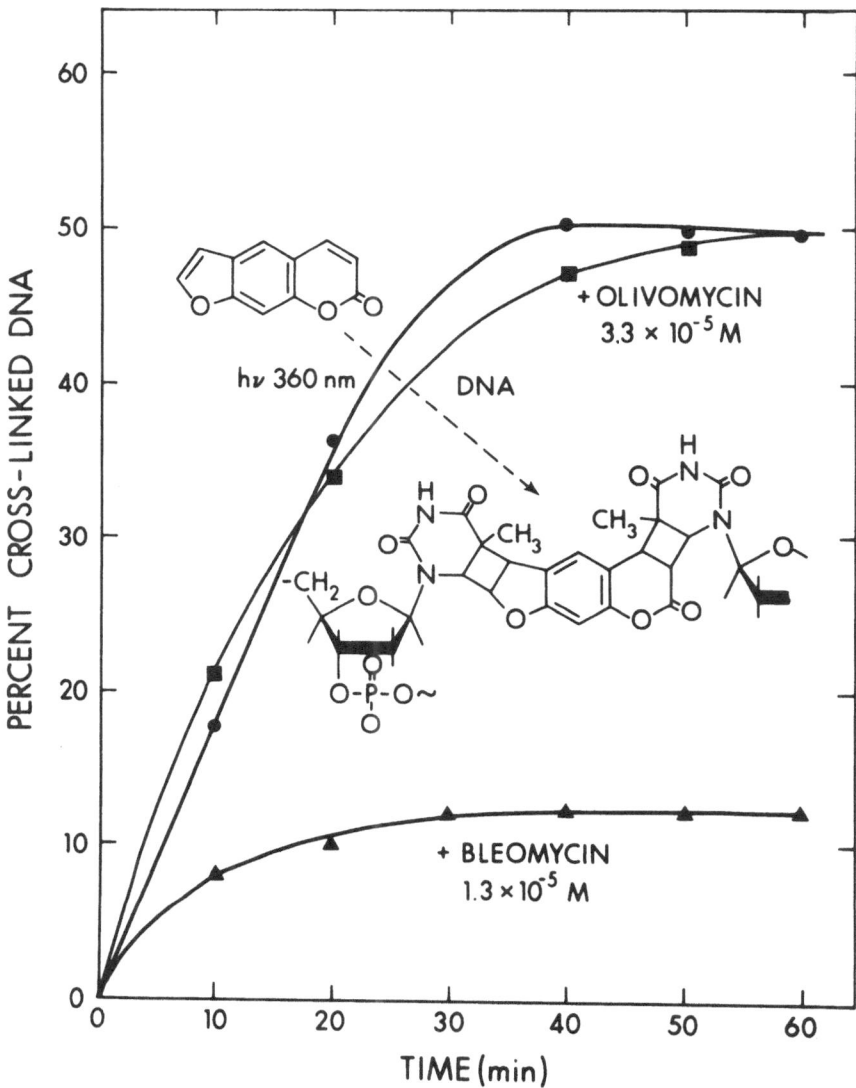

Fig. 5. Inhibition of psoralen photo-cross-linking of DNA by BLM.

10^{-5} M CuSO$_4$ with NADH nicks 70% PM2–DNA in 20 min. The mechanism is evidently a Fenton-like process since it is suppressed by superoxide dismutase and catalase. In contrast to the behavior of Fe, the nicking by reduced Cu is suppressed by sequestering with BLM. This is in agreement with the stronger binding of Cu to the antibiotic[31] where it is evidently more difficult to reduce. Umezawa and co-workers have obtained evidence for the intracellular reduction

$$BLM + Fe^{2+} \rightleftharpoons BLM \cdot Fe^{2+} \qquad\qquad 1$$

$$BLM \cdot Fe^{2+} + O_2 \longrightarrow BLM \cdot Fe^{3+} + O_2^{\bar{\cdot}} \qquad 2$$

$$BLM \cdot Fe^{3+} + NADPH \xrightarrow{-H^+} BLM \cdot Fe^{2+} + NADP \qquad 3$$

$$BLM \cdot Fe^{2+} + O_2^{\bar{\cdot}} \xrightarrow{2H^+} BLM \cdot Fe^{3+} + H_2O_2 \qquad 4$$

$$2O_2^{\bar{\cdot}} + 2H^+ \xrightarrow[\text{DISMUTASE}]{\text{SUPEROXIDE}} H_2O_2 + O_2 \qquad 5$$

$$2H_2O_2 \xrightarrow{\text{CATALASE}} 2H_2O + O_2 \qquad 6$$

$$BLM \cdot Fe^{2+} + H_2O_2 \longrightarrow BLM \cdot Fe^{3+} + \overset{\cdot}{O}H + \overset{\cdot}{O}H \qquad 7$$

$$OH^{\cdot} + DNA \longrightarrow STRAND\ CLEAVAGE \qquad 8$$

HABER-WEISS REACTION

$$H_2O_2 + O_2^{\bar{\cdot}} \longrightarrow \overset{\cdot}{O}H + \overset{\bar{}}{O}H + O_2 \qquad 9$$

of the Cu(II) in the BLM–Cu complex and the transfer of the Cu(I) to a cellular protein in the process by which BLM is activated[32]. The results of the DNA nicking with the selective inhibitors confirm the intermediacy of $O_2^{\bar{\cdot}}$, H_2O_2 and implicate $OH^{\cdot}$ and allow us to formulate a chemical mechanism for the free radical component of the cleavage of DNA by BLM.

The fact that the cleavage of DNA by the oxygen-labile BLM–Fe complex [for which spectroscopic evidence has been obtained[12]] is not repairable by ligase[6] is in accord with the suggested nicking by $OH^{\cdot}$, although alternative explanations are possible. Further proof for the generation of $OH^{\cdot}$ from BLM is afforded by spin-trapping experiments with phenyl-*tert*-butylnitrone (PBN). Incubation of 10^{-5} M BLM with PBN slowly gives rise to the PBN·OH adduct nitroxide radical detected and confirmed by its six line EPR spectrum with characteristic hyperfine couplings (see Fig. 6).

The Haber–Weiss reaction 9 commonly regarded as a source of $OH^{\cdot}$ radicals has been shown to be quite slow with $k < 0.3$ M^{-1} sec^{-1}[33,34] and therefore is unlikely to compete with steps 4 and 6. It has been suggested rather that *in vivo* the $OH^{\cdot}$ radical is produced by reaction of $O_2^{\bar{\cdot}}$ with Fe(III) complexed with protein or ATP[35]. A factor which may contribute to the selectivity of action of the antibiotic is that it has been found that superoxide dismutase levels are suppressed in tumor tissue[36].

Hydroxyl radicals are known to degrade DNA and to release free bases and aldehyde sugars of the type isolated from BLM degradation[37] (see Fig. 7). The

$$BLM \cdot Fe^{2+} + O_2 \longrightarrow BLM \cdot Fe^{3+} + O_2^{\overset{\cdot}{-}}$$

$$BLM \cdot Fe^{2+} + O_2^{\overset{\cdot}{-}} \xrightarrow{\ 2H^+\ } BLM \cdot Fe^{3+} + H_2O_2$$

$$BLM \cdot Fe^{2+} + H_2O_2 \longrightarrow BLM \cdot Fe^{3+} + \overset{\cdot}{O}H + \overset{\cdot}{O}H$$

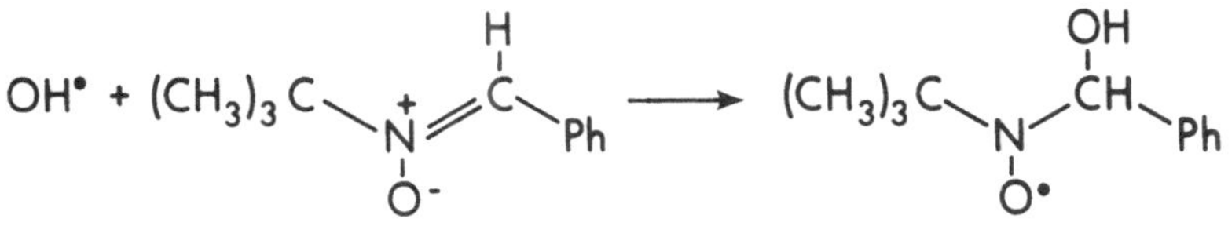

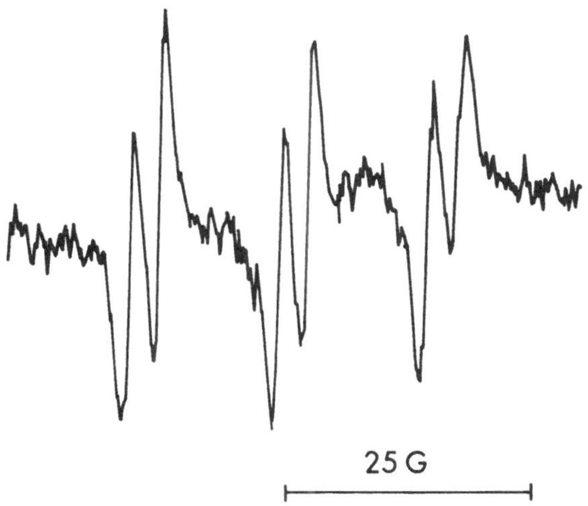

$$OH^\bullet + (CH_3)_3 C \overset{\overset{\displaystyle H}{|}}{\underset{\underset{\displaystyle O^-}{|}}{N^+}} = C \diagdown_{Ph} \longrightarrow (CH_3)_3 C \overset{\overset{\displaystyle OH}{|}}{\underset{\underset{\displaystyle O^\bullet}{|}}{N}} CH \diagdown_{Ph}$$

25 G

Fig. 6. Spin trapping of hydroxyl radical from oxidation of BLM.

cleavage of DNA by BLM increases markedly in the pH range 7–8[38]. For a given concentration of Fe(II) the radical induced DNA cleavage proceeds more slowly at pH 9.0 than at pH 8.0 or 7.0. Recent spectral evidence by Dabrowiak indicates that metals are bonded in BLM to the basic NH_2 group of the antibiotic (pK_a = 7.3), the imidazole moiety, the pyrimidine ring, and sugar carbamate group[39]. Thus it appears plausible that the higher pH is necessary to permit binding of the $-NH_2$ group to the Fe(II) and of the resulting complex to the DNA and, once bound, the sequestered Fe(II) generates radicals close to

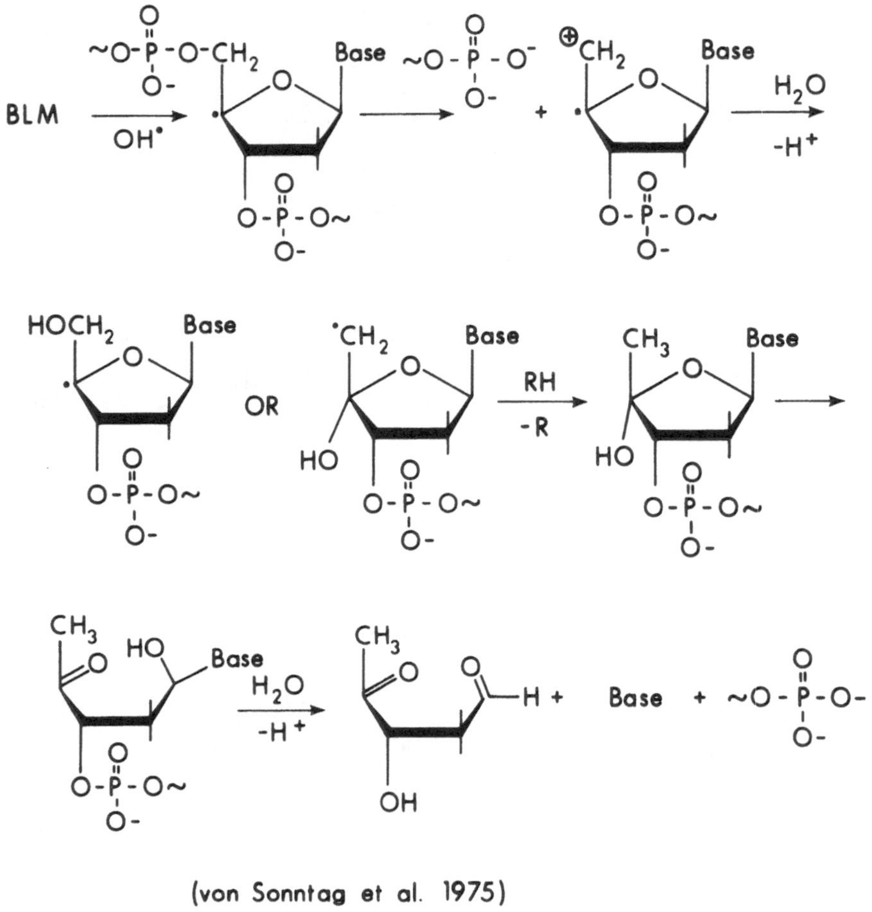

(von Sonntag et al. 1975)

Fig. 7. DNA strand breakage by hydroxyl radical (37).

the DNA albeit at less than optimum efficiency for the Fe(II). Several other clinically useful antitumor antibiotics also degrade DNA by the superoxide anion-hydroxyl radical pathway. These include mitomycin C[21], mitomycin B[40], the anthracyclines[12], streptonigrin[23], carminic acid[41], and neocarzinostatin[42]. In the case of streptonigrin and carminic acid this appears to be the principal mechanism of action and is enhanced in the case of streptonigrin by the inactivation of superoxide dismutase by the antibiotic.

Thus we may conclude that a free radical pathway may well make a significant contribution to the breakage of DNA by bleomycin and in turn to its antineoplastic action.

References

1. H. Umezawa, K. Maeda, T. Takeuchi, and Y. Okami, *J. Antibiot. (Tokyo)*, **19**, 200 (1966).
2. H. Umezawa, In *Antibiotics II* (J. W. Corcoran and F. E. Hahn, Eds.), pp. 21–23. Springer-Verlag, Berlin (1975).
3. H. Suzuki, K. Nagai, H. Yamaki, N. Tanaka, and H. Umezawa, *J. Antibiot. (Tokyo)*, **21**, 379 (1968).
4. T. Terasima. M. Yasukawa, and H. Umezawa, *Gann*, **61**, 513 (1970).
5. M. T. Kuo and T. C. Hsu, *Nature (London)*, **271**, 83 (1978).
6. M. Miyaki, T. Ono, and H. Umezawa, *J. Antibiot. (Tokyo)*, **24**, 587 (1971).
7. H. Yamaki, H. Suzuki, K. Nagai, N. Tanaka, and H. Umezawa, *J. Antibiot. (Tokyo)*, **24**, 178 (1971).
8. T. Ono, M. Miyaki, and J. Kuroda, Report of the Annual Meeting of the Japan Cancer Society, Nagoya, November 1972.
9. H. Suzuki, K. Nagai, H. Yamaki, N. Tanaka, and H. Umezawa, *J. Antibiot. (Tokyo)*, **23**, 473 (1970).
10. G. F. Saunders, C. W. Haidle, P. P. Saunders, and M. T. Kuo, In *Pharmacological Basis of Cancer Chemotherapy*, pp. 507–529. Williams and Wilkins, Baltimore (1975).
11. C. W. Haidle, *Mol. Pharmacol.* 7, 645 (1971).
12. E. A. Sausville, J. Peisach, and S. B. Horwitz, *Biochem. Biophys. Res. Commun.*, **73**, 814 (1976).
13. R. Ishida and T. Takahashi, *Biochem. Biophys. Res. Commun.*, **65**, 1432 (1975).
14. M. Takeshita, A. P. Grollman, and S. B. Horwitz, *Virology*, **69**, 453 (1976).
15. R. L. Wilson, *Iron Metabolism, Ciba Foundation Symposium 51*, pp. 331–354. Elsevier, New York (1977).
16. J. W. Lown and S. K. Sim, *Biochem. Biophys. Res. Commun.*, **77**, 1150 (1977).
17. V. C. Bode, *J. Mol. Biol.*, **26**, 125 (1967).
18. H. S. Rosenkrantz and S. Rosenkrantz, *Arch. Biochem. Biophys.*, **146**, 483 (1971).
19. A. R. Morgan, R. L. Cone, and T. M. Elgart, *Nucleic Acid Res.*, **3**, 1139 (1975).
20. A. R. Morgan and D. E. Pulleyblank, *Biochem. Biophys. Res. Commun.*, **61**, 396 (1974).
21. J. W. Lown, A. Begleiter, D. Johnson, and A. R. Morgan, *Can. J. Biochem.*, **54**, 110 (1976).
22. J. W. Lown, S. K. Sim, K. C. Majumdar, and R.-Y. Chang, *Biochem. Biophys. Res. Commun.*, **76**, 705 (1977).
23. R. Cone, S. K. Hasan, J. W. Lown, and A. R. Morgan, *Can. J. Biochem.*, **54**, 219 (1976).
24. J. W. Lown, K. C. Majumdar, A. I. Meyers, and A. Hecht, *Bioorg. Chem.*, **6**, 453 (1977).
25. J. W. Lown and K. C. Majumdar, *Can. J. Biochem.*, **55**, 630 (1977).
26. W. E. G. Muller, Z. Yamazaki, H. Breter, and R. K. Zahn, *Eur. J. Biochem.*, **31**, 518 (1972).
27. C. W. Haidle, K. K. Weiss, and M. T. Kuo, *Mol. Pharmacol.*, **8**, 531 (1972).
28. R. S. Cole, *Biochim. Biophys. Acta*, **27**, 30 (1970).
29. J. W. Lown and S. K. Sim, *Bioorg. Chem.*, **7**, 85 (1978).

30. D. E. Pulleyblank and A. R. Morgan, *Biochemistry*, **14**, 5205 (1975).
31. A. D. Nunn, *J. Antibiot. (Tokyo)*, **29**, 1102 (1976).
32. K. Takahashi, O. Yoshioka, A. Matsuda, and H. Umezawa, *J. Antibiot (Tokyo)*, **30**, 861 (1977).
33. J. J. van Hemmen and W. J. A. Meuling, *Arch. Biochem. Biophys.*, **182**, 743 (1977).
34. G. J. McClune and J. A. Fee, *FEBS Lett.*, **67**, 294 (1976).
35. G. Czapski and Y. A. Ilan, International Conference on Singlet Oxygen and Related Species in Chemistry and Biology, August 21–26, 1977, Pinawa, Manitoba, Abstract C-11.
36. L. W. Oberley, I. B. Bize, and S. K. Sahu, International Conference on Singlet Oxygen and Related Species in Chemistry and Biology, August 21–26, 1977, Pinawa, Manitoba, Abstract O-2.
37. M. Dizdaroglu, C. von Sonntag, and D. Schulte-Frohlinde, *J. Am. Chem. Soc.*, **97**, 2277 (1975).
38. H. Umezawa, H. Abakura, K. Oda, S. Hori, and M. Hori, *J. Antibiot. (Tokyo)*, **26**, 521 (1973).
39. J. C. Dabrowiak, F. T. Greenaway, W. E. Longo, M. Van Husen, and S. T. Crooke, *Biochim. Biophys. Acta*, **517**, 517 (1978).
40. J. W. Lown and G. Weir, *Can. J. Biochem.*, **56**, 249 (1978).
41. J. W. Lown, H. H. Chen, S. K. Sim, and J. A. Plambeck, *Bioorg. Chem.*, **8**, 17 (1979).
42. S. K. Sim and J. W. Lown, *Biochem. Biophys. Res. Commun.*, **81**, 99 (1978).

Interaction of Bleomycin with DNA

MAKOTO HORI

The activity of bleomycin in causing strand scission of superhelical SV40 DNA was used as the basis for a quantitative assay. Strand scission was inhibited by simultaneous addition of DNA from other species (protection by competitive binding) but not by the addition of RNA. Based on the protective effect of the added DNA, the preferred DNA structure for interaction with bleomycin was determined. The results indicate that bleomycin binds preferentially to certain purine/pyrimidine alternating sequences of double-stranded DNA.

Those structural features of bleomycin essential for activity were also studied: (i) terminal amines of complex structures can be replaced by NH_2; (ii) hydrogenation of a thiazole ring does not affect the activity; (iii) deamidation of the β-aminoalanine moiety markedly lowers activity; and (iv) Cu^{2+}-chelated bleomycin is totally inactive.

The fluorescence of the bithiazole group of bleomycin, having an emission maximum at 353 nm, quenches in the presence of various nucleic acids. The fluorescence studies gave the following results: (i) ribopolynucleotides are as effective as deoxyribopolynucleotides; (ii) guanine is a required structure; (iii) double-stranded structure is not necessary; and (iv) Cu^{2+}-chelated bleomycin, having no DNA-cleaving activity, shows stronger interaction with nucleic acids than does Cu^{2+}-free bleomycin.

Bleomycin (BLM) causes strand scission of DNA *in vitro* as well as *in vivo*[1,2]. For quantitative analysis of the activity, we used [^{3}H]SV40 DNA as a substrate[3], the native form of which is double-stranded, covalently closed circular, and superhelical. It sediments at 53 S at alkaline pH, and if bleomycin causes a single scission on either strand of the native form, the products sediment at 18 S (single-stranded, closed circular DNA) and 16 S (single-stranded, linear DNA) under the same conditions. The reaction was terminated at alkaline pH. Most experiments were conducted under restricted conditions where 16 S and 18 S DNA were the major products, while the amounts of slower sedimenting DNA fragments or acid-soluble materials were negligible. The assay conditions utilized were such that the reaction velocity, the rate of disappearance of the characteristic form of DNA, depended on the concentration of bleomycin as well as on the reaction time. This method permitted the quantitative measurement and characterization of bleomycin activity which was found to be largely independent of temperature, being somewhat greater at 0 than at 37°C. On the other hand,

195

the activity was fairly pH dependent (cleavage occurring between pH 6 and 13), with an optimum at pH 9.1. Cu(II) (1 mM) almost completely inhibited the reaction while 2-mercaptoethanol at 1mM stimulated the reaction by about 20-fold. No sulfydryl compound was included in our standard assay system. Ross and Moses recently reported that bleomycin, in the absence of a sulfhydryl compound, could not cleave DNA but produced only some alkali-labile structures in the DNA[4]. To examine this possibility, we omitted the use of alkali for termination of the reaction and analyzed the reaction products by electrophoresis at neutral pH; the results showed that bleomycin did cause strand scission of DNA[5]. Ethidium bromide (5 μg/ml) inhibited the reaction while actinomycin-D (2 μg/ml) enhanced it[6]. Fe(II) (0.1 mM), which by itself had no effect on SV40 DNA, markedly enhanced the reaction. The activity of Fe(II)–bleomycin was inhibited by EDTA, while that of bleomycin was not.

The relative ratio between the concentrations of DNA and bleomycin was critical to the sensitivity of the assay. Upon addition of unlabeled DNA, either from SV40 or other species, to a reaction mixture which contained a fixed amount of [^{3}H]SV40 DNA, the apparent effect of bleomycin diminished. As shown in Fig. 1, additional DNA from *Pseudomonas fluorescens* largely protected SV40 DNA from bleomycin action. On the other hand, tRNA from *Escherichia coli* had no effect on the sensitivity of the assay system even at high concentrations. The effect of the additional DNA was designated "protection by competitive binding," since the DNA should compete with [^{3}H]SV40 DNA for bleomycin, making less antibiotic available. In contrast, bleomycin does not bind to RNA, thus there is no effect. Various nucleic acids, either natural or synthetic, were tested for possible "protection by competitive binding" to reveal the structural features of DNA that promote binding of bleomycin. The results are shown in Table 1. Calf thymus DNA, either native or denatured, and most double-stranded copolymers effectively protected SV40 DNA. In contrast, *E. coli* tRNA, apurinic acid, and homopolymers afforded little protection. There is no doubt that denatured calf thymus DNA contains some double-stranded regions and that denatured poly(dA–dT) anneals readily. CPK models of bleomycin showed that bleomycin B$_1'$, the active member of the bleomycin family with the smallest terminal amine, is as large as a trinucleotide. Hence, some protection by oligonucleotides was expected. However, as shown in Table 1, no protection was observed.

From the results of Experiment 1 in Table 2[7], it is clear that bleomycin showed extensive binding with every purine-pyrimidine alternating copolymer (double-stranded), some binding with poly(dG)· poly(dC), and little or no binding with any of the four homopolymers or poly(dA)· poly(dT). These results suggest that for efficient bleomycin binding (i) the DNA must be double-stranded, (ii) some purine–pyrimidine neighboring sequences on each strand are desirable, although a G strand paired with a C strand is acceptable, and (iii) thymine is not required for binding, although its selective release is thought to be a cause of strand scission[1]. It should be noted that there was no extensive strand scission

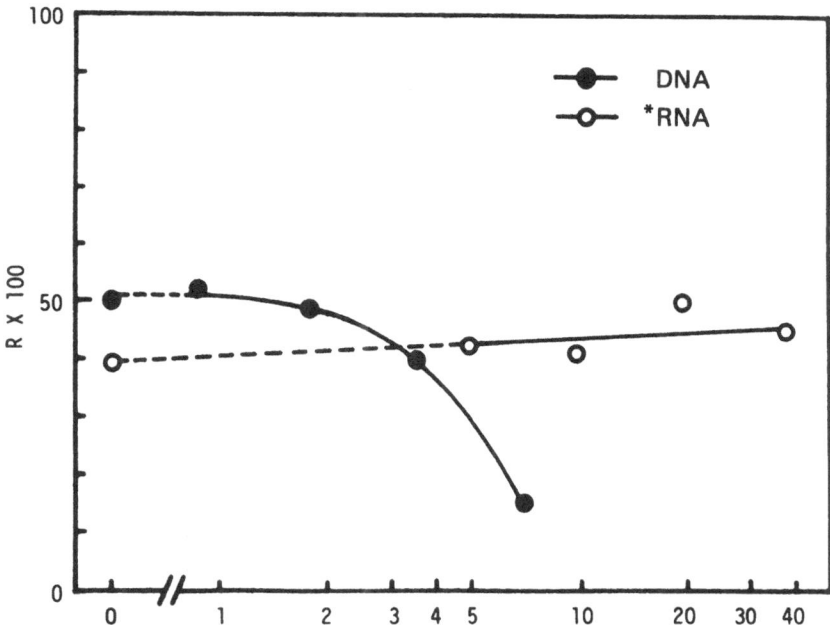

Fig. 1. Effect of DNA or RNA concentration on bleomycin action [3]. The ordinate indicates the extent of strand scission of [^{3}H]SV40 DNA. R is the ratio of radioactivity in the 16–18 S DNA fraction to the total radioactivity in the 53 S DNA and 16–18 S DNA fractions. Zero and 100 in the ordinate represent R of the control run which had no BLM and the R in the total loss of the 53 S DNA fraction, respectively [3]. DNA from *Pseudomonas fluorescens* KW4032 or tRNA from *E. coli* K12W6 was added to the reaction mixture at the desired concentration and the reaction was initiated by addition of bleomycin B$_2$ (1 μg/ml) and allowed to proceed at 0°C for 30 min.

of DNA under our assay conditions because of the low concentration of bleomycin employed and the absence of any enhancing agent such as dithiothreitol. With respect to the second prerequisite, it should be noted that poly(dA–dC)· poly(dG–dT) showed the strongest protection followed by poly(dG–dC) and poly(dA–dT) in that order and that these were even more efficient than calf thymus DNA. The preferred structure, therefore, seems to involve alternating G and T on one strand and alternating A and C on the other strand. In this structure, a G·C pair(s) may be more important than an A·T pair(s). This assumption rests on the observation that poly(dG)·poly(dC) [but not poly(dA)· poly(dT)] showed some protection in this experiment and that poly(dG,dC) gave stronger protection than poly(dA,dT) in Experiment 2 (see below). It was also interesting to find that a copolymer of unnatural bases, poly [dI–(5–Br· dC)], afforded good protection.

Table 1: Protection of SV40 DNA from Bleomycin Action by Simultaneous
Addition of Various Nucleic Acids[6] (Protection by Competitive Binding)

Additional nucleic acid	Amount added ($\times 10^{-3} A_{260}$ units)	Rate of protection[a]
None: control		0^b
DNA (calf thymus)	1.2	69.7
Heat-denatured DNA	1.3	61.0
tRNA (*E. coli*)	100	8.9
Apurinic acid	20	0
poly(dA–dT)	0.5	45.2
	2	75.5
Alkaline-denatured poly (dA–dT)	0.5	32.5
	2	62.6
poly(dG–dC)	0.5	62.2
	2	88.5
poly(dA)	100	7.3
poly(dT)	100	12.6
d(pA–pT)$_3$	37.4	0
d(pG–pT)$_2$	33.6	0
d(pC–pA)$_2$	39.0	0

[a]Rate of protection = $\left(1 - \dfrac{\text{rate of strand scission in a test run}}{\text{rate of strand scission in a control run}}\right) \times 100$ (%).

[b]Rate of conversion from superhelical form to nicked open circular form (% conversion) by 0.39 μM BLM B$_2$ was 60%.

The requirement for double-stranded structures and the preference for G·C relative to A·T were demonstrated again in Experiment 2. Here, copolymers with random sequences of two bases were examined. Protection was significant only with poly(dG,dC) and poly(dA,dT) and both can form double-stranded structures. Poly(dG,dC) was over 10 times more effective than poly(dA,dT).

In the hope of identifying the minimum structure to which bleomycin can bind, we tested a pair of oligonucleotides with complementary sequences, d(pT–pG)$_x$ and d(pC–pA)$_y$, in various combinations (Experiment 3). No protection was observed with either d(pT–pG)$_{6-9}$ alone or d(pC–pA)$_{6-9}$ alone. Since both oligonucleotides are single stranded, this result is reasonable. In contrast, there was considerable protection if both were present simultaneously, even at very low concentrations. In other combinations, we found that one oligonucleotide can be as small as a tetranucleotide, provided that the other is somewhat larger. This result seems very reasonable considering the molecular size of bleomycin.

As described above, at low concentrations Fe(II) enhanced the ability of bleomycin to cleave DNA. At high concentrations Fe(II) alone also cleaves DNA strands. "Protection by competitive binding" was observed with Fe(II)–bleomycin but not with Fe(II) alone. The nonspecific nature of the degradation of DNA by Fe(II) was suggested also by the fact that Fe(II), acting on SV40

Table 2: Protection of SV40 DNA from Bleomycin Action by Simultaneous Addition of Various Poly- and Oligonucleotides[7]

Additional nucleic acid	Amount added ($\times 10^{-3} A_{260}$ units)	Percentage protection
Expt: 1: Homo- and copolynucleotides		
Calf thymus DNA		52.7
poly(dG–dC)	2.5	77.2
poly(dA–dT)	2	53.2
poly(dA–dC)·poly(dG–dT)	1	82.9
poly[dI-(5-Br·dC)]	1	69.3
poly(dA)·poly(dT)	100	0
poly(dG)·poly(dC)	10	25.3
	100	97.3
poly(dC)	100	0.7
poly(dG)	100	3.8
poly(dT)	100	0
poly(dA)	100	0
Expt 2: Copolymers with random sequences		
poly(dC,dT)	100	0
poly(dI,dT)	100	0
poly(dA,dC)	100	22.8
poly(dC,dG)	1	76.2
poly(dA,dT)	1	6.7
	10	52.0
Expt 3: Oligonucleotides		
d(pT–pG)$_{6-9}$	100	0
d(pC–pA)$_{6-9}$	100	3.3
d(pT–pG)$_{6-9}$ + d(pC–pA)$_{6-9}$	1	34.1
	10	96.6
d(pT–pG)$_4$ + d(pC–pA)$_4$	10	30.0
	100	100.0
d(pT–pG)$_3$ + d(pC–pA)$_3$	10	17.7
	100	100.0
d(pT–pG)$_2$ + d(pC–pA)$_2$	10	0
	100	13.8
d(pT–pG)$_{6-9}$ + d(pC–pA)$_2$	10	8.5
	100	86.7
d(pT–pG)$_{6-9}$ + d(pC–pA)	100	4.0
d(pT–pG)$_{6-9}$ + d(CpA)	100	1.1
d(pT–pG)$_2$ + d(pC–pA)$_{6-9}$	10	0
	100	60.0

DNA, gave no detectable amount of form III as a degradation intermediate as opposed to bleomycin alone or bleomycin in combination with Fe(II) or a sulfhydryl compound (data not shown).

Form I of SV40 DNA appeared somewhat more susceptible to bleomycin-induced cleavage than form II of the DNA[3]. We wondered if the first nick might take place at a specific point in form I DNA. To examine this possibility, bleomycin-induced form II DNA (labeled with ^{32}P) was treated with *EcoRI*, heat denatured, and submitted to electrophoresis through an agarose gel. The products were found to be a mixture of single-stranded DNAs of various lengths, including the one having full length as the major component. This result indicated that the first nick was not at a specific point and that bleomycin recognized only a short sequence in form I DNA, if any (data not shown).

Another approach to characterization of the action of bleomycin on DNA involves a description of the group or groups in bleomycin required for activity. For this purpose, various bleomycins and their derivatives were tested for possible activity in the cleavage of SV40 DNA (Table 3). For each compound, the antimicrobial activity is also indicated for comparison. As far as the *in vitro* activity is concerned, the terminal amines can be replaced by NH_2 (BLM B_1') or an alcohol (bleomycin ester). In contrast, bleomycinic acid, having a free carboxyl group, was inactive. It is possible that the negative charge on the free carboxyl group may interfere with the binding of an active group or groups within the same molecule. These results indicate that the terminal amine of a complexed structure is not essential for the activity but may primarily influence permeability.

Bleomycin is inactivated in various organs; an inactivated product was isolated and its structure was studied. The enzymically inactivated bleomycin also showed greatly reduced activity in our assay system. There is considerable evidence that bleomycin is inactivated by an aminopeptidase which hydrolyzes the amide group (* in Fig. 2) of the 2,3-diaminopropionic acid moiety and that no other reaction is involved. This modification is thought to strengthen the basicity of the neighboring amino group (** in Fig. 2) which originally has pK_a' 7.3 in bleomycin. In the inactivated bleomycin, the corresponding pK_a' is raised to 9.4. Therefore, the amino group of the inactivated bleomycin is predominantly protonated under our assay conditions (pH 9.0 at 0°C) which seems to be the reason for the lack of activity of this compound in view of the following: (i) bleomycin is inactive at pHs below 6, where the amino group is protonated, and (ii) bleomycin becomes inactive upon chelation with Cu^{2+}, in which the lone-pair electrons are thought to be involved, as judged from the simultaneous loss of the ionizable group with pK_a' 7.3. These observations suggested the importance of the lone-pair electrons. Cu(II)-chelated bleomycin did not interfere with the action of bleomycin on DNA, suggesting that the complex did not even bind to DNA. The *in vivo* activity of the complex may be exerted after Cu(II) ion is removed in tissues.

Phleomycin differs from bleomycin only in (formal) hydrogenation of a thiazole ring (*** in Fig. 2). This structurally related antibiotic was as active as bleomycin in our assay system, suggesting that the thiazole ring is not important

Table 3: Structure–Activity Relationship of Bleomycin[a](6)

Compounds tested		Strand scission of DNA	Antimicrobial activity (u/mg)
Bleomycins with various terminal amine parts			
$-NH-(CH_2)_4-NH-C-NH_2$ with $=NH$	(BLM B₂)	100	3,385
$-NH-(CH_2)_3-N$ (morpholine)	(MOP–BLM)	113	592
$-NH-(CH_2)_3-N<^{CH_3}_{CH_3}$	(DMP–BLM)	100	810
$-NH-CH_2-CH-CH_3$ with $=NH$	(DAP–BLM)	83	2,178
$-NH-(CH_2)_3-NH-(CH_2)_3-NH-$ (cyclohexyl)	(CHPP–BLM)	28	12,335
$-NH_2$	(BLM-B₁′)	≥100	685
$-OH$	(Bleomycinic acid)	5	159
$-O-(CH_2)_3-NH_2$	(BLM ester)	≥100	1,335
$-NH_2(CH_2)_3\overset{+}{S}-(CH_3)_2$	(BLM A₂)	≥100	938
Other bleomycins			
Iso-BLM A₂		5	305
Phleomycin A₂		87	595
Phleomycin A₂ (Cu²⁺-chelated)		<2	595
Epi-BLM B₂		≥100	584
Enzymically inactivated BLM B₂		5	<300
BLM B₂ (Cu²⁺-chelated)		<2	3,480
BLM B₂ + BLM B₂ (Cu²⁺-chelated)		≥100	

[a]The activity of each compound is expressed relative to the activity of bleomycin B₂ (100%). The antimicrobial activities were determined using Mycobacterium 607 as a test organism.

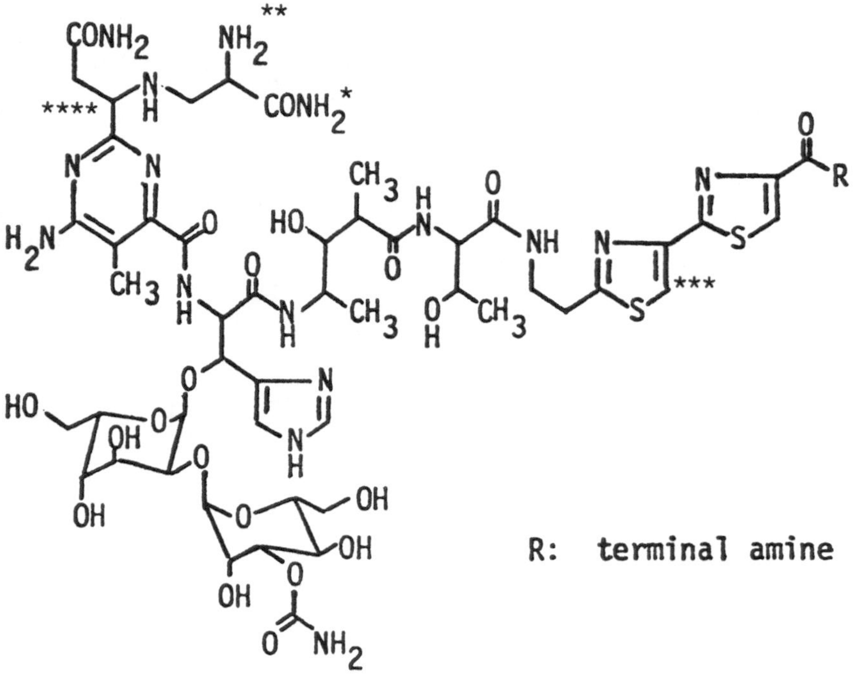

Fig. 2. Generalized bleomycin structure showing the carboxamide hydrolyzed by an aminopeptidase (*), the amino group normally involved in metal binding, whose pK_a is affected by the aminopeptidase hydrolysis (**), the site at which bleomycin and phleomycin differ structurally (***), and the site of epimerization when Cu(II) BLM is treated with triethylamine (****).

for this activity and that a possible conformational change due to the hydrogenation does not affect the function of the molecule. Epibleomycin, an epimer at the α-methine carbon of the pyrimidine moiety (**** in Fig. 2), was more active than bleomycin *in vitro*, in contrast to its reduced antimicrobial activity. Therefore the conformational change seems to affect only the permeability of bleomycin to micrococcal cells but not the activity causing strand scissions of DNA. Isobleomycin, an isomer containing 2-0-carbamoyl-D-mannose instead of 3-0-carbamoyl mannose, was inactive *in vitro*, although it had considerable antimicrobial activity. In micrococcal cells, the carbamoyl group of isobleomycin may shift from 3-0- to 2-0- of the mannose moiety. The proper location of the carbamoyl group seems to be another structural feature essential for bleomycin activity.

No chemical fragments of bleomycin were active in causing strand scission of SV40 DNA at 5–6 μM concentration. The fragments tested were heptapeptide, tetrapeptide A, tripeptide S, compound II, compound IV, compound VI, and methyl 3-0-carbamoyl-α-D-mannopyranoside. They also did not interfere with the action of bleomycin on DNA. It is suggested, therefore, that none of these fragments even binds to DNA (data not shown).

The fluorescence emission spectrum of bleomycin at 353 nm, which is attributed to the bithiazole moiety of the molecule, quenches when bleomycin is mixed with DNA[8]. We examined the possible correlation between the apparent interaction of bleomycin with DNA based on fluorescence spectroscopy[9] and the above results based on the observed strand scission of SV40 DNA. As shown in Fig. 3, the fluorescence spectra of various bleomycins and related structural components were determined at a concentration of 7.5×10^{-6} M. Compound

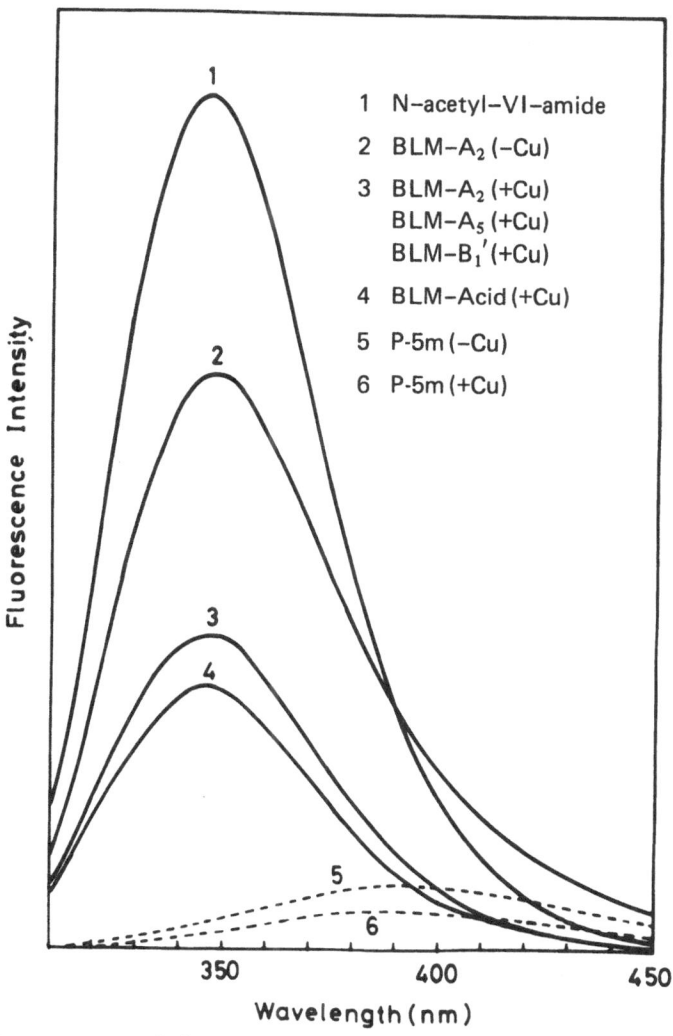

Fig. 3. Fluorescence emission spectra of bleomycins in 1.5 mM EDTA, 15 mM NaCl, and 15 mM Tris–HCl (pH 7.6); excitation at 290 nm.

1 is a chemical fragment that includes the bithiazole group. Compounds **5** and **6** lack a bithiazole group and hence are without emission maxima at 353 nm.

When calf thymus DNA was added at increasing concentrations, increasing quenching of fluorescence resulted, as shown in Fig. 4. The marked quenching observed with bleomycin A_5 seems to indicate that the strongly cationic terminal amine of this bleomycin enhanced the interaction between the bithiazole group and DNA. This is consistent with the much weaker quenching of fluorescence obtained with bleomycinic acid, which has no terminal amine. It is interesting to note that chelation with Cu^{2+} strengthens this interaction, although the chelation leads to loss of DNA-cleaving activity.

Various poly-, oligo-, and mononucleotides were examined for possible inter-action with the bithiazole moiety of bleomycin; the results are shown in Table 4.

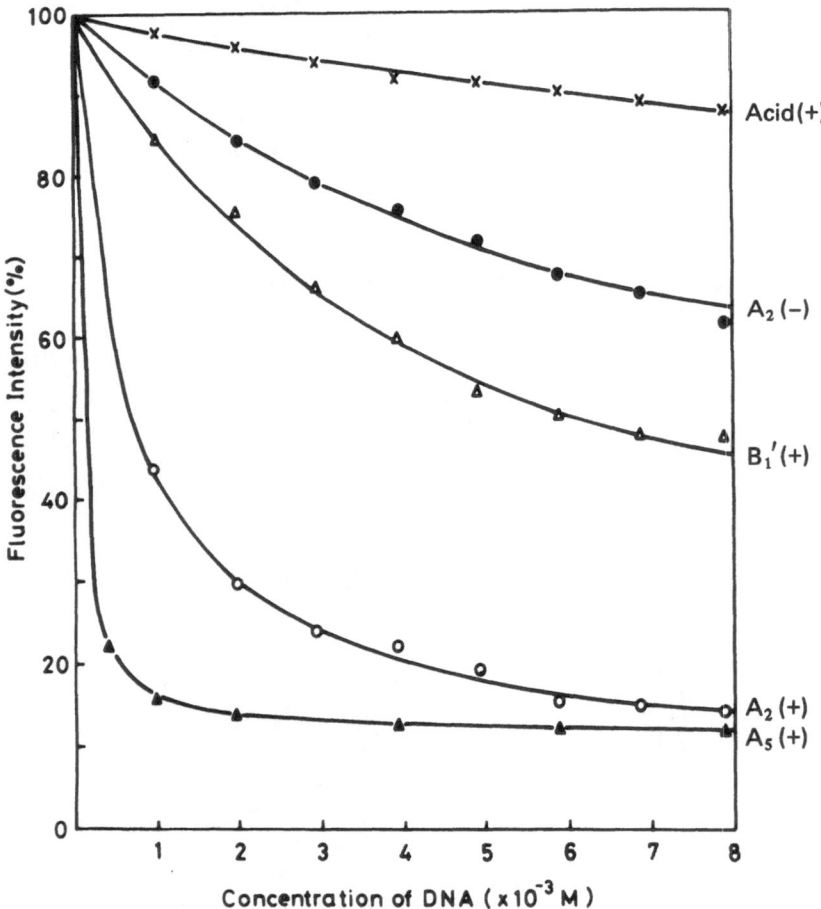

Fig. 4. Fluorescence quenching of bleomycin on addition of various amounts of DNA; concentration of bleomycin, 0.075×10^{-4} M.

Table 4: Fluorescence Quenching of BLM-A$_5$
(+Cu) on Addition of Nucleic Acids^a

Nucleic acids	Quenching (%)
5′ -pA	0
pU	0
pG	4
pC	0
Calf thymus DNA	56
Denatured calf thymus DNA^b	65
poly(dG–dC)	64
poly(dG)·poly(dC)	61
poly(dA–dT)	31
poly (dA)·poly(dT)	0
poly(A)·poly(U)	35
poly(I)·poly(C)	18
d(pT–pG)$_{6-9}$	29
d(pC–pA)$_{6-9}$	13
d(pT–pG)$_{6-9}$ + d(pC–pA)$_{6-9}$	32
poly(A)	39
poly(U)	14
poly(G)	77
poly(C)	0
poly(dA)	18
poly(dT)	20
poly(dG)	80
poly(dC)	0

^{a}Concentrations of BLM A$_5$ (+Cu) and nucleic acids
were 0.075 × 10^{-4} and 1.22 × 10^{-4} M, respectively.
^{b}Hyperchromicity, 31%.

All the mononucleotides tested were inactive. Denatured DNA was more active than native DNA. Synthetic double-stranded DNAs consisting of G·C pairs were somewhat more active than those consisting of A·T pairs. Ribohomopolymers were as active as deoxyribopolymers. The guanine moiety seemed to be important for maximal interaction. These fluorescence studies have thus given various results which conflict with those based on strand scission of DNA. Therefore, the function of the bithiazole group in bleomycin action on DNA has yet to be established.

Acknowledgment

This work was partly supported by a Grant-in-Aid for Cancer Research from the ministry of Education, Science and Culture.

References

1. W. E. G. Müller and R. K. Zahn, *Prog. Nucleic Acid Res. Mol. Biol.*, **20**, 20 (1977).
2. H. Umezawa, *Lloydia*, **40**, 67 (1977).
3. H. Umezawa, H. Asakura, K. Oda, S. Hori, and M. Hori, *J. Antibiot. (Tokyo)*, **26**, 521 (1973).
4. S. L. Ross and R. E. Moses, *Biochemistry*, **17**, 581 (1978).
5. H. Asakura *et al.*, in preparation.
6. H. Asakura, M. Hori, and H. Umezawa, *J. Antiobiot. (Tokyo)*, **28**, 537 (1975).
7. H. Asakura, H. Umezawa, and M. Hori, *J. Antibiot. (Tokyo)*, **31**, 156 (1978).
8. M. Chien, A. P. Grollman, and S. B. Horwitz, *Biochemistry*, **16**, 364 (1977).
9. H. Kasai, H. Naganawa, T. Takita, and H. Umezawa, *J. Antibiot. (Tokyo)*, **31**, 1316 (1978).

A Molecular Basis for the Interaction of Bleomycin with DNA

Masaru Takeshita and Arthur P. Grollman

The bleomycin antibiotics form complexes with DNA (cf. 1) and with certain metal ions[2,3]. Binding parameters for these complexes have been reported[4]. Nuclear magnetic resonance studies[4] suggest that the bithiazole and terminal amine moieties of the bleomycins bind to DNA; certain other functional groups of the antibiotic serve as ligands for the metal ion[3,5]. Some of the structural requirements for biological activity of bleomycin have been established[1,6]. Various DNAs have been compared for their ability to bind these antibiotics[7].

In the presence of small amounts of ferrous ion and molecular oxygen, bleomycin induces strand scissions in DNA[8-11], accompanied by release of thymine and other bases[12-14] and the formation of a low-molecular-weight product that produces a chromophore when reacted with thiobarbituric acid[15]. Fragmentation of DNA occurs when eukaryotic or prokaryotic cells are treated with bleomycin; this effect may be responsible for the antimicrobial[16], antiviral[17], and antitumor[16] activities of this antibiotic as well as for some of its toxic[18] actions.

In this chapter, we apply a new method for determining the base sequence of DNA, developed by Maxam and Gilbert[19], which has enabled us to detect precisely nucleotide sequences that are preferentially cleaved by bleomycin. Taken together with other data concerning the properties of the cleavage products, we propose a model for a DNA–iron–oxygen–bleomycin complex which explains, on a molecular basis, most reported observations relating to the interaction of bleomycin with DNA, including the role of ferrous ion and molecular oxygen in the fragmentation reaction.

Methods

Preparation of DNA Restriction Fragments

Bacteriophage ϕX 174 was grown by infecting *Escherichia coli* HF4704 in the presence of chloramphenicol[20]. DNA was extracted from the bacteriophage and purified by the method of Godson and Vapnek[21]. The alkaline sucrose density gradient step in this method was replaced by centrifugation on cesium chloride density gradients prepared in the presence of ethidium bromide (350

μg/ml). Following extraction with isopropanol and precipitation with ethanol, the DNA was digested with *Hae*III[22], yielding the double-stranded DNA restriction fragment designated Z_8 in the map reported by Sanger *et al.*[22]. This fragment was phosphorylated at the 5'-terminus with $[\gamma$-^{32}P] ATP ($\sim$ 2000 Ci/mmole) and polynucleotide kinase, then digested with *Hinf*I. A portion was treated with alkali to obtain single-stranded DNA[19]. Single- and double-stranded DNA fragments obtained by these procedures were purified further by electrophoresis on 4% polyacrylamide gels and their sequence (Fig. 1) was determined as described by Maxam and Gilbert[19]. In a similar fashion, restriction fragments were prepared from pSM1 and pSM15 plasmid DNA[23].

Fragmentation of DNA with Bleomycin

The standard reaction mixture contained 50 mM Tris–HCl, pH 8.5, 10 mM 2-mercaptoethanol, ^{32}P-labeled restriction fragments, 5 μg/ml calf thymus DNA, and 25 μg/ml bleomycin A_2 in a total volume of 40 μl. After incubating for varying periods of time at 37°C, ethylenediaminetetraacetic acid (EDTA) was added to a final concentration of 10 mM and the solution was rapidly frozen and lyophilized. Each sample was dissolved in 20 μl of 0.1 M NaOH containing 1 mM EDTA, to which was added 20 μl of 10 M urea containing 0.05% bromphenol blue–xylene cyanol. The solution was heated at 90°C for 15 sec and aliquots of 5–10 μl were transferred to 20% polyacrylamide slab gels for analysis of the nucleotide sequence of the reaction products. Autoradiography of polyacrylamide gels was carried out at –20°C.

```
5'
   ³²p C C C C T T A C T T G A G G A T A A A T T A T G T C T A A T -
3'    G G G G A A T G A A C T C C T A T T T A A T A C A G A T T A -

      A T T C A A A C T G G C G C C G A G C G T A T G C C G C A T -
      T A A G T T T G A C C G C G G C T C G C A T A C G G C G T A -

      G A C C T T T C C C A T C T T G G C T T C C T T G C T G G T -
      C T G G A A A G G G T A G A A C C G A A G G A A C G A C C A -

      C A G A T T G G T C G T C T T A T T A C C A T T T C A A C T -
      G T C T A A C C A G C A G A A T A A T G G T A A A G T T G A -

      A C T C C G G T T A T C G C T G G C G
      T G A G G C C A A T A G C G A C C G C T G A p
```

Fig. 1. Base sequence of a ^{32}P-labeled double-stranded restriction fragment prepared from ϕX 174.

Preparation of [^{14}C]-labeled HeLa Cell DNA

HeLa S$_3$ cells were grown overnight in Eagle's minimal essential media containing (A) 0.1 μCi [*methyl*-^{14}C]thymidine (53 mCi/m mole), 2.5 μCi [8-^{14}C]deoxyadenosine (39 mCi/ml), and 2.5 μCi [8-^{14}C]deoxyguanosine (56 mCi/m mole) or (B) 4 μCi uniformly labeled [^{14}C]cytidine (378 mCi/m mole). [^{14}C]DNA was purified from isolated nuclei[24] by methods previously described[25]. [^{14}C]DNA, isolated from culture (A), contained 22% of the total radioactivity in the thymine moiety, 37% as adenine, and 40% as guanine; [^{14}C]DNA, isolated from culture (B), contained 26% of the total radioactivity as thymine, 22% as cytosine, and 52% as deoxyribose. Radioactivity in the deoxyribose moiety was determined as follows: DNA was degraded enzymatically to deoxynucleosides which were separated by paper chromatography. Deoxycytidine (containing 44% of the total radioactivity) was hydrolyzed further to cytosine and deoxyribose by heating with 0.01 *M* HCl for 2 hr at 100°C. The deoxyribose of thymidine is assumed to have the same specific activity as that derived from deoxycytidine.

Results

Fragmentation of DNA Produced by Bleomycin

Cleavage of double-stranded and single-stranded restriction fragments of φX DNA, 139 nucleotides in length, is illustrated in Figs. 2 and 3. Oligonucleotides, ranging in length from 2 to 60 bases, were identified as reaction products of the double-stranded fragment. Longer oligonucleotides were excluded from the analysis of sequence specificity since the position at which they were cleaved could not be precisely distinguished. The electrophoretic mobility of oligonucleotides produced by the action of bleomycin is slightly greater than that of comparable markers; this difference is most clearly seen when comparing fragments of less than 10 nucleotides in length (cf. Fig. 4).

Bases released by treatment with bleomycin and the relative quantity of each oligonucleotide produced are shown in Fig. 3. Under the experimental conditions used, approximately half of all pyrimidines in this part of the restriction fragment were released. Seven of the eight oligonucleotides present in largest quantity (i.e., the more intense bands) are those from which a pyrimidine was released to form a new 3'-terminus. Pyrimidines were released three times as often as purines. Analysis of data for the double-stranded fragment reveals that the base located to the 3' side of *every* guanine molecule was released (Table 1). In contrast, bases located to the 3' side of cytosine, thymine, and adenine were released 6, 12, and 31% of the time, respectively. Pyrimidines, but not purines, were released with a frequency of approximately 50% when located on the 3' side of adenine. Our experiments do not permit double-stranded breaks to be distinguished from single-stranded breaks. Thus, the preferred "recognition site" for bleomycin in DNA involves GT or GC sequences. Sequences with the reverse polarity, for example, TG and CG, are not recognized (Table 1).

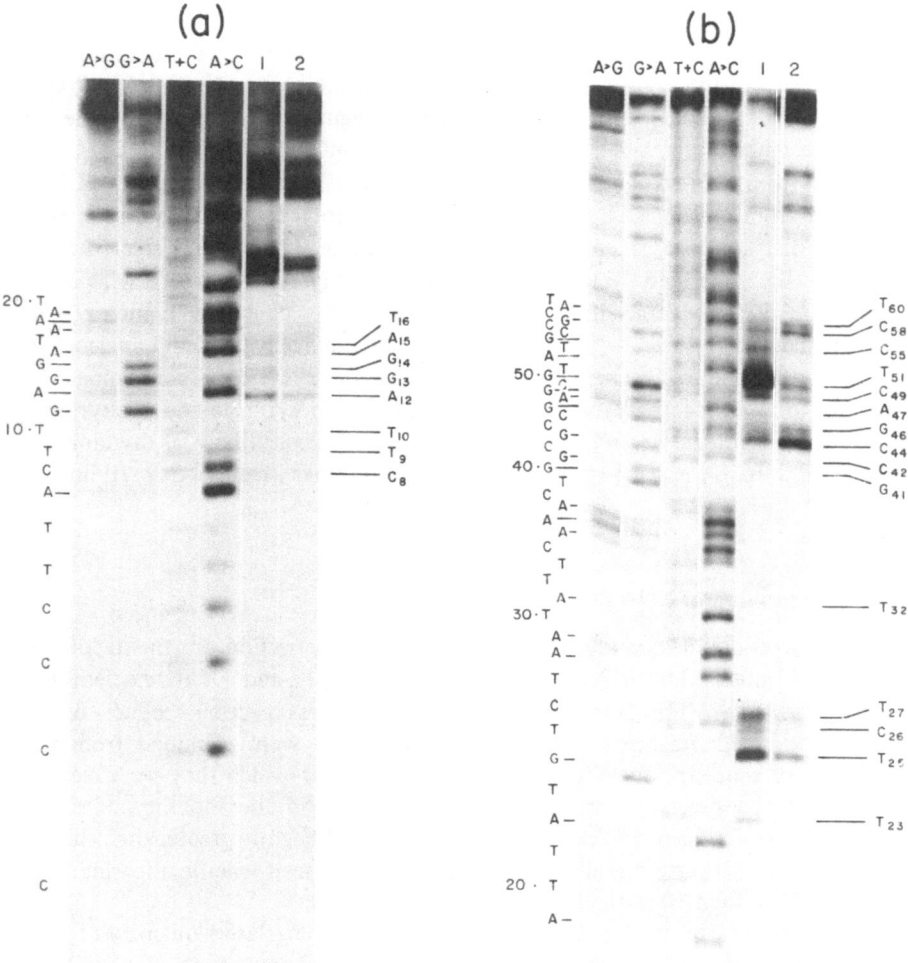

Fig. 2. Autoradiogram of a polyacrylamide gel used for sequence analysis of restriction fragments and to determine the products of bleomycin digests of ϕX DNA. The standard reaction mixture, containing ^{32}P-labeled double- or single-stranded restriction fragments, was incubated for 5 min at 37°C then treated as described under Methods. Electrophoresis was for 4 hr (a) or 22 hr (b). A > G, G > A, T + C, and A > C represent standard markers prepared as described by Maxam and Gilbert [19] in which cleavage predominates at the bases indicated. Double-stranded fragments are in lane 1; single-stranded fragments are in lane 2. Bases corresponding to the markers are identified on the left side of each gel and some of the bases released in the presence of bleomycin are indicated on the right side.

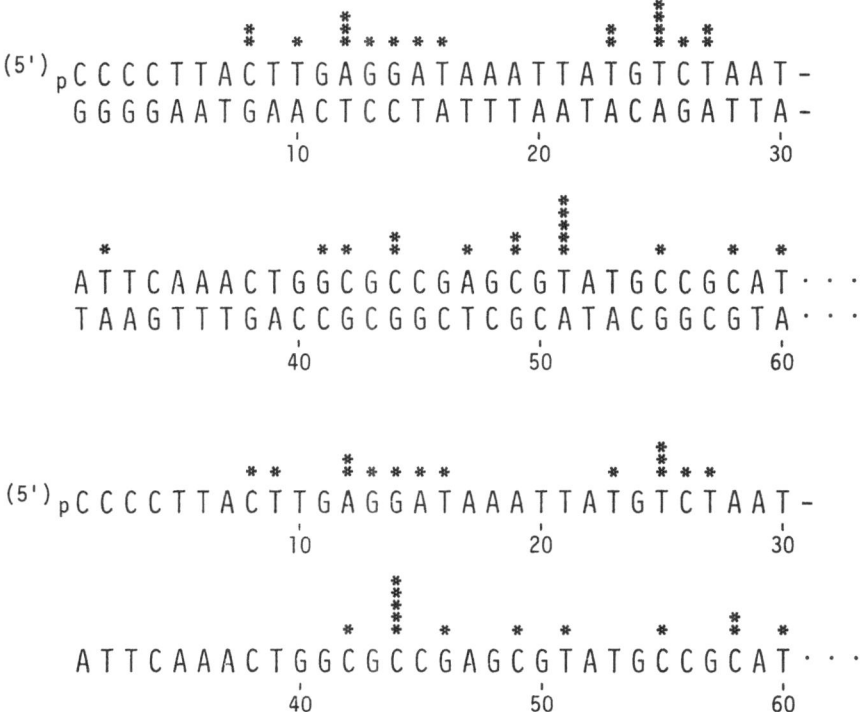

Fig. 3. Release of bases from double- and single-stranded restriction fragments of φX DNA. The sequences shown represent a portion of the double- and single-stranded restriction fragment used for analysis of base specificity. Data are taken from the experiment described in the legend to Fig. 2; bases released are indicated by asterisks and the relative intensities of the bands are indicated by the number of symbols. The thymine located at position 10 corresponds to position 989 in Sanger's map [22].

The sequence analysis described above was also conducted using single-stranded fragments. The rate of reaction of bleomycin with the single-stranded fragment was slower; two of the bases cleaved in the single-stranded form were not released from the double-stranded material and four bases cleaved in the double-stranded form were not released from single-stranded material. There was also a qualitative difference in the intensity of bands produced from the two forms, for example, at positions T_{51}, C_{44}, and T_{25}. Overall, however, bleomycin shows similar base specificity toward single-stranded and double-stranded DNA in that pyrimidines adjacent to guanine or adenine are preferentially released.

The addition of mercaptoethanol, H_2O_2, or NaBH₄ to the standard reaction mixture facilitated cleavage of restriction fragments without altering sequence specificity (data not shown). Effects on plasmid DNA are similar to those on φX DNA (Table 2).

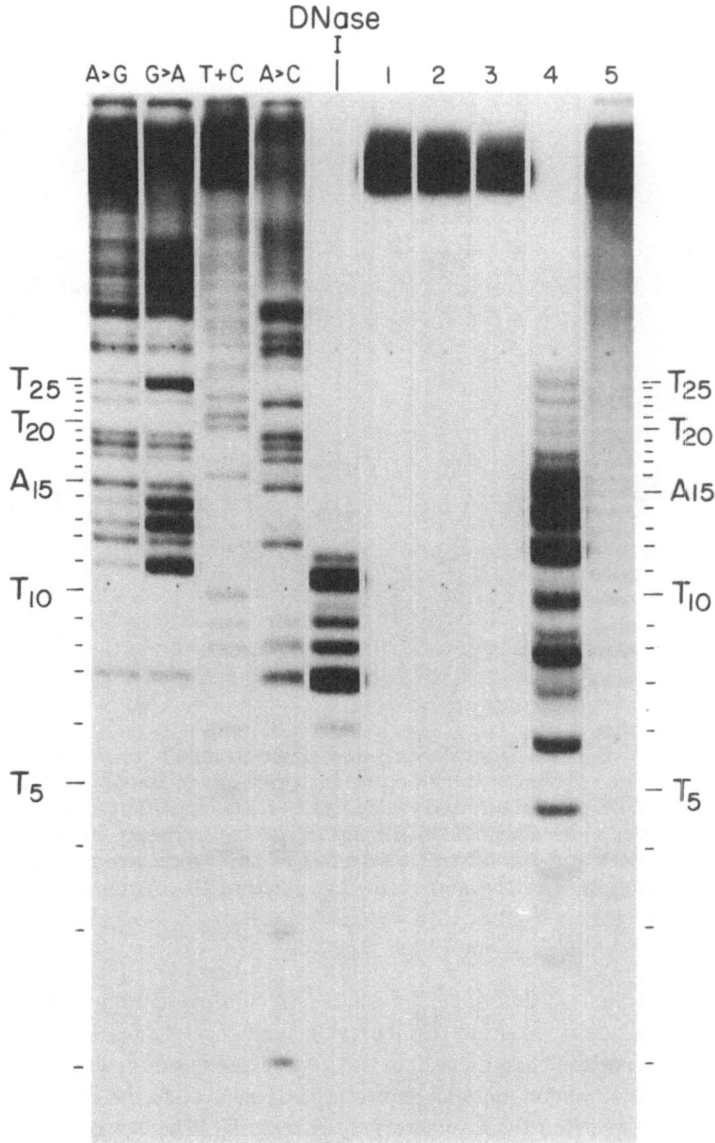

Fig. 4. Autoradiogram of a polyacrylamide gel used to determine the effects of ferrous ion and bleomycin on cleavage of DNA. The double-stranded restriction fragment of ϕX 174 was incubated with bleomycin as described in the legend to Fig. 2 with the following modifications: 2-mercaptoethanol was omitted from all reaction mixtures and bleomycin was omitted from reactions analyzed in lanes (1), (3) and (5). Fe(NH$_4$)$_2$(SO$_4$)$_2$ was added as follows: (1) none, (2) none, (3) 0.15 mM, (4) 0.15 mM, and (5) 1.5 mM. Electrophoresis was for 4 hr. "DNase I" represents an enzymatic digest of the restriction fragment.

Table 1:   Relation of Base Released to the Neighboring Nucleotide[a]

Base sequence	Number present	Frequency of release			
		from dsDNA		from ssDNA	
		No.	%	No.	%
GT*[b,c]	2	2	100	2	100
AT*	7	4	57	3	43
CT*	4	1	25	2	50
TT*	4	1	25	0	0
GC*	5	5	100	5	100
AC*	2	1	50	1	50
CC*	5	0	0	0	0
TC*	2	1	50	1	50
GA*	3	3	100	2	67
AA*	5	0	0	0	0
CA*	2	0	0	0	0
TA*	6	0	0	0	0
GG*	2	2	100	1	50
AG*	2	1	50	1	50
CG*	4	0	0	1	25
TG*	4	0	0	0	0
*TG	4	2	50	1	25
*TA	6	3	50	3	50
*TC	2	1	50	1	50
*TT	4	1	25	1	25
*CG	4	2	50	2	50
*CA	2	1	50	1	50
*CC	5	2	40	2	40
*CT	4	2	50	2	50
*AG	2	2	100	1	50
*AA	5	0	0	0	0
*AC	2	0	0	0	0
*AT	7	1	14	1	14
*GG	2	1	50	1	50
*GA	3	1	33	2	67
*GC	5	1	20	0	0
*GT	2	0	0	0	0

[a]Data taken from Fig. 3.
[b]Written as a 5'-phosphoguanylyl (3'-5')-thymidine sequence.
[c]Asterisk indicates base released.

Effect of Ferrous Ion

The effect of ferrous ion on the restriction fragment was tested in the presence and absence of bleomycin (Fig. 4). Addition of 0.15 mM Fe(NH$_4$)$_2$(SO$_4$)$_2$, a concentration 10 times that of the DNA bases, produced minimal cleavage

Table 2:   Bases Released from ϕX and Plasmid DNAs

DNA	Base released (%)			
	T	C	A	G
ϕX (ds)[a]	47	46	18	25
ϕX (ss)[a]	41	40	13	25
Plasmid (ss)[b]	31	27	7	4

[a]Data taken from experiment shown in Fig. 3.
[b]Three single-stranded fragments of *E. coli* plasmid DNA [319–420, 416–319, -61'–73, of the Ohtsubos' map[23]] were digested in a reaction mixture[24] containing 50 mM Tris–HCl, pH 8.5, 50 μg/ml calf thymus DNA, 10 mM 2-mercaptoethanol, 10 mM ATP, 10 mM Mg/Cl$_2$, and 250 μg/ml bleomycin A$_2$. Aliquots were removed after 3 min and after 2 hr, then combined.

under the experimental conditions employed. If the concentration of ferrous ion was increased to 1.5 mM, fragmentation occurred at *all* positions; the bands corresponding to oligonucleotides produced were equal in intensity. The electrophoretic mobility of these oligonucleotides was identical to those of the chemically degraded markers.

In the presence of 0.015 mM bleomycin and 0.15 mM ferrous ion, strand scission occurred with the same sequence specificity as when ferrous ion was omitted from the reaction and mercaptoethanol was present. Oligonucleotides produced under these conditions did not have the same electrophoretic mobility as the markers nor did they correspond to oligonucleotides with a 3'-hydroxyl terminus, produced by treating the restriction fragment with DNase. These differences were most clearly seen when the time of electrophoresis was extended to 8 hr.

Base Release

In separate experiments, HeLa cell DNA, labeled in thymine, adenine, and guanine or, uniformly, in deoxycytidine and thymidine, was treated with bleomycin and the reaction mixtures were subjected to paper chromatography. In both experiments, identical amounts of thymine were released and the data are combined in Table 3 to illustrate the overall pattern of base release from DNA. As shown, thymine and cytosine were preferentially released; some adenine and lesser amounts of guanine were also formed. When expressed as a percentage of total bases present in DNA, the relative order of bases released from HeLa cell DNA was similar to that observed in comparable sequencing experiments which analyzed the effects of bleomycin on restriction fragments prepared from plasmid and ϕX DNA (Table 2).

Fragmentation of Deoxyribose

[^{14}C]HeLa cell DNA, labeled in thymidine and deoxycytidine, was digested with bleomycin and subjected to column chromatography on Sephadex G-10

Table 3:   Bases Released following Treatment of HeLa Cell DNA with Bleomycin[a]

	Amount released (%)		
Base	Expt 1	Expt 2	Expts 1 and 2
Native DNA			
T	14.4	14.5	14.5
C	–	10.0	10.0
A	4.3	–	4.3
G	1.5	–	1.5
Denatured DNA[b]			
T	11.1	10.9	11.0
C	–	8.5	8.5
A	2.9	–	2.9
G	1.2	–	1.2

[a][^{14}C]HeLa cell DNA, obtained from culture A (cf. Methods) in Expt 1 and from culture B in Expt 2 was incubated for 3 h with bleomycin as described for plasmid DNA in Table 2 and the bases released were determined by paper chromatography in butanol–ammonia–water (86:1:13).
[b]Denatured DNA was prepared by heating HeLa cell DNA (100 μg/ml) in 0.02 M Tris–HCl, pH 8.5, at 100°C for 10 min followed by rapid cooling.

(Fig. 5). Calf thymus DNA, treated similarly with bleomycin, was added as carrier. Fractions were analyzed for radioactivity and for reactivity with thiobarbituric acid. Ten percent of the total radioactivity present was released as thymine or cytosine; 2% (30–40% of the malondialdehyde that could, theoretically, be formed from deoxyribose) was recovered in fractions containing thiobarbituric acid-reactive material.

Discussion

Based on the experimental data presented in this paper and observations from other laboratories, we have formulated a model (Fig. 6) for the interaction of bleomycin with DNA[26]. The principal features of this model and supporting evidence for this proposal follow:

Intercalation of the Bithiazole Group

We have shown[4] that tripeptide S, a cleavage product of bleomycin A$_2$, binds to DNA with approximately the same affinity as the parent compound. The bithiazole and dimethylsulfonium groups, both of which are present in tripeptide S, are involved in binding to DNA[4]. The work of Povirk et al.[27] shows that binding of bleomycin and tripeptide S relaxes supercoiled DNA and lengthens linear DNA, suggesting that part of the drug intercalates between base pairs. The planar bithiazole rings are logical candidates for such an intercalation process.

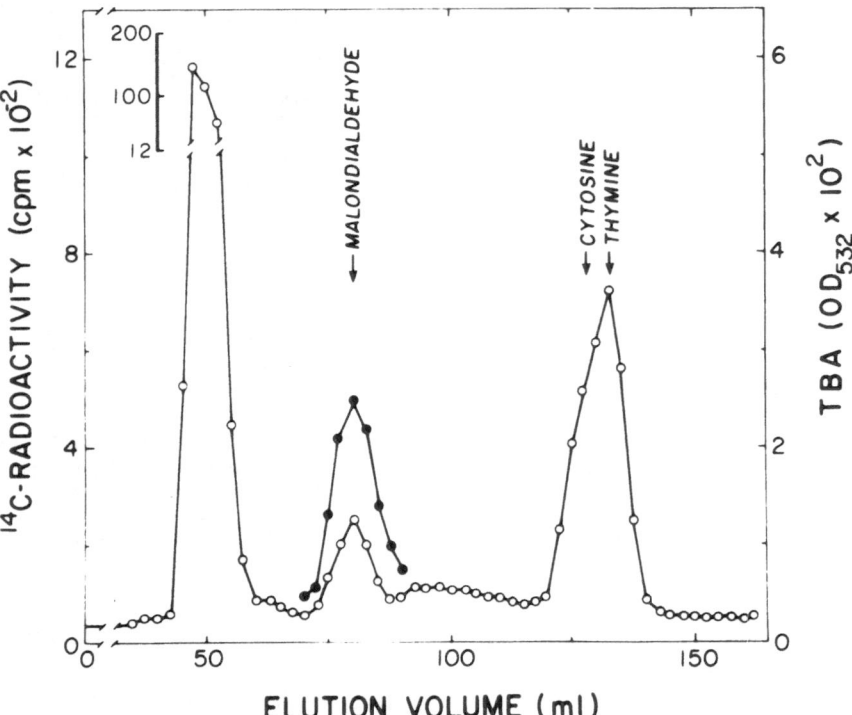

Fig. 5. Column chromatography of products obtained from digestion of HeLa cell DNA with bleomycin. HeLa cell DNA (0.15 mg), labeled in thymidine and deoxycytidine as described under Methods, and calf thymus DNA (1 mg) were digested with bleomycin in separate reactions under conditions described in the footnotes to Table 3. Reaction mixtures were combined and subjected to gel filtration on Sephadex G-10 equilibrated with 0.1% ammonium carbonate, pH 8.5. Radioactivity in each fraction (1 ml) was measured and the OD_{532} was determined after heating for 20 min at $100°C$ at pH 2, with 0.4% thiobarbituric acid. The arrows indicate the peak absorbancy of standard markers. (○ – ○), cpm; (● – ●), OD_{532}.

Specificity for GT or GC Base Sequences

Pyrimidine bases are released preferentially from DNA by the action of bleomycin, thymine being slightly favored over cytosine[26]. Pyrimidines released were almost always located to the 3′ side of a purine base. In most instances, the purine base was guanosine; thus, when bleomycin binds to and cleaves DNA, the base sequence most frequently recognized is GT or GC. No such specificity is observed when guanine is located to the 3′ side of the pyrimidine released.

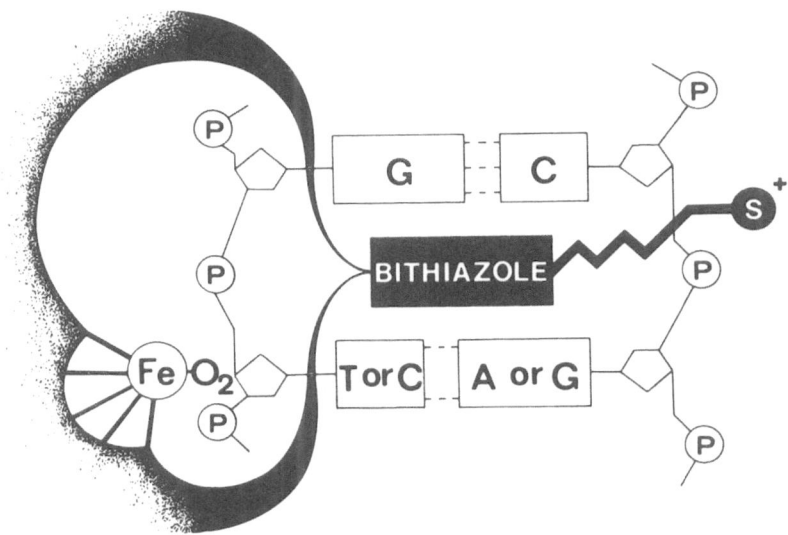

Fig. 6. A proposed model for the bleomycin–DNA–ferrous ion–oxygen complex.

Preferential release of bases, deduced by sequence analysis, is also demonstrated by quantitative measurements of bases released from labeled HeLa cell DNA.

Electrostatic Interactions

Bleomycin B_2 demonstrates the same base specificity as bleomycin A_2 in cleaving DNA (Takeshita and Grollman, unpublished data), indicating that the terminal amine moiety of bleomycin is not a critical determinant of base specificity. However, virtually all biologically active bleomycin analogs possess a positively charged group on the side chain. Binding to DNA could be partially stabilized through an electrostatic interaction between these cationic functions and negatively charged phosphates on the DNA backbone. Presence of a cation cannot be an absolute requirement for complex formation since bleomycin B_1', in which the side chain is electronically neutral, is biologically active[1]. However, related compounds with a negatively charged side chain, such as bleomycinic acid, are inactive.

Bleomycin-Fe^{2+}-O$_2$ Complex

Metals, such as Cu^{2+} and Zn^{2+}, form stable complexes with bleomycin[1,5,28]. Bleomycin also binds ferrous ion[11,14] although the geometry of this complex and specific binding sites for the metal have not been established. Since ferrous ion serves as cofactor in the cleavage of DNA and iron is routinely found in commercial preparations of bleomycin[29] and polynucleotides, it seems likely that strand scission reported by previous investigators[8,12,13,15,25] was mediated by ferrous ion. If this were the case, the demonstrated requirement for reducing agents, such as mercaptoethanol[8], would reflect a requirement for iron to be in the ferrous(II) rather than ferric(III) form.

High concentrations of ferrous ion are capable of cleaving DNA in the absence of bleomycin[30]; however, in this instance, the fragmentation pattern is non-specific and all base pairs are attacked. In the presence of lower concentrations of ferrous ion, strand scission occurs only when bleomycin is also present. Under such conditions, cleavage is base-specific.

Presence of a Free Electron Pair on Fe^{2+}

If ferrous ion is liganded to five of the atoms on bleomycin that appear to bind copper and other metals[3,5,28], one coordinating position on the metal remains free, allowing molecular oxygen or water to bind to the iron–bleomycin complex. Molecular model building studies show that such molecules could be positioned near the deoxyribose of the adjacent pyrimidine.

A Reaction Mechanism to Account for Products of Strand Scission

In the presence of ferrous ion, molecular oxygen, and bleomycin, DNA is degraded to acid-insoluble oligonucleotides, free bases, and a low-molecular-weight compound which forms a chromophore with thiobarbituric acid[12–14]. The latter has been tentatively identified as malondialdehyde. Our experiments, using DNA labeled in the deoxyribose moiety, suggest that the sugar is the source of the thiobarbituric acid-reacting material.

A plausible mechanism (Fig. 7) for this process involves generation of a radical species [1] at C-4 by means of an oxygen radical anion which, in turn, arises from reduction of oxygen by the ferrous ion–bleomycin complex [Eq. (1)]. Sausville et al.[11,14] have shown that molecular oxygen is required for the action of bleomycin; evidence suggesting formation of superoxide radicals during the reaction has also been reported[29,31].

$$Fe(II)\text{–bleomycin} + O_2 \qquad Fe(III)\text{–bleomycin} + {}^{-}O\text{–}O^{\cdot} \qquad (1)$$

The initial radical [1] reacts rapidly[32] with a second molecule of oxygen to give a new radical [2] which undergoes further reduction (by ferrous ion or reducing agents such as mercaptoethanol) to the hydroperoxy species [3]. Subsequent cleavage of the bond between carbons $3'$ and $4'$ would be very facile under neutral to slightly acidic conditions and hydrolysis of the acyl hemiaminacetal [4] yields the fragmentation products depicted: malondialdehyde and the free pyrimidines, leaving a free $5'$-phosphate ester [5] and a $3'$-phosphate ester of glycolic acid [6] as end-terminal groups on DNA.

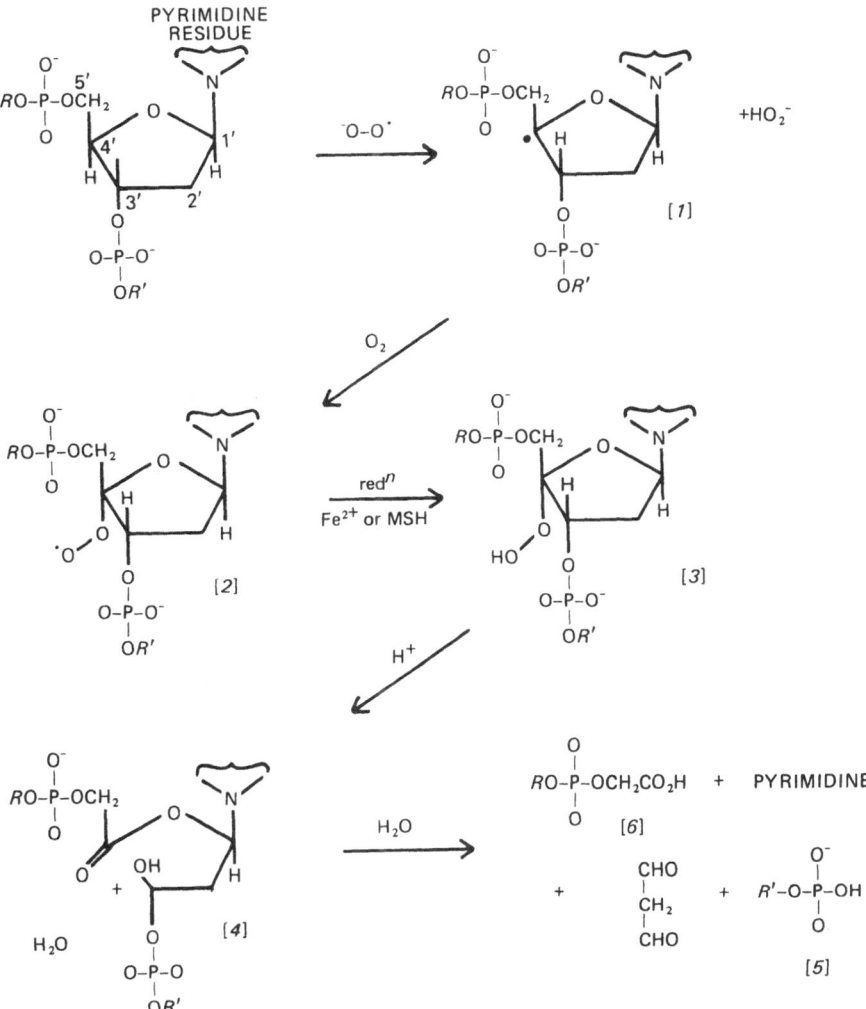

Fig. 7. Postulated reaction mechanism for cleavage of DNA in the presence of ferrous ion and molecular oxygen.

The structure of oligonucleotides formed following treatment with bleomycin contrasts with those produced by treatment of DNA with neocarcinostatin[33]. In this case, malondialdehyde is also produced but the electrophoretic mobilities of the oligonucleotides are identical to that of the standard markers and the 3'- and 5'-termini remain intact[34]. The observed difference in electrophoretic mobility between oligonucleotides produced by chemical degradation of restriction fragments (standard markers) and those produced by reaction with bleomycin suggests that some alteration of the 3'-phosphate terminus has occurred.

Acknowledgments

We are grateful to Professors Francis Johnson and Albert Haim for their help in formulating the chemical mechanism described in this paper. We thank the editors of the *Proceedings of the National Academy of Sciences* for permission to reproduce Fig. 5 and certain other data published in Reference[26].

Editor's Note. Subsequent to the presentation of this contribution, a related set of experiments was described by another laboratory (A. D. D'Andrea and W. A. Haseltine, *Proc. Nat. Acad. Sci. USA,* **75**, 3608 (1978).

References

1. H. Umezawa, *Lloydia*, **40**, 67 (1977).
2. H. Umezawa, K. Maeda, T. Takeuchi, and Y. Okami, *J. Antibiot. (Tokyo)*, **19A**, 200 (1966).
3. J. Dabrowiak, F. T. Greenaway, W. E. Longo, M. Van Husen, and S. T. Crooke, *Biochim. Biophys. Acta,* **517**, 517 (1978).
4. M. Chien, A. P. Grollman, and S. B. Horwitz, *Biochemistry,* **16**, 3641 (1977).
5. T. Takita, This volume, p. 156.
6. H. Umezawa, *Biomedicine*, **18**, 459 (1973).
7. H. Asakura, H. Umezawa, and M. Hori, *J. Antibiot. (Tokyo)*, **31**, 156 (1978).
8. H. Suzuki, K. Nagai, E. Akutsu, H. Tamaki, H. Tanaka, and H. Umezawa, *J. Antibiot. (Tokyo)*, **23**, 473 (1970).
9. R. Ishida and T. Takahashi, *Biochem. Biophys. Res. Commun.*, **66**, 1432 (1975).
10. T. Onishi, H. Iwata, and Y. Takagi, *J. Biochem.*, **77**, 745 (1975). (1978).
12. C. W. Haidle, K. K. Weiss, and M. T. Kuo, *Mol. Pharmacol.*, **8**, 531 (1972).
13. W. E. G. Müller, Z. Yamazaki, H. J. Breter, and R. K. Zahn, *Eur. J. Biochem.*, **31**, 518 (1972).
14. E. A. Sausville, R. Stein, J. Peisach, and S. B. Horwitz, *Biochemistry*, **17**, 2746 (1978).
15. M. T. Kuo, C. W. Haidle, and L. D. Inners, *Biophys. J.*, **13**, 1296 (1973).
16. T. Ishizuka, H. Takayama, T. Takeuchi, and H. Umezawa, *J. Antibiot. (Tokyo)*, Ser. A, **20**, 15 (1967).
17. M. Takeshita, S. B. Horwitz, and A. P. Grollman, *Virology*, **60**, 455 (1974).
18. A. Yagoda, B. Mukherji, C. Young, E. Etcubamas, C. Lamonte, J. R. Smith, C. T. C. Tan, and I. H. Krakoff, *Ann. Int. Med.*, **77**, 861 (1972).
19. A. Maxam and W. Gilbert, *Proc. Nat. Acad. Sci. USA*, **74**, 560 (1977).
20. T. Komano and R. L. Sinsheimer, *Biochim. Biophys. Acta*, **155**, 295 (1968).
21. G. N. Godson and D. Vapnek, *Biochim. Biophys. Acta*, **299**, 516 (1973).
22. F. Sanger, G. M. Air, B. G. Barrell, N. L. Brown, A. R. Coulson, J. C. Fiddes, C. A. Hutchison III, P. M. Slocombe, and M. Smith, *Nature (London)*, **265**, 687 (1977).
23. H. Ohtsubo and E. Ohtsubo, *Proc. Nat. Acad. Sci. USA*, **75**, 615 (1978).
24. T. W. Borun, M. D. Scharff, and E. Robbins, *Biochim. Biophys. Acta*, **149**, 302 (1967).

25. M. Takeshita, A. P. Grollman, and S. B. Horwitz, *Virology*, **69**, 453 (1976).
26. M. Takeshita, A. P. Grollman, E. Ohtsubo, and H. Ohtsubo, *Proc. Nat. Acad. Sci. USA*, **75**, 598 (1978).
27. L. F. Povirk, M. Hogan, and N. Dattagupta, *Biochemistry*, **18**, 96 (1979).
28. J. Dabrowiak, F. T. Greenaway, and R. Grulich, *Biochemistry*, **17**, 4090 (1978).
29. J. W. Lown and S. Sim, *Biochem. Biophys. Res. Commun.*, **77**, 1150 (1977).
30. S. Zamenhof, G. Griboff, and N. Marullo, *Biochim. Biophys. Acta*, **13**, 459 (1954).
31. R. Ishida and T. Takahashi, *Biochem. Biophys. Res. Commun.*, **66**, 1432 (1975).
32. C. von Sonntag and D. Schulte-Frohlinde, in *Effects of Ionizing Radiation on DNA* (J. Hüttermann, W. Köhnlein, and R. Teoule, Eds.), pp. 204–251. Springer-Verlag, New York (1978).
33. T. Hatayama, I. H. Goldberg, M. Takeshita, and A. P. Grollman, *Proc. Nat. Acad. Sci. USA*, **75**, 3603 (1978).
34. L. S. Kappen and I. H. Goldberg, *Biochemistry*, **17**, 729 (1978).

Molecular Aspects of Bleomycin-Promoted Damage of Covalently Closed Circular DNA

Charles W. Haidle, R. Stephen Lloyd,
and Donald L. Robberson

An extensive amount of information has been accumulated concerning the extracellular degradation of DNA by bleomycin since its discovery in 1966 by Umezawa et al.[1] It has been established that the drug causes the release of free bases, ruptures the phosphodiester backbone, and damages the deoxyribose moiety of DNA. There are no detectable damaging effects on RNA, nor does the drug alter the ability of mRNA to direct cell-free protein synthesis. For a complete review of both the intra- and extracellular effects of bleomycin, the reader is referred to several recent review articles[2–4]. Most of the previous studies on the extracellular DNA fragmentation reaction have relied heavily on the use of velocity sedimentation analyses (usually sucrose gradients) to assay for the introduction of strand scissions into DNA. Additionally, most of these studies employed linear duplex DNA as the substrate for the reaction. The use of linear DNA in conjunction with velocity sedimentation analyses requires the introduction of many single- or double-strand breaks into the DNA before an appreciable (and measurable) decrease in the sedimentation velocity occurs. The use of such a limited technology precludes the possibility of asking some of the more fundamental questions concerning molecular mechanisms which are operative in this bleomycin–DNA reaction at low drug concentrations. To answer some of these basic questions requires the use of a defined DNA substrate that is homogeneous with respect to size and base sequence content. Additionally, the substrate must allow for the detection of very low levels of phosphodiester bond scission. Umezawa et al.[5] have described the use of closed circular SV40 DNA in velocity sedimentation assays of bleomycin-induced damage. This procedure relies on the difference in sedimentation behavior between a DNA molecule (form I), which exists as a superhelical, covalently closed circle, and a molecule (form II), with one or more single-strand breaks resulting in a non-superhelical nicked circle.

We have chosen to use the superhelical covalently closed circular DNA from the bacteriophage PM2[6,7], which has a molecular length of approximately 10,250 base pairs (D. L. Robberson, unpublished results), and can be prepared in very large quantities[8,9]. Damage to this DNA can be assayed very precisely using an agarose gel electrophoresis technique, rather than by velocity sedimentation analyses.

Analysis of PM2 DNA Breakage by Agarose Gel Electrophoresis

The three topological forms of PM2 DNA, shown in Fig. 1, are completely resolved after electrophoresis in agarose gels prepared as described by Sharp et al.[10] In 0.9% gels, form I DNA (superhelical covalently closed circular native PM2 DNA) migrates faster than form III DNA (full-length linear duplex PM2 DNA) which in turn migrates faster than form II DNA (nicked circular PM2 DNA). In 1.4% gels, form III DNA migrates faster than form I DNA which in turn migrates faster than form II DNA. Form I° DNA (covalently closed circles without superhelical turns), when complexed with ethidium bromide in 1.4% gels, migrates faster than form II DNA, but slower than form III DNA.

The mass of PM2 DNA at each position in the gel was determined, after staining with ethidium bromide, by measuring the enhanced fluorescence of intercalated ethidium bromide using an Aminco–Bowman spectrofluorometer, equipped with a scanning device for cylindrical gels[11]. The excitation wavelength was 470 nm, and the emission wavelength was 590 nm. The integral fluorescence intensity of forms I and II DNA was found to be a linear function of DNA mass in the range of 0.25–1.50 μg per gel (determined spectrophotometrically prior to dilution for electrophoresis)[12].

Since form I DNA is topologically restricted with respect to its ability to intercalate ethidium bromide, it is necessary to correct the integral fluorescence

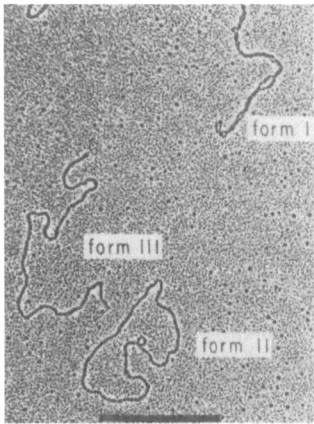

Fig. 1. Electron micrograph of the three topological forms of PM2 DNA. Samples for electron microscopy were prepared by the aqueous basic protein Kleinschmidt technique described by Davis et al. [28] The grids were rotary-shadowed with Pt–Pd (80:20) and examined in a Philips 300 electron microscope. Bar length equals 1 μm.

intensity of form I DNA. This multiplicative correction factor has been determined as follows[12]: nine identical samples of PM2 DNA containing a high percentage of form I molecules were analyzed by measuring fluorescence enhancement after electrophoresis as described above. An additional nine identical samples containing the same mass of DNA were converted to a mixture of forms II and III DNA by DNase I treatment[13] and the fluorescence enhancement after electrophoresis was likewise determined. Since the total DNA mass was the same after conversion of form I DNA to forms II and III DNA, one can calculate a multiplicative factor by which the integral fluorescence intensity of form I DNA is reduced by the topologically restricted intercalation of ethidium bromide. Thus the correction factor (x) was determined from the following relationship:

$$x (A_I) + A_{II} = A^t_{II} + A^t_{III}$$

where A_I and A_{II} are the respective areas under the peaks of forms I and II DNA prior to treatment with DNase I, and A^t_{II} and A^t_{III} are the respective areas under the peaks for forms II and III DNA after DNase I treatment. The factor determined by the DNase I method was 1.41 ± 0.14. A correction factor of 1.42 ± 0.09 was also calculated after conversion of form I DNA to forms II and III DNA by nicking the form I DNA with short wavelength (254 nm) ultraviolet light followed by restaining and rescanning the gels. In addition, a correction factor for form I DNA of 1.42 ± 0.005 was determined independently using uniformly ^{32}P-labeled PM2 DNA[8] by comparing the areas under the peaks of the three DNA forms with the radioactivity associated with each band determined by scintillation spectrometry. It can be seen that these factors determined by different methods are in excellent agreement with each other. Finally, since form I° DNA (relaxed, covalently closed circular DNA) is also topologically restricted with respect to its ability to intercalate ethidium bromide, a correction factor of 2.04 ± 0.07 was determined using the short wavelength ultraviolet nicking procedure.

Bleomycin-Mediated Akaline-Labile Damage

Since bleomycin is known to cause not only single-strand and possibly double-strand breaks[14], but also the release of free bases from DNA[15], the question arises as to what proportion of this latter damage results in alkaline-labile sites that lead to strand breakage. Povirk et al.[16] also reported the occurrence of bleomycin-promoted alkaline-labile sites and double-strand breaks in DNA.

To analyze alkaline-labile damage effectively, it was necessary to determine the conditions under which alkaline-labile sites could be hydrolyzed without irreversibly denaturing the DNA. A method similar to that described by Lindahl and Nyberg[17] using acid-heat treatment of the DNA to remove purines selectively was used to generate an alkaline-labile DNA population. Such a DNA population under neutral conditions (pH 7.6) showed a very slow rate of introduction of single-strand breaks as measured by the loss of form I DNA. However, when this DNA was maintained at pH 11.7 for 2.5 hr, neutralized, and analyzed on agarose gels, a significant loss of form I DNA was observed with a concomitant increase in form III DNA (not shown). This demonstrates that alkaline-labile damage in the form of purine base removal can be assayed effectively using PM2 DNA at pH 11.7 without irreversibly denaturing the DNA. Continued alkaline treatment, beyond 2.5 hr, showed no further alkaline hydrolysis, indicating complete hydrolysis of alkaline-labile sites.

In Fig. 2, a composite of the time course of reaction using 0.5 μg/ml of bleomycin, 25 mM 2-mercaptoethanol and 50 μg/ml of PM2 DNA analyzed under both neutral and nondenaturing alkaline conditions is presented. It is evident that low doses of bleomycin (0.5 μg/ml) cause extensive damage to the DNA in a relatively short reaction time. Under netural conditions, the mass fraction of form I DNA (open circles) decreased to less than 0.05 after 20 min of reaction with bleomycin. In contrast, analysis by nondenaturing alkaline conditions in

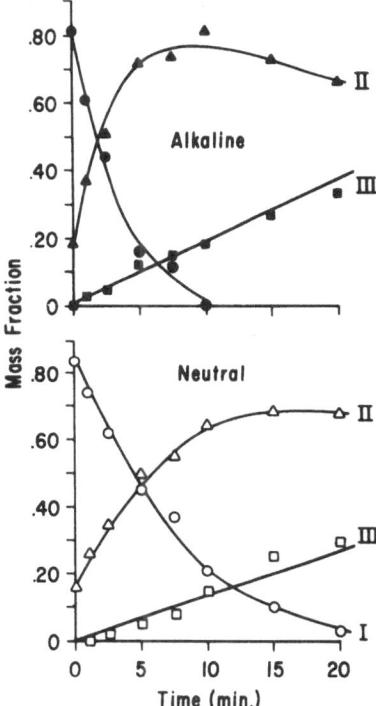

Fig. 2. Effect of bleomycin on PM2 DNA with increasing time. The bottom panel shows the mass fraction of DNA in each form as a function of reaction time analyzed at neutral pH: ○, form I; △, form II; □, form III. The top panel shows the same reaction analyzed at pH 11.7: ●, form I; ▲, form II; ■, form III.

which both breaks and hydrolyzed alkaline-labile sites contribute to the loss of form I DNA (closed circles), indicated that all detectable superhelical DNA was absent after only 10 min of incubation with bleomycin. The mass fraction of form II DNA (open triangles) increased fairly rapidly to approximately 0.65 after 10 min and remained fairly constant throughout the remainder of the experiment at neutral pH. However, form II DNA when analyzed with nondenaturing alkali (closed triangles) reached a fairly constant mass fraction after 5 min of reaction but began to gradually diminish toward the end of the experiment. Form III DNA increased linearly with time in both cases but somewhat more slowly under neutral conditions. Determination of the average rate at which damage was introduced into isolated DNA in the form of single-strand breaks and alkaline-labile sites showed that the reaction proceeds 1.9 times faster when analyzed under alkaline compared with neutral conditions. Therefore, in the bleomycin–DNA fragmentation reaction, for every single-strand break directly introduced by the drug, a second alkaline-labile site is formed.

The rate of appearance of form III DNA molecules was also studied under the neutral and nondenaturing alkaline conditions. Over a wide range of bleomycin concentrations and reaction temperatures, approximately one and a half to two linear DNA molecules were produced when analyzed under alkaline conditions for every one produced under neutral conditions (not shown).

Direct Double-Strand Breaks Produced by Bleomycin

Examination of the data presented in Figure 2 (bottom panel) reveals a fairly rapid rate of accumulation of form III DNA in the presence of low drug concentrations. Since the average number of breaks per molecule is approximately two to three after 20 min of reaction, it does not appear possible to account for the large number of linear molecules present by a mechanism involving the accumulation of random single-strand scissions.

If these single-strand scissions accumulate so that two such breaks in the complementary strands lie within approximately 16 base pairs, at 37°C under the ionic strength conditions utilized here, then one would expect the DNA helix to be destabilized with probable dissociation of base pairs and formation of linear duplexes. Alternatively, bleomycin treatment could specifically produce double-strand breaks in either form I or form II DNA in addition to single-strand scissions. To distinguish between these two alternatives, we have established a test which involves measurement of the observed frequencies of formation of form III DNA and loss of form I DNA, respectively.

If single-strand scissions are introduced at random, we estimate the average number of strand scissions per molecule, $<S>$, as a function of time, will be

$$\left[\sum_{n}^{\infty} n P_n \right] \left[f_{II}(t) + f_{III}(t) \right] + 2 f_{III}(t) \qquad (1)$$

where n is the number of strand scissions, P_n is the Poisson probability that a given molecule will be found to contain n scissions, and $f(t)$ is the mass fraction of the indicated forms of DNA as a function of reaction time with bleomycin. By definition, form I DNA contains zero single-strand scissions and form III DNA contains at least two single-strand scissions. We also note that $f_{III}(0) = 0$. For a Poisson distribution, $n=0$ nP_n is equal to $<S>$. Substituting this last term into Eq. (1) with the requirement that

$$f_I(t) + f_{II}(t) + f_{III}(t) = 1 \tag{2}$$

we obtain

$$<S> = \frac{2 f_{III}(t)}{f_I(t)} \tag{3}$$

The observed ratio of form III DNA to form I DNA at corresponding times of bleomycin reaction permits calculation of $<S>$. If $f_I(t)$ is plotted as a function of $<S>$, the curve obtained when specific double-strand scissions are produced is expected to differ significantly from the curve that predicts $f_I(t)$ as a function of $<S>$ for only random accumulation of single-strand scissions. The random accumulation of single-strand scissions without introduction of specific double-strand breaks predicts that $f_I(t)$ will decrease as a function of $<S>$ according to a simple Poisson distribution

$$f_I(t) = \frac{e^{-<S>} <S>^0}{0!} = e^{-<S>} \tag{4}$$

Using Eq. (4) we plot $f_I(t)$ as a function of $<S>$ (Fig. 3) and compare this function with a plot of the experimentally observed $f_I(t)$ as a function of $<S>$ [calculated from the ratio $f_{III}(t)/f_I(t)$ observed experimentally as in Eq. (3)]. It is clear that the observed time-dependent loss of form I DNA through bleomycin treatment as a function of the average number of breaks per molecule is significantly different from that expected by simple accumulation of single-strand scissions at random. In particular, the observed loss of form I DNA from the reaction is more rapid than that expected on the basis of random accumulation of single-strand scissions. We infer from this result that the increased efficiency in conversion of form I DNA to forms II and III DNA is a direct result of bleomycin-specific double-strand cleavages.

The analysis above is rather indirect but provides a test for direct double-strand breaks by actual measurement of both $f_I(t)$ and $f_{III}(t)$. A more straightforward analysis to test for double-strand breaks expected to arise by random accumulation of single-strand scissions has been developed by Freifelder[18]. In this latter approach, the formation of full-length linear duplexes (form III DNA) is plotted as a function of reaction time with bleomycin (Fig. 4). The dashed line

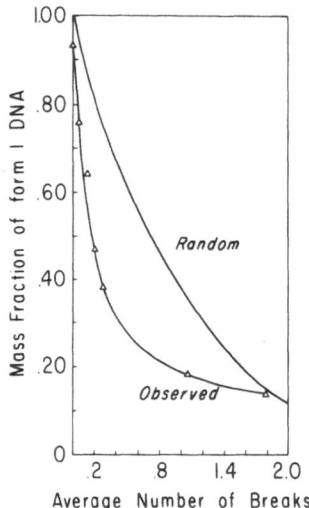

Fig. 3. Loss of bleomycin-treated form I PM2 DNA as a function of increasing average number of breaks per molecule. The experimentally observed time-dependent mass fraction of form I PM2 DNA is compared with the fraction expected for random accumulation of single-strand scissions produced by bleomycin treatment.

in Fig. 4 represents the mass fraction of form III DNA expected to be produced as a function of time resulting from the accumulation of single-strand scissions introduced at random sites in the PM2 DNA molecule. This was calculated using the method of Freifelder[18] for random accumulation of single-strand scissions separated on opposing strands by 16 base pairs; after 25 min of reaction, the expected average number of double-strand breaks is 2×10^{-3}. Thus by this analysis as well, the double-strand breaks have not arisen from single-strand scissions accumulated at random sites on the PM2 genome. It is possible, however, that the double-strand breaks result from single-strand scissions introduced randomly at closely spaced sites on opposing strands in specific regions of the genome.

Further investigation of the bases of the double-strand scissions produced by bleomycin treatment has revealed that negative superhelical turns in the reactant DNA are not required. Essentially the same rate of production of double-strand scissions was observed using form I° DNA (a relaxed covalently closed circular DNA) as reactant.

Although the presence of superhelical turns in the reactant DNA is not required for bleomycin-promoted double-strand scissions, it is possible that a site of depurination may be required. This requirement of the reaction might be expected on the basis of the alkaline sensitivity of PM2 DNA strands observed after

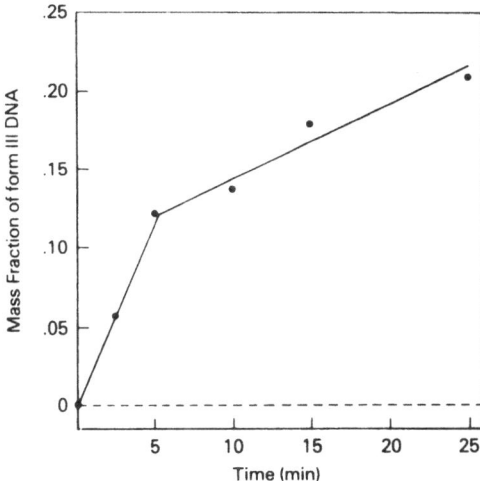

Fig. 4. Time course of bleomycin fragmentation of PM2 DNA. The mass fraction of the total PM2 DNA appearing as full-length linear duplexes is plotted as
a function of reaction time with bleomycin (solid line). Native superhelical
PM2 DNA (50 μg/ml) was treated with 0.5 μg/ml of bleomycin in 20 mM Tris
buffer, 0.27 mM CaCl$_2$, 25 mM 2-mercaptoethanol, pH 8.0. Reactions were
incubated at 37°C and stopped with 20 mM EDTA. Samples at the indicated
times were subjected to agarose gel (1.4%) electrophoresis to separate forms I,
II, and III PM2 DNAs and stained with ethidium bromide, and the mass of DNA
in each band was determined by scanning spectrofluorometry of the tube gels
as described [12]. The dashed line represents the mass fraction of form III DNA
expected to be produced as a function of time resulting from accumulation of
single-strand scissions introduced at random sites in the PM2 DNA molecule
(see text). After 25 min of reaction, the expected average number of double-
strand breaks was 2×10^{-3}.

bleomycin treatment[19]. We therefore prepared form I PM2 DNA containing an
average of at least three depurinations per molecule by thermal treatment under
acidic conditions by methods previously described[17]. When the depurinated
form I DNA was treated with bleomycin, it was found that the rate of conversion of form I DNA to forms II or III DNAs was essentially the same as that
observed for form I DNA without depurination. Furthermore, the rates of
formation of form II and form III DNAs, respectively, were indistinguishable
from the corresponding rates of formation when the reactant PM2 DNA had not
been subjected to depurination prior to bleomycin treatment (data not shown).
We conclude that sites of depurination produced prior to reaction with bleomycin are not required for bleomycin-promoted double-strand or single-strand
scissions of form I PM2 DNA at neutral pH. This result does not, however,
rule out the possibility that removal of a base from the DNA may accompany
the breakage reactions.

Since the bleomycin used in this study is the clinically used mixture of bleomycins, the question arises as to whether the different components of the bleomycin mixture may be separately promoting either single- or double-strand scissions. If a single component of the bleomycin mixture promotes double-strand scissions, then dilution of the bleomycin mixture could lead to an assessment of the kinetics of breakage in the formation of form III DNA analogous to "single-hit" kinetic analysis as would be applied, for example, in assaying ultraviolet inactivation of bacteriophage[20]. It was found that the mass fraction of form III DNA that was formed after 1 min of reaction with bleomycin decreased exponentially as the concentration of bleomycin was reduced (Fig. 5A). When the data are plotted logarithmically, the least-squares line obtained has a slope of 0.92 ± 0.19 (± standard deviation) which is seen to parallel approximately a line describing single-hit double-strand breakage of PM2 DNA. This result is compatible with there being a single component of the mixture which promotes double-strand scissions. Similarly, a plot of the log of the mass fraction of form II DNA produced at a fixed reaction time versus log of relative bleomycin concentration (less than 0.3) produces a least-squares line with a slope of 0.95 ± 0.31 which also approximately parallels single-hit events in the production of form II DNA (Fig. 5B). The initial lag in mass fraction of form II DNA produced upon small dilutions (Fig. 5B) probably represents saturation of the reaction at the higher bleomycin concentrations.

If two separate components are individually responsible for the formation of forms II and III DNA, respectively, then one would expect that a plot of the log of the average number of breaks per molecule as a function of the log of relative bleomycin concentration will follow a line that parallels a two-hit reaction, inferring the existence of two molecular components of the bleomycin mixture which independently contribute to the average number of breaks or strand scissions per molecule. This result was indeed found in that the data were fit by a least-squares line with a slope of 1.80 ± 0.20 (Fig. 5C) and suggests that separate components of the bleomycin mixture are promoting single-strand and double-strand scissions, respectively. Alternatively, a single component could account for both single- and double-strand scissions, but these analyses require independent molecular encounters of the component with the DNA substrate to produce different structural alterations.

If separate bleomycin components predominate in the formation of single-strand and double-strand scissions, respectively, one might expect to find an experimental condition which differentially alters these two activities. We have found that increasing the NaCl concentration in the reaction reduces the rate of nicking of form I PM2 DNA. The initial rate of formation of form III DNA was markedly reduced as the NaCl concentration was increased from 0.17 to 0.8 M. By way of contrast, the initial rate of formation of form II DNA was much less affected by increasing salt concentration. This differential effect on the initial rates of formation of the two forms of DNA is apparent when their ratios (initial rate of form III produced divided by the initial rate of form II produced) are

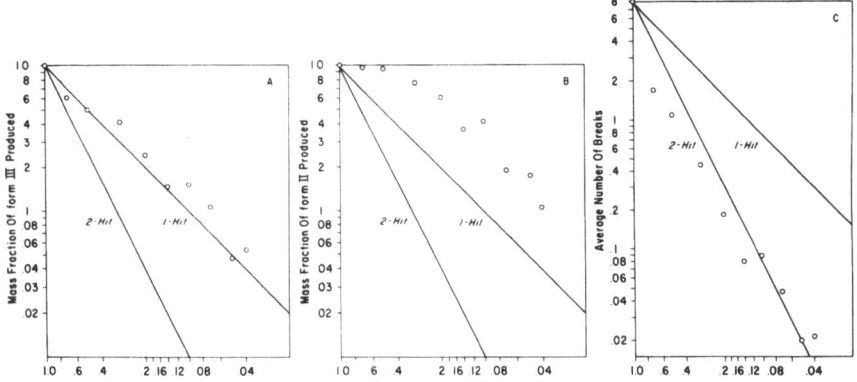

Fig. 5. Analysis of bleomycin-promoted strand scissions produced by one or two components of the reaction mixture. The logarithms of the mass fractions of total DNA as form III (A) or form II (B) PM2 DNAs produced by treatment with bleomycin for a fixed reaction time (1 min) and temperature (37°C) are plotted as a function of the bleomycin concentration relative to a starting concentration of 5.0 μg/ml (taken as 1.0). In (C), the logarithm of the average number of breaks per molecule is plotted as a function of the same relative bleomycin concentrations. Solid lines in each panel depict patterns expected to parallel one-hit (single reaction component) or two-hit (reaction requiring two components).

plotted as a function of ionic strength of the reaction (Fig. 6). These results are consistent with the previous conclusion that two components, or at least one of two activities in each of two components of bleomycin, are acting individually in the production of single- and double-strand scissions, respectively. In the case of double-strand scissions that produce form III DNA, one also observes that after 30 min of reaction the mass fraction of total DNA as form III DNA was dramatically reduced from 18%, when the reaction was performed in 0.017 M NaCl to 2% when the reaction was performed in 0.8 M NaCl solutions.

Site Preference and Mapping of Bleomycin-Promoted Double-Strand Scissions on the PM2 Genome

Bleomycin-promoted double-strand scissions could occur either at specific sites on the genome or could be introduced at randomly selected sites throughout the genome. To examine the spectrum of possible site-specific double-strand scissions in bleomycin-treated PM2 DNA, the pattern of fragmentation produced by restriction endonuclease digestions in combination with bleomycin fragmentation was determined. When PM2 DNA was digested with HindIII restriction endonuclease as described by Brack et al.[21,22], the limit products were seven

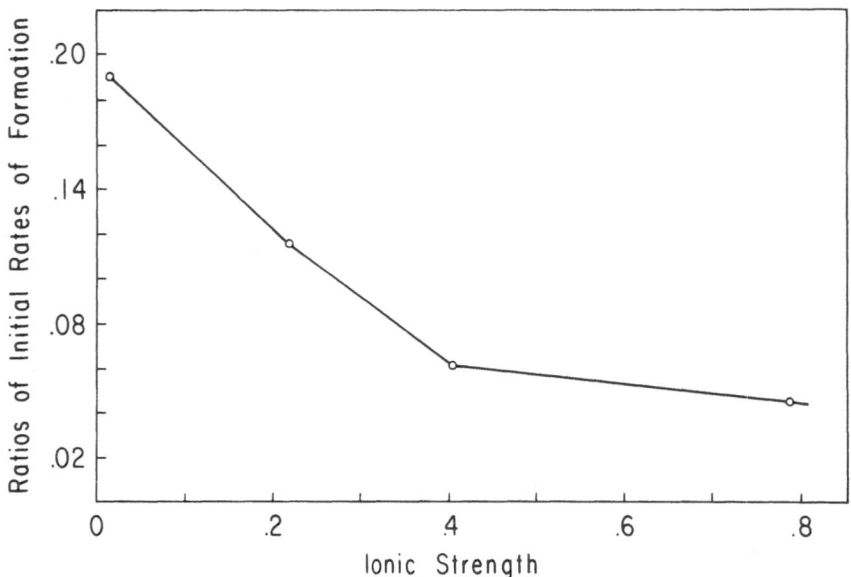

Fig. 6. Preferential inhibition of bleomycin-promoted double-strand scissions of PM2 DNA. The ratios of initial rates of formation of form III PM2 DNA to form II PM2 DNA produced by reaction with bleomycin are plotted as a function of the ionic strength of the solution.

linear duplex fragments which were resolved by agarose gel electrophoresis (Fig. 7, A–G). These fragments ranged in molecular weight from 3.4×10^6 (fragment A) to 4.8×10^4 (fragment G) (Fig. 7, track a). When the HindIII PM2 DNA fragments were subsequently treated with 2.5 μg/ml of bleomycin for reaction times which, under equivalent conditions, would progressively convert a population of form I and II DNA molecules to mixtures of forms II and III DNAs, a series of discrete DNA fragments with intermediate sizes became apparent. The intermediate size fragments migrated to positions in the agarose gel primarily between HindIII fragments A and B and between fragments B and C. Ten discrete molecular species are indicated in Fig. 7 (at the right) and a total of 15 species have been detected by visual inspection of the agarose gels after staining with ethidium bromide and illumination with short wavelength ultraviolet light. This observation requires that bleomycin-promoted double-strand scissions occur at specific sites on the PM2 genome and that many such sites are available for reaction.

The restriction endonuclease, HpaII, is known to introduce one double-strand cleavage at a specific site in PM2 DNA[21]. When PM2 DNA was digested with HpaII and subsequently treated with bleomycin as described above, we again observed a spectrum of molecular species with discrete fragment sizes apparent in agarose gel electrophoresis (Fig. 8, at left); however, the majority of fragments

Hind III then Bleomycin

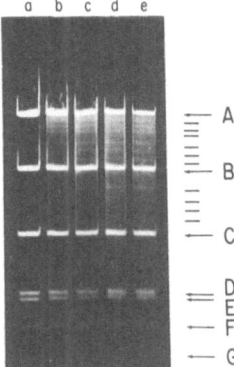

Fig. 7. Bleomycin fragmentation at discrete sites on PM2 DNA. PM2 DNA was digested with the HindIII restriction endonuclease followed by bleomycin (2.5 μg/ml) treatment for increasing reaction times to give fragments of discrete sizes which were resolved by agarose gel electrophoresis. Tracks a–e correspond to 0, 15, 30, 60, and 180 min of reaction with bleomycin. The HindIII fragments of PM2 DNA are indicated in order of decreasing sizes, A–G, by arrows at the right. Predominant intermediate size fragments produced by HindIII digestion followed by bleomycin treatment are indicated by lines at the right.

migrated to positions in the gel corresponding to near full-length linear duplexes of PM2 DNA (Fig. 8, position marked by arrow). When purified form III PM2 DNA, derived from bleomycin treatment, was subsequently digested with HpaII restriction endonuclease, discrete fragment sizes were again detected (Fig. 8, at right). As before, the majority of fragment mass remains near the position of full-length PM2 linear duplex DNA. These combined observations suggest that although bleomycin-promoted double-strand scissions occur at many discrete sites on the PM2 genome, a preponderance of these breaks occur near a site corresponding to the HpaII restriction endonuclease cleavage site.

The availability of a second restriction enzyme (PstI) in addition to HpaII that cleaves PM2 DNA at a single unique site provides the basis for mapping the sites of bleomycin-promoted double-strand scissions. We therefore isolated full-length linear duplex PM2 DNA after digestion with PstI or HpaII[21]. These linear duplex molecules were treated with bleomycin under conditions that would correspond to the introduction of less than one double-strand scission for every full-length molecule[23]. This limited fragmentation favors breakage at only one of several possible bleomycin-sensitive sites. This was done to provide pairs of linearly permuted fragments derived by cleavage at a unique site by the HpaII or PstI (R. J. Legerski, J. L. Hodnett, and H. B. Gray, Jr., unpublished results) restriction enzymes followed by bleomycin fragmentation at a single site. This approach to mapping of cleavage sites is conceptually the same as the method introduced by Parker et al.[24] in which pairs of fragments are detected whose combined molecular weights are equal to the full-length genome.

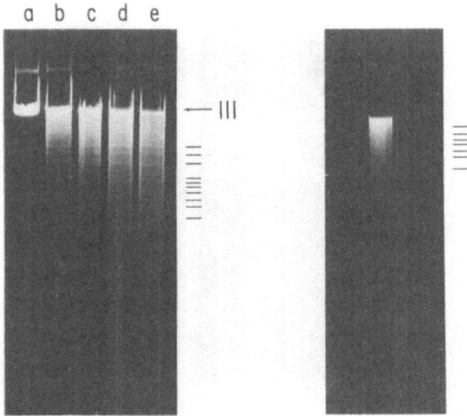

Fig. 8. Bleomycin fragmentation of PM2 DNA at discrete sites near the *Hpa*II restriction enzyme cleavage site. PM2 form I DNA was exhaustively digested with *Hpa*II restriction endonuclease followed by bleomycin (2.5 μg/ml) treatment for increasing times of incubation and the reaction products were resolved by agarose slab gel electrophoresis (at left, tracks a–e corresponding to 0, 15, 30, 60, and 180 min). The position of form III PM2 DNA is indicated by an arrow. Form III PM2 DNA derived from bleomycin treatment of form I DNA was purified and then exhaustively digested (12 hr) with *Hpa*II restriction endonuclease to give discrete fragments resolved by agarose slab gel electrophoresis (at right). The positions of predominant fragment sizes in both agarose slab gels are indicated by lines to the right of each photograph.

When full-length linear duplex PM2 DNA from *Pst*I restriction endonuclease cleavage is subsequently treated with bleomycin for increased reaction times, a set of discrete fragments is detected which migrate more rapidly than the full-length genome in agarose gel electrophoresis (Fig. 9, left panel). The molecular weights of these fragments are estimated from the gel positions relative to the *Hin*dIII fragments of PM2 DNA included as molecular weight standards. The molecular weights of all fragments were interpolated from plots of the logarithm of the molecular weight of *Hin*dIII fragments versus distance of migration, and are presented in Table 1. The results of a similar molecular weight analysis of discrete fragments produced by bleomycin treatment of full-length linear duplex PM2 DNA derived from *Hpa*II restriction enzyme digestion[21] are also summarized in Table 1.

We have shown that *Hin*dIII digestion of PM2 DNA followed by bleomycin treatment also results in fragments of discrete size whose molecular weights are intermediate between those of the *Hin*dIII fragments (Fig. 7). To provide another pattern of fragmentation, we purified the *Hin*dIII fragments A, B, and C (the three largest fragments resulting from digestion of PM2 DNA) and treated these with bleomycin, with the results shown in Table 1. The pattern of bleomycin

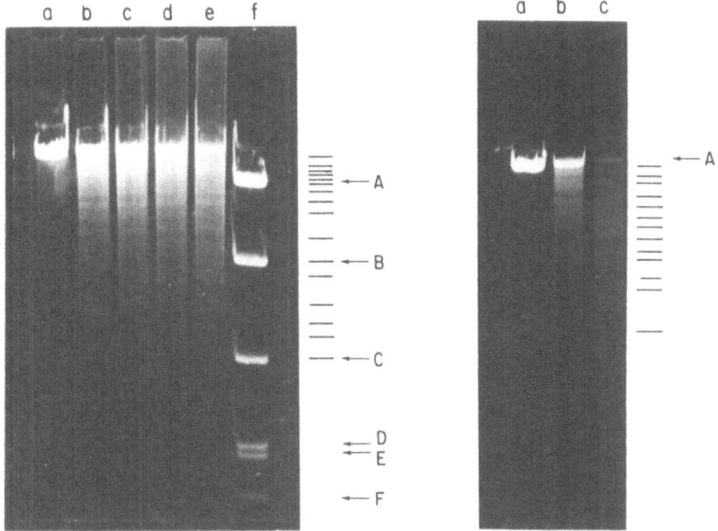

Fig. 9. Bleomycin fragmentation at discrete sites on PM2 DNA. PM2 DNA was digested with the *Pst*I restriction endonuclease followed by bleomycin (2.5 µg/ml) treatment for increasing reaction times to give fragments of discrete sizes which were resolved by agarose gel electrophoresis (left). Tracks a–e correspond to reaction times of 0, 15, 30, 60, and 120 min, respectively, and track f shows a complete *Hind*III digest of PM2 DNA. The fragments are indicated in order of decreasing sizes A–F by arrows. Predominant bleomycin-mediated fragments are indicated by lines at the right. Purified *Hind*III fragment A was treated with bleomycin (2.5 µg/ml) for increasing reaction times. Tracks a–c correspond to 0, 15, and 60 min, respectively; the arrow at A indicates the position of electrophoretic migration of *Hind*III fragment A and the lines at the right indicate the major bleomycin-mediated fragments.

fragmentation for fragment A for increasing reaction times is shown in Fig. 9, right panel).

The rationale for mapping the positions of bleomycin-promoted double-strand breaks is illustrated in Fig. 10. If one considers the middle circle, the largest fragment from a combined digestion with *Pst*I followed by bleomycin has a molecular weight of 5.26×10^6, designated *Pst*–BLM 1. When the molecular weight is added to the molecular weight of 0.73×10^6, determined for the fragment designated *Pst*–BLM 13, one obtains a combined molecular weight of 5.99×10^6. This combined molecular weight is, allowing for experimental error, equal to the molecular weight of full-length PM2 DNA (6×10^6). Ambiguity in the exact position of this site of breakage can be resolved by an examination of the *Hpa*II–bleomycin fragmentation pattern shown in the outer circle of Fig. 10. In this case the fragment designated *Hpa*II–BLM 3 having a molecular weight of

Table 1: Molecular Weights of Fragments Derived from Bleomycin Treatment of PM2 DNA Restriction Nuclease Fragments

*Hpa*II then bleomycin		*Pst*I then bleomycin		*Hind*III fragment A then bleomycin	
Fragment designation[a]	Molecular weight $\times 10^{-6}$ a.u.[b]	Fragment designation[a]	Molecular weight $\times 10^{-6}$ a.u.[b]	Fragment designation[a]	Molecular weight $\times 10^{-6}$ a.u.[b]
*Hpa*II–BLM		*Pst*I–BLM		*Hind*III A–BLM	
1	5.31	1	5.26	1	3.13
2	4.98	2	4.73	2	2.87
3	4.50	3	4.16	3	2.60
4	4.24	4	3.78	4	2.12
5	3.91	5	3.45	5	1.86
6	3.54	6	3.24	6	1.59
7	3.16	7	2.86	7	1.42
8	2.91	8	2.19	8	1.27
9	2.52	9	1.92	9	1.11
10	2.06	10	1.58	10	0.89
11	1.68	11	1.12	11	0.84
12	1.54	12	0.88	12	0.74
13	1.43	13	0.73	13	0.50
14	1.28	14	0.67		
15	1.07			*Hind*III fragment B then bleomycin	
16	0.89			*Hind*III B–BLM	
17	0.80			1	1.23
				2	1.03
				3	0.88
				4	0.67
				5	0.55
				6	0.50

HindIII C–BLM	HindIII fragment C then bleomycin
1	0.54
2	0.48
3	0.45
4	0.37
5	0.29

[a]Fragments have been designated according to the restriction enzyme used to produce the fragment upon which bleomycin (BLM) has acted. Fragments are presented in order of decreasing size with the largest designated 1.
[b]Molecular weights of fragments were determined relative to the HindIII fragments of PM2 DNA included for internal calibrations in agarose gel electrophoresis. The molecular weights are expressed as atomic units (a.u.).

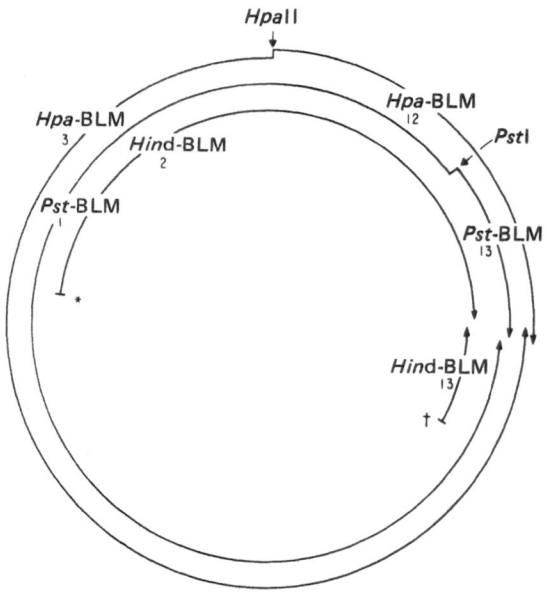

Fig. 10. Rationale for mapping of sites of preferential cleavage by bleomycin. See text for details. *, HindIII fragment A–D junction; †, HindIII fragment A–B junction.

4.5×10^6 when added to the molecular weight determined for fragment HpaII-BLM 12 of 1.54×10^6 yields a combined molecular weight of 6.04×10^6. Similarly, this position can be further confirmed by reacting HindIII fragment A with bleomycin (inner circle) and looking for fragments whose combined molecular weights equal that of the intact HindIII fragment A (3.4×10^6).

The combined molecular weights of fragment pairs from the restriction enzyme-bleomycin treatment, whose sum is approximately equal to the parental DNA, are presented in Table 2. The corresponding map positions for sites of bleomycin breakage also are presented in Table 2. These positions were measured as percentage of the circular PM2 genome length clockwise from the HpaII cleavage site (taken as 0.00%) with the PstI cleavage site located clockwise from this origin at 14% of the genome length. Ambiguities in positions of bleomycin breakage of PstI or HpaII linear duplex PM2 DNA were resolved by assignment of map positions for bleomycin fragmentation of the HindIII fragments A, B, or C (Tables 1 and 2) as described above.

The results of this analysis are summarized in the physical map of 11 bleomycin fragmentation sites presented in Fig. 11. These sites are distributed about the PM2 genome at positions (measured clockwise from the HpaII cleavage site) of 3 ± 1.1, 7 ± 1.1, 12 ± 1.0, 21 ± 2.2, 25 ± 1.9, 34 ± 0.8, 41 ± 1.8, 50 ± 1.8, 67 ± 0.5, 71 ± 2.5, and $82 \pm 2.0\%$ of the genome.

Table 2:   Physical Map Sites of Bleomycin Fragmentation of PM2 DNA

Fragment pairs		Molecular weight of fragment pairs $\times 10^{-6}$ a.u.	Fragment sizes (% of genome)[a]
HpaII–BLM 1	+ HpaII–BLM 17	6.11	88.5, 13.3
HpaII–BLM 2	+ HpaII–BLM 15	6.05	83.0, 17.8
HpaII–BLM 3	+ HpaII–BLM 12	6.04	75.0, 25.7
HpaII–BLM 4	+ HpaII–BLM 11	5.92	70.6, 28.0
HpaII–BLM 5	+ HpaII–BLM 10	5.98	65.2, 34.3
HpaII–BLM 6	+ HpaII–BLM 9	6.06	59.0, 42.0
HpaII–BLM 7	+ HpaII–BLM 8	6.07	42.7, 48.5
PstI–BLM 1	+ PstI–BLM 13	5.99	87.7, 12.2
PstI–BLM 2	+ PstI–BLM 11	5.85	78.8, 18.7
PstI–BLM 3	+ PstI–BLM 9	6.08	69.3, 32.0
PstI–BLM 4	+ PstI–BLM 8	5.97	63.0, 36.5
PstI–BLM 5	+ PstI–BLM 7	6.13	57.5, 44.7
PstI–BLM 6	+ PstI–BLM 7	5.91	54.0, 44.7
HindIII A–BLM 2	+ HindIII A–BLM 13	3.37	47.8, 8.3
HindIII A–BLM 3	+ HindIII A–BLM 11	3.44	43.3, 14.0
HindIII A–BLM 4	+ HindIII A–BLM 8	3.39	35.3, 21.2
HindIII A–BLM 5	+ HindIII A–BLM 6	3.45	31.4, 26.5
HindIII B–BLM 3	+ HindIII B–BLM 6	1.38	14.8, 8.3
HindIII B–BLM 4	+ HindIII B–BLM 5	1.22	11.2, 9.2
HindIII C–BLM 4	+ HindIII C–BLM 5	0.63	6.2, 4.8

[a]Distances in percent of the PM2 DNA genome length measured clockwise from the HpaII restriction enzyme site; the PstI cleavage site is positioned at 14%.

Discussion

Bleomycin-produced damage to DNA structure consists of a spectrum of chemical alterations. These experimentally observed alterations include: removal of bases[15]; damage to the deoxyribose moiety[25]; introduction of alkaline-labile sites[19]; and direct single- and double-strand scissions[12]. Modifications that result in depurination or depyrimidation, as well as single-strand scissions of the phosphodiester backbone, are expected to be repaired and would not necessarily constitute lethal cellular damage. By way of contrast, bleomycin-promoted double-strand scissions are expected to be largely irreparable alterations. In fact, cytological studies have demonstrated that fragmentation of metaphase chromosomes occurs when cultured mouse fibroblasts are exposed to high concentrations of bleomycin[26]. Thus bleomycin-specific double-strand breaks in DNA may represent important physiological aspects for effective clinical application of the drug to control cellular growth in certain human malignancies.

In the present study, it is demonstrated that bleomycin reaction with PM2 DNA results directly in double-strand scissions rather than accumulation of single-strand scissions that lead to the formation of linear form III DNA. These

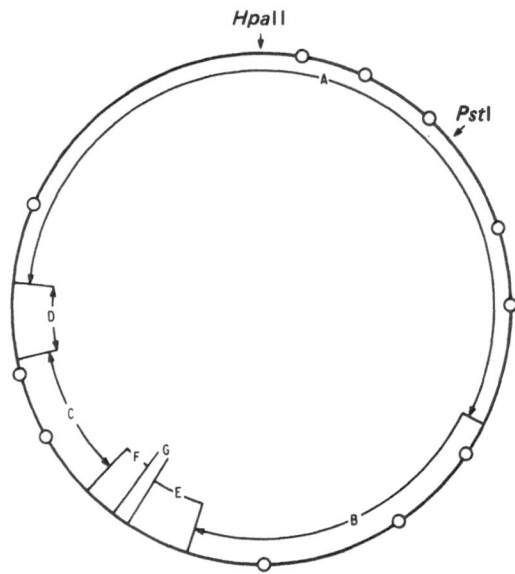

Fig. 11. Map of regions on the PM2 DNA genome which are preferentially cleaved by bleomycin (open circles). A–G indicate the fragments formed by HindIII action on PM2 DNA. HpaII and PstI sites are indicated by arrows.

studies have been facilitated considerably by the use of a covalently closed circular DNA as substrate for the reaction. Although bleomycin-promoted double-strand scissions do not require superhelical turns or sites of depurination, it is clear that the breaks are introduced in PM2 DNA at several specific sites which are most frequently located near the HpaII cleavage site (Fig. 11). This region of the PM2 genome is not particularly distinguished in base composition[22]. However, the small number of uniquely positioned bleomycin fragmentation sites suggests the existence of unique sequences of nucleotides in the PM2 genome that specify either the binding of bleomycin or hydrolysis of phosphodiester bonds. We consider it remarkable that a small molecular weight glycopeptide antibiotic can produce a specific fragmentation of duplex DNA resembling that obtained with sequence-specific restriction endonucleases. Specific fragmentation by bleomycin in the PM2 genome may, nevertheless, reflect the presence of particular deoxyribonucleotide sequences that permit binding of the drug in a configuration that results in scission of complementary strands at closely spaced sites. For a random sequence of deoxyribonucleotides, the probability of finding a specific sequence n nucleotides in length is $(1/4)^n$. For a 10-fold repetition of a particular unique sequence in the PM2 genome (containing 10,250 nucleotide pairs), one would expect the sequence to contain four to six nucleotide pairs. It is also possible that 10 separate unique sequences each comprising four to six nucleotide pairs could account for the observed frequency of double-strand scissions. In either case, the lengths of PM2 DNA sequences that are calculated to account for the observed frequency of specific fragmentation produced by bleomycin are approximately one-half the number of nucleotides that are thought to be involved in, or at least blocked by, the binding of bleomycin such that a break will result[27]. Thus it is possible to account for bleomycin-promoted fragmentation of PM2 DNA on the basis of sequence-specific strand breakage even for a random arrangement of deoxyribonucleotides in the genome.

By assessing the frequencies of double-strand and single-strand scissions for a fixed reaction time and temperature as a function of bleomycin dilution (Fig. 5), it was possible to demonstrate that these breakage events each follow single-hit kinetics. Interestingly, the same dilution analysis of the average number of breaks per molecule followed two-hit kinetics, a result that suggests that separate molecular components of the bleomycin mixture used in this study are responsible for single- and double-strand scissions, respectively. In this regard, it is particularly relevant that our unpublished observations with bleomycin A_2 purified to homogeneity indicate that both single-strand and double-strand scissions also occur with approximately the same yield and kinetics as was found for the bleomycin mixture. The bleomycin mixture (Blenoxane) contains two major components, bleomycins A_2 and B_2, and at least nine minor components. It is therefore necessary to resolve the paradox that a purified component of bleomycin reacts with PM2 DNA in a manner that is essentially indistinguishable from the bleomycin mixture, which is shown by this study to contain two separate reactive species. One explanation to resolve this paradox would be to postulate the existence of a monomer molecular species of bleomycin A_2 which reacts with DNA to produce a single-strand scission whereas a dimer species of bleomycin A_2 reacts with DNA to produce double-strand scissions[12]. Such a dimer species necessarily requires an asymmetric arrangement of the two bleomycin molecules that would interact with double-stranded DNA as illustrated diagrammatically in Fig. 12. In the double-strand breakage reaction, two bleomycin A_2 molecules would dimerize and bind to double-stranded DNA such that

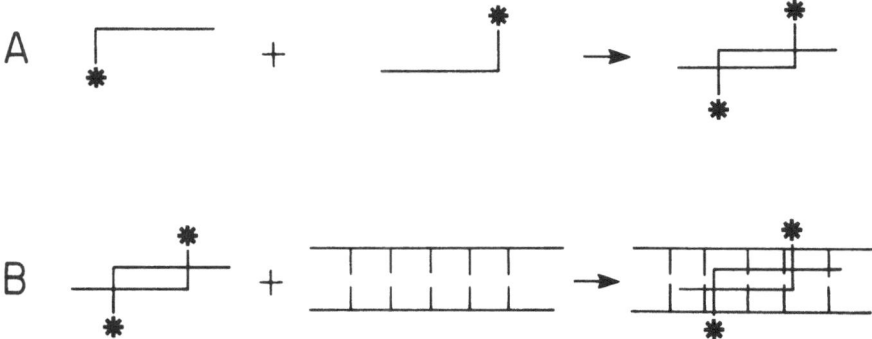

Fig. 12. Model of bleomycin binding to DNA leading to strand scissions.

the reactive groups which lead to phosphodiester bond scissions are juxtaposed on the complementary strands. If the region of juxtaposition encompasses less than approximately four nucleotide pairs, then strand scissions would result in a double-strand break upon dissociation of the intervening base pairs under the reaction conditions used. We consider this scheme to be the simplest in explaining the spectrum of properties for bleomycin-specific fragmentation of double-stranded DNA, although other, more complicated schemes may be found to apply to these results.

The possibility of different chemical reactions with DNA mediated by dimer or even higher oligomeric species of bleomycin components is an intriguing though speculative concept and emphasizes the need to define experimentally the spectrum of chemical damage to DNA promoted by a mixture of pharmacologically active agents that are routinely applied in chemotherapy. The methods of analyses applied in this study provide a first step in defining the different modes of action expressed by one such mixture of bleomycins and provide insight into the possible basis for cytotoxicity of drug action.

Acknowledgments

The authors wish to express their appreciation to Dr. Roger R. Hewitt for his generous gift of the PM2 DNA. The authors also gratefully acknowledge the assistance of Dr. Hamao Umezawa and Dr. Akira Matsuda for their kind assistance in obtaining certain bleomycin derivatives. The Blenoxane used in this research was the generous gift of Bristol Laboratories, Syracuse, New York.

This research was supported in part by Grant Nos. CA-13246 and CA-16527 awarded by the National Cancer Institute, DHEW, and by Grant No. G-441 awarded by the Robert A. Welch Foundation.

References

1. H. Umezawa, K. Maeda, T. Takeuchi, and Y. Okami, *J. Antibiot. (Tokyo)*, **19**, 200 (1966).
2. C. W. Haidle and R. S. Lloyd, In *Antibiotics. Mechanism of Action of Antimicrobial and Antitumor Agents* (J. W. Corcoran and F. E. Hahn, Eds.), Vol. V. Springer, New York–Heidelberg–Berlin, in press.
3. H. Umezawa, In *Gann Monograph on Cancer Research No. 19. Fundamental and Clinical Studies of Bleomycin* (S. K. Carter, T. Ichikawa, G. Mathé, and H. Umezawa, Eds.), pp. 3–36. University Park Press, Baltimore–London–New York (1976).
4. W. E. G. Müller and R. K. Zahn, *Prog. Nucleic Acid Res. Mol. Biol.*, **40**, 21 (1977).
5. H. Umezawa, H. Asakura, K. Oda, S. Hori, and M. Hori, *J. Antibiot. (Tokyo)*, **26**, 521 (1973).
6. R. T. Espejo and E. S. Canelo, *Virology*, **34**, 738 (1968).
7. R. T. Espejo, E. S. Canelo, and R. L. Sinsheimer, *Proc. Nat. Acad. Sci. USA.*, **63**, 1164 (1969).
8. J. E. Strong and R. R. Hewitt, In *Isozymes, Vol. III: Developmental Biology* (C. Markert, Ed.), pp. 473–483. Academic Press, New York (1975).

9. M. Salditt, S. N. Braunstein, R. D. Camerini-Otero, and R. M. Franklin, *Virology*, **48**, 259 (1972).
10. P. A. Sharp, B. Sugden, and J. Sambrook, *Biochemistry*, **12**, 3055 (1973).
11. D. J. Grdina, P. H. M. Lohman, and R. R. Hewitt, *Anal. Biochemistry*, **51**, 255 (1973).
12. R. S. Lloyd, C. W. Haidle, and D. L. Robberson, *Biochemistry*, **17**, 1890 (1978).
13. H. Sierakowsha and D. Shugar, *Prog. Nucleic Acid Res. Mol. Biol.*, **20**, 59 (1977).
14. C. W. Haidle, *Mol. Pharmacol.*, **7**, 645 (1971).
15. C. W. Haidle, K. K. Weiss, and M. T. Kuo, *Mol. Pharmacol.*, **8**, 531 (1972).
16. L. F. Povirk, W. Wübker, W. Köhnlein, and F. Hutchinson, *Nucleic Acid Res.*, **4**, 3573 (1977).
17. T. Lindahl and B. Nyberg, *Biochemistry*, **11**, 3610 (1972).
18. D. Freifelder, *Biopolymers*, **7**, 681 (1969).
19. R. S. Lloyd, R. R. Hewitt, and C. W. Haidle, *Fed. Proc.*, **36**, 694 (1977).
20. S. E. Luria and R. Dulbecco, *Genetics*, **34**, 93 (1949).
21. C. Brack, H. Eberle, T. A. Bickle, and R. Yuan, *J. Mol. Biol.*, **104**, 305 (1976).
22. C. Brack, H. Eberle, T. A. Bickle, and R. Yuan, *J. Mol. Biol.*, **108**, 583 (1976).
23. R. S. Lloyd, C. W. Haidle, D. L. Robberson, and M. L. Dodson, Jr., *Current Microbiol.*, **1**, 45 (1978).
24. R. C. Parker, R. M. Watson, and J. Vinograd, *Proc. Nat. Acad. Sci. USA*, **74**, 851 (1977).
25. M. T. Kuo and C. W. Haidle, *Biochim. Biophys. Acta*, **385**, 109 (1973).
26. K. D. Paika and A. Krishan, *Cancer Res.*, **33**, 961 (1973).
27. M. T. Kuo, C. W. Haidle, and L. D. Inners, *Biophys. J.*, **13**, 1296 (1973).
28. R. W. Davis, M. Simon, and N. Davidson, In *Methods in Enzymology* (L. Grossman and K. Moldave, Eds.), pp. 413–428. Academic Press, New York (1971).

The Use of Covalently Closed Circular DNA to Investigate Properties of Bleomycin and Its Analogs

JAMES E. STRONG AND STANLEY T. CROOKE

The mechanism of action for bleomycin appears to involve production of both single- and double-strand DNA breakage[1-3]. A variety of methods have been used to investigate this phenomenon. Several investigators have used covalently closed circular DNA (CCC DNA) because of its unique properties in determining conformational alterations[4-7]. Bleomycin-induced single-strand breakage in superhelical CCC DNA (form I) produces a relaxed open circular conformation (form II) and renders the DNA susceptible to alkali or heat denaturation. The product of this denaturation (single-strand DNA) can be detected by velocity sedimentation in alkaline sucrose gradients or another method depending on the loss of DNA intercalative binding by the fluorescent dye, ethidium bromide. Since the quantum yield of fluorescence by ethidium bromide is enhanced upon intercalative binding to double-stranded DNA[8] a decrease in fluorescence results upon alkaline denaturation of broken CCC DNA. This provides the basis for a fluorescent assay of DNA breakage (Fig. 1). Bleomycin-induced DNA fragmentation can also be detected using electrophoresis of CCC DNA. This is because the products of broken superhelical CCC DNA (form I), relaxed circular DNA (form II), and linear duplex DNA (form III) are electrophoretically distinguishable under appropriate conditions (Fig. 2)[9]. Electrophoresis also may be used to quantitate the double-strand breakage by bleomycin since production of linear duplex DNA (form III) requires a double-strand break.

The fluorescence assay and electrophoresis of DNA-degradation products provide methods for investigating *in vitro* bleomycin-induced degradative effects. The purpose of this investigation is to describe development and use of CCC DNA breakage systems to compare the mechanism of action of bleomycin and its third generation analog, tallysomycin.

Materials and Methods

Fluorescence Assay for PM-2 DNA Degradation

Bacteriophage PM-2 was isolated following lytic infection of *Pseudomonas* BAI-31 host as previously described[10]. DNA was extracted from purified PM-2

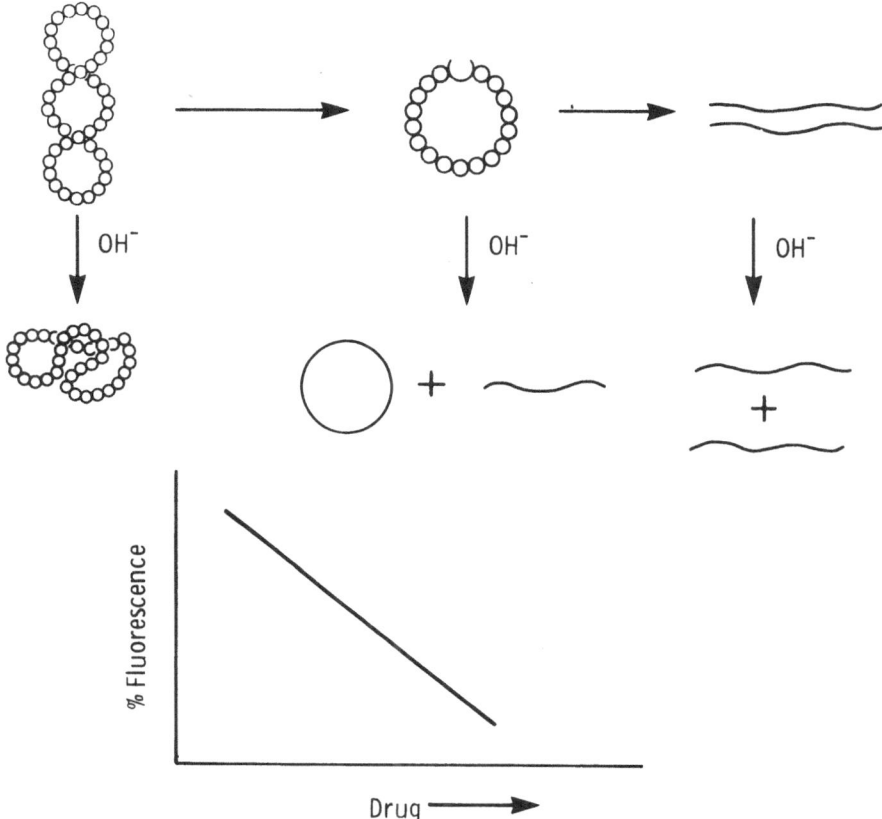

Fig. 1. Fluorescence assay for DNA breakage. Superhelical PM-2 DNA produces a relaxed circular conformation upon receipt of a single-strand break or linear duplex DNA as the result of a double-strand break. Alkali denaturation produces single-stranded conformations in single- or double-strand broken DNA. However, covalently closed circular DNA remains at least partially double stranded. The percentage fluorescence, relative to fluorescence without drug, decreases as a result of DNA breakage and concomitant loss of ethidium bromide binding.

by lysis with sarkosyl NL97 (ICN Pharmaceuticals Inc., Cleveland, Ohio) (0.5%, 10 min at 60°C) and repeated extraction with chloroform: isoamyl alcohol (24:1). The DNA was precipitated in ethanol (-20°C) and purified CCC PM-2 DNA was isolated in cesium chloride–ethidium bromide gradients[11]. The analysis was performed by adding 25 μg PM-2 DNA in 0.5 ml buffer (0.015 M NaCl, 0.05 M sodium borate, pH 9.5) containing 25 mM freshly added 2-mercaptoethanol. Incubation was performed at room temperature for 30 min unless otherwise indicated. Triplicate aliquots (100 μl) of the assay solution were placed in denaturation buffer (900 μl) (0.04 M Na$_2$PO$_4$, 0.01 M EDTA, 0.1 M NaCl, pH adjusted to 12.1 with 0.15 M NaOH) followed by addition of 0.1 ml ethidium bromide (Sigma Chemical Co., St. Louis, Missouri) (22 μg/ml in

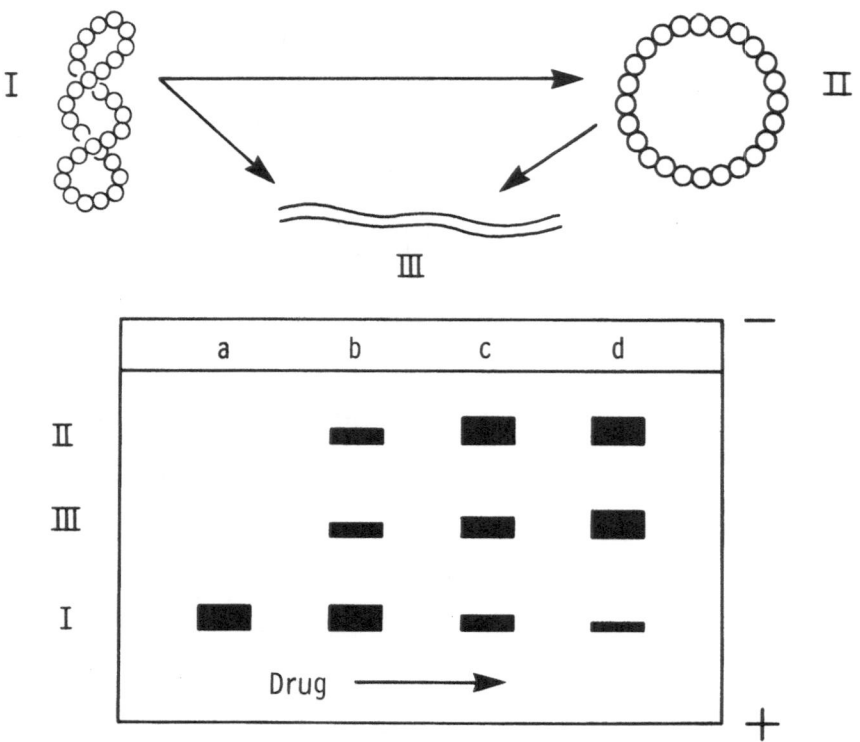

Fig. 2. Agarose electrophoresis of circular PM-2 DNA following DNA breakage. The electrophoretically separable conformations of circular DNA are identifiable following incubation with increasing amounts of drug.

denaturation buffer). Fluorescence of the ethidium bromide–DNA mixture was determined using a spectrophotofluorometer (Aminco–Bowman) at 530 mM excitation and 590 nM emission. The change in concentration of covalently closed, circular PM-2 DNA was determined by the percentage decrease in fluorescence, after subtraction of background fluorescence, relative to control reactions not containing drug. Control reactions were also performed without 2-mercaptoethanol present and no decrease in fluorescence was detected because of 2-mercaptoethanol addition. A similar technique using heat treatment of PM-2 DNA for denaturation has been previously reported[12].

Agarose Gel Electrophoresis

The products of PM-2 DNA reaction with either bleomycin or tallysomycin were electrophoretically separated on agarose gels. The reaction was terminated by addition of 50 μl of the reaction mixture described above to 10 μl of 50 mM EDTA. Bromphenol blue (0.1%) in 75% glycerol (20 μl) was then added and 30 μl of the resultant mixture was electrophoresed on a 1.0% agarose gel. Electrophoresis was performed in 40 mM Tris, 5 mM sodium acetate, and 1 mM EDTA, pH 7.8, at 5 V/cm for 8 hr. The agarose (agarose ME) was obtained from Miles Laboratories (Elkhart, Illinois) and was adjusted to 1% concentration by addition of water, following agarose dissolution in 100°C electrophoretic buffer.

Drugs

Bleomycin A$_2$ (lot U-18A$_2$S) and tallysomycin A (lot 34-F-12) were obtained from Bristol Laboratories, Inc. (Syracuse, New York).

Results

The amount of PM-2 DNA added to denaturation buffer in the fluorescence assay was adjusted so that changes in fluorescence quantitatively reflected changes in PM-2 DNA concentration. When increasing quantities of PM-2 DNA were added to denaturation buffer containing 2 μg/ml ethidium bromide, changes in fluorescence directly corresponded to changes in PM-2 DNA concentration below 7.5 μg/ml (Fig. 3). Therefore, 5 μg/ml PM-2 DNA plus 2 μg/ml

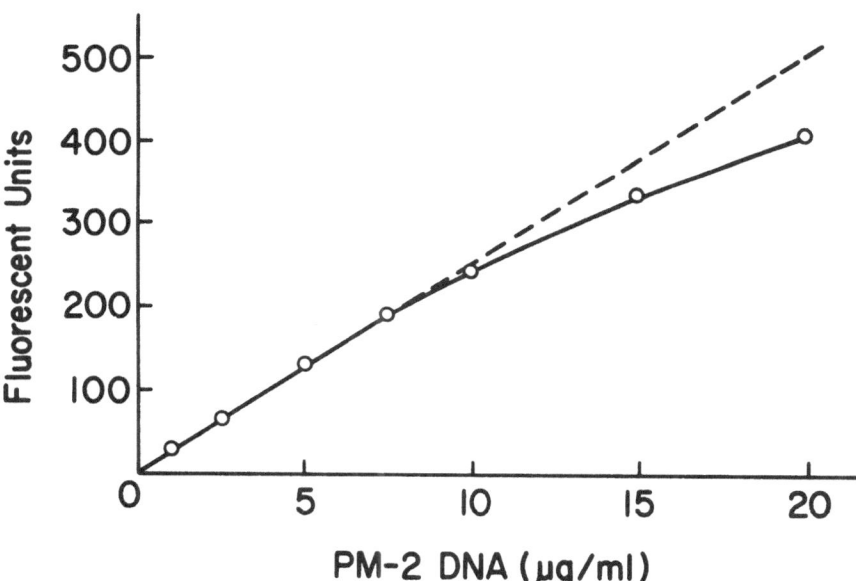

Fig. 3. Fluorescence increase caused by addition of PM-2 DNA. The PM-2 DNA was added to denaturation buffer containing 2 μg/ml ethidium bromide.

ethidium bromide was selected for use in the assay. The efficiency of linear duplex DNA denaturation in this system was confirmed by adding up to 1 mg/ml calf thymus DNA without increase in fluorescence. Thus, the fluorescence caused by ethidium bromide binding to intact covalently closed circular PM-2 DNA was responsible for the fluorescence greater than background.

Optimal conditions for bleomycin-induced DNA degradation in this assay were investigated. Maximum degradation of PM-2 DNA occurred in the presence of 2.5 mM β-mercaptoethanol at pH 9.5. The decrease in fluorescence was essentially a first order reaction and rate constants determined from the time necessary to produce a 50% decrease in fluorescence were directly proportional to bleomycin concentration (Fig. 4). The presence of bleomycin-induced alkaline labile

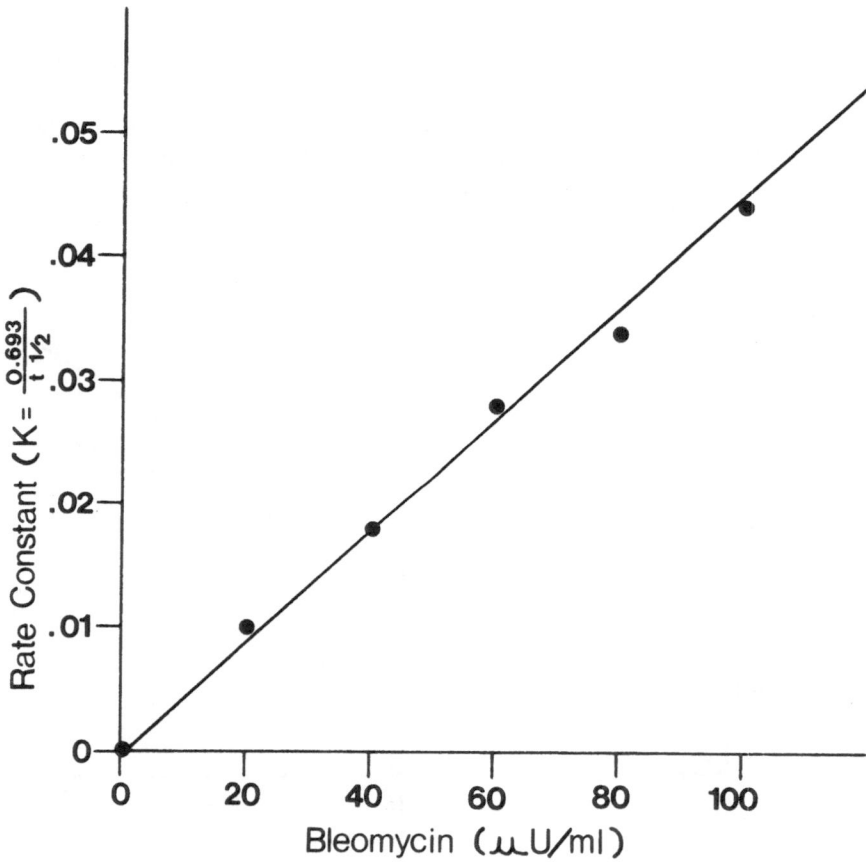

Fig. 4. Rate constants for DNA breakage. The rate constants were determined from the time necessary to reduce fluorescence by 50% after addition of the indicated concentration of bleomycin.

damage was investigated by determining the DNA breakage induced prior to addition of denaturation buffer. This was performed using the relative increase in fluorescence produced by breakage of superhelical CCC PM-2 DNA and addition of ethidium bromide under nondenaturing conditions[8]. The rate of DNA breakage before and after addition of denaturation buffer was compared (Fig. 5). Approximately two-fold greater concentration of bleomycin under neutral conditions was required to produce an equivalent amount of DNA breakage as that following alkaline denaturation, Thus, the fluorescence assay detected alkaline-labile damage in addition to DNA breakage.

The fluorescence assay was used to compare DNA breakage produced by bleomycin and its third-generation analog tallysomycin (Fig. 6). Approximately four-fold greater concentration of tallysomycin was required to produce a 50% decrease in relative fluorescence. The DNA breakage produced by these compounds was also compared using agarose gel electrophoresis (Fig. 7). Both double- and single-strand breaks were produced in PM-2 DNA by bleomycin A_2 and tallysomycin. Bleomycin-induced DNA breakage was also greater than tallysomycin in this system.

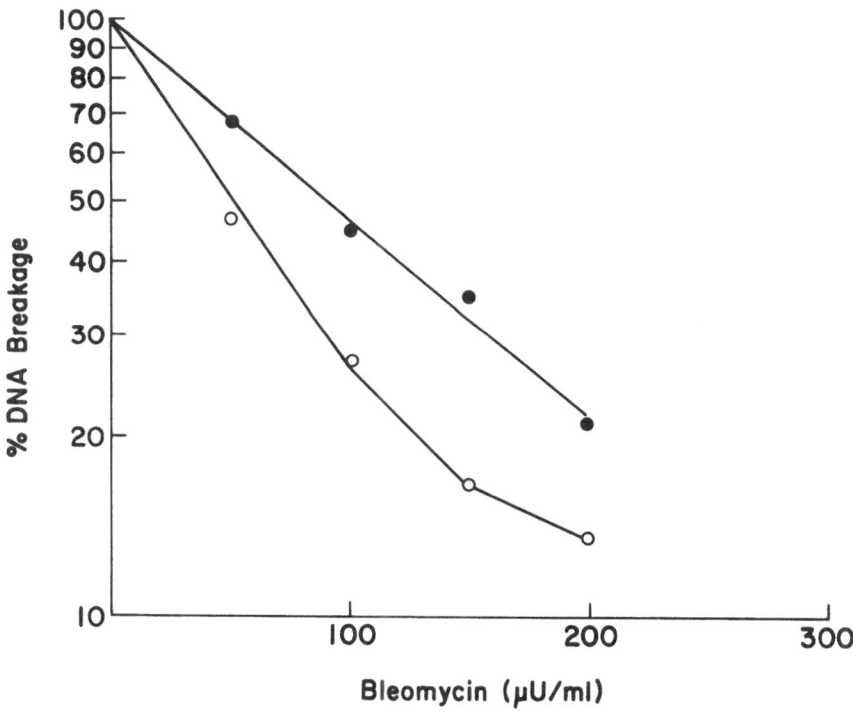

Fig. 5. Comparison of DNA breakage before and after alkaline denaturation. The relative percentage DNA breakage was determined before addition of denaturation buffer (●) and after addition of denaturation buffer (○).

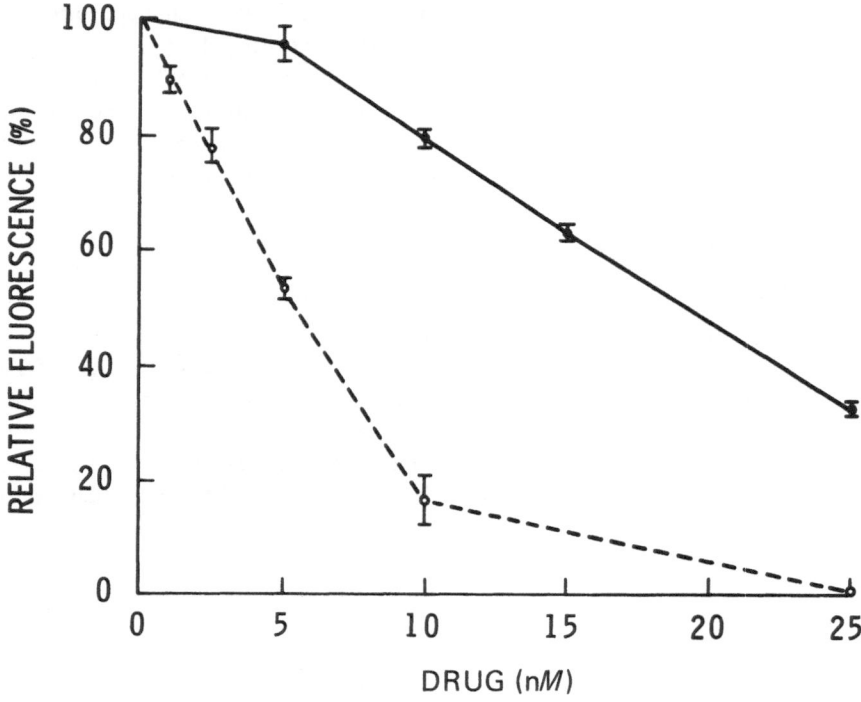

Fig. 6. Fluorescence assay comparison of DNA breakage. The indicated concentrations of tallysomycin (●) or bleomycin (○) were incubated 30 min in the fluorescence assay. Results are shown with indicated standard deviations for triplicate aliquots.

A role for ferrous ion in the degradation of DNA by bleomycin has recently been proposed[12,13]. One possible explanation for the enhanced DNA breakage by bleomycin, compared to tallysomycin, could be that bleomycin contained greater amounts of ferrous ion. However, atomic absorption spectrometry of the drug lots revealed less iron in bleomycin (10.1 μg Fe/g) than tallysomycin (12.4 μg Fe/g) (Bristol Laboratories, personal communication). Thus, the trace metal concentration of Fe ion in tallysomycin did not appear to be a predominant factor in the drug's reduced DNA breakage activity compared with bleomycin A_2. The effect of iron on DNA breakage in the fluorescence assay and on agarose gels was further investigated. Both drugs were incubated at 10 nM concentration and various amounts of $FeCl_2$ were added. Table 1 depicts the effect of $FeCl_2$ addition on bleomycin and tallysomycin activity. The rate of DNA breakage was doubled by addition of 100 nM $FeCl_2$ for both tallysomycin and bleomycin. However, tallysomycin with 100 nM $FeCl_2$ had less than half the activity of bleomycin A_2 without additional ferrous ion. This result was confirmed with agarose gel electrophoresis of the reaction products (Fig. 8). The

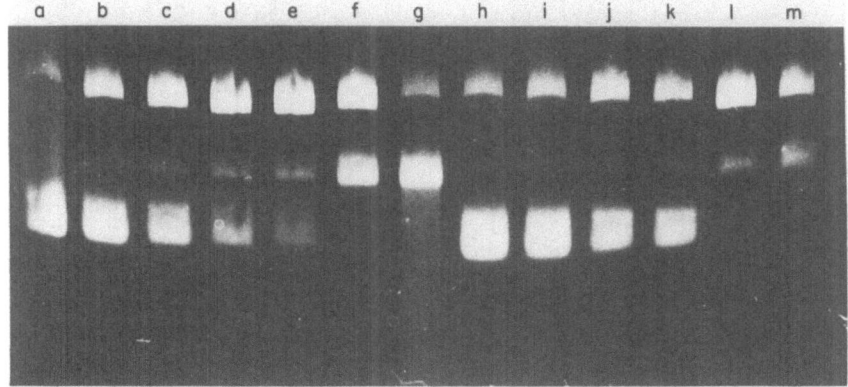

Fig. 7. Agarose gel electrophoresis of PM-2 DNA. Following incubation with bleomycin or tallysomycin, aliquots of the reaction mixture were electrophoresed on a horizontal agarose gel. The samples represent (a) control, (b) bleomycin-5 nM, (c) bleomycin-10 nM, (d) bleomycin-25 nM, (e) bleomycin-50 nM, (f) bleomycin-100 nM, (g) bleomycin-500 nM, (h) tallysomycin-5 nM, (i) tallysomycin-10 nM, (j) tallysomycin-25 nM, (k) tallysomycin-50 nM, (l) tallysomycin-100 nM, (m) tallysomycin-500 nM.

amount of iron present as trace contaminant in 10 nM (14.8 ng/ml) bleomycin A_2 was 0.265 pM ($\sim$ 15 fg/ml). Thus, addition of greater than 10^5-fold concentration of trace iron present in bleomycin did not increase the DNA breakage of tallysomycin to the level of activity found in bleomycin. No DNA breakage activity was detected by up to 10 μM $FeCl_2$ without bleomycin or tallysomycin present. DNA breakage activity by both bleomycin and tallysomycin was completely inhibited with addition of ethylenediaminetetraacetic acid (EDTA, 5 mM) or by reduction of dissolved oxygen in the incubation buffer through nitrogen bubbling.

Discussion

The third-generation bleomycin analog, tallysomycin, differs from bleomycin in two of the amino acids found upon hydrolysis and in a unique amino-sugar: 4-amino-4,6-dideoxy-L-talose[14]. Tallysomycin A contains a lysine molecule

Table 1: Rate Constants for DNA Breakage in the Presence of Iron

$FeCl_2$ (nM)	BLM (K min^{-1})	TLM (K min^{-1})
0	0.0433	0.00714
1	0.0513	0.00888
10	0.0578	0.0126
100	0.0866	0.0151

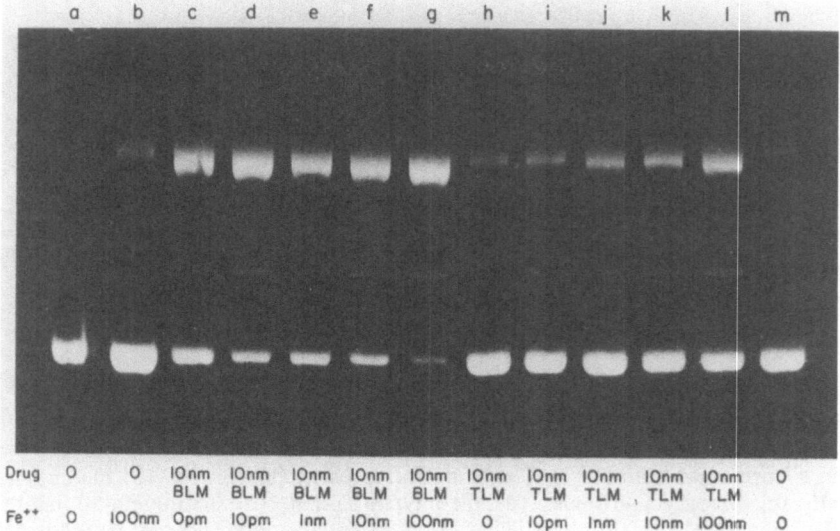

Fig. 8. Agarose gel electrophoresis of PM-2 DNA plus iron. Following incubation with bleomycin or tallysomycin, aliquots of the reaction mixture were electrophoresed on a horizontal agarose gel. The samples contained the indicated amounts of $FeCl_2$ and drug.

plus spermidine as the terminal side chain[14]. Some of the biological properties of tallysomycin include[15]: (a) antimicrobial activity up to 100-fold greater than bleomycin, (b) lysogenic bacteriophage induction in *Escherichia coli* W-1709 (lambda) 30-fold more potent than bleomycin, (c) more active than bleomycin on P-388 Leukemia (10-fold) and Walker 256 carcinosarcoma (14-fold), (d) slightly less active than bleomycin in B-16 melanoma, and (e) a lower incidence of pneumonitis and fibrosis in chronically treated mice than bleomycin when given at equivalent LD_{50} fractions. Thus, tallysomycin appears to be generally more potent than bleomycin in bactericidal and antineoplastic activity.

The mechanism of action for tallysomycin-increased biological potency was investigated using covalently closed circular PM-2 DNA. However, tallysomycin-induced PM-2 DNA breakge measured in the fluorescence assay system and by agarose gel electrophoresis was less than bleomycin. This reduced DNA breakage activity was not caused by differences in iron concentration since tallysomycin contained a greater amount of trace iron. Moreover, addition of exogenous iron at up to 10^5-fold greater than trace concentration did not cause tallysomycin–DNA breakage activity equivalent to bleomycin. The difference in DNA break-activity of the two compounds did not appear to be confined to PM-2 DNA as a substrate. Experiments involving alkaline sucrose sedimentation of DNA from bleomycin- and tallysomycin-treated Novikoff hepatoma ascites cells also showed that more DNA breakage occurred in bleomycin-treated cells[16]. However, colony survival studies on bleomycin- and tallysomycin-treated Novikoff hepatoma cells cultured *in vitro* indicated no significant difference in cellular lethality (unpublished results). Thus, innate DNA breakage activity did not appear to be the mechanism of enhanced biological activity for tallysomycin. Although DNA breakage by tallysomycin has not been compared to bleomycin in systems where tallysomycin is more potent, various members of the bleomycin complex and their derivatives were examined for DNA breakage activity on SV40 DNA by Asakura *et al.*[17]. They found little correlation between antimicrobial potency and *in vitro* DNA breakage activity.

A summary of results from a previous study[16] comparing the properties of bleomycin (BLM) and tallysomycin (TLM) interactions with DNA is presented in Table 2. One property of tallysomycin that correlated to its increased biological activity was tallysomycin affinity for DNA. Tallysomycin affinity for salmon sperm DNA was approximately 25-fold greater than bleomycin, determined using a fluorescence quenching technique[16]. This enhanced affinity corresponded to the increased bacteriophage induction potency of tallysomycin. Tallysomycin DNA affinity may play a significant role in the drug's antitumor effectiveness. Other pharmacologic factors may be involved in determining *in vivo* effectiveness of the drug as an antitumor agent. These factors include: susceptibility of tallysomycin to inactivation by bleomycin-inactivating enzymes and differences in pharmacologic disposition between tallysomycin and bleomycin. Studies are currently in progress in our laboratory to examine these and other questions involving tallysomycin mechanism of action.

Table 2: Comparison of the Interactions of Bleomycin and Tallysomycin with DNA

	BLM	TLM
pH optimum	9.5	9.5
2-MEO optimum	25 mM	25 mM
Temperature optimum	30–37°C	30–37°C
Energy of activation	13,909 cal/mole	33,547 cal/mole
Single- and double-strand breaks	Yes	Yes
Alkali-labile bonds	Yes	Yes
Degradation of cellular DNA	+++	++
Nucleotides/drug	3.39 ± 0.51	4.04 ± 0.13
Affinity constant for DNA	3.41 ± 0.42 × 10^4	8.47 ± 1.06 × 10^5

Acknowledgments

The authors wish to express appreciation to Dr. Roger Hewitt of the M.D. Anderson Hospital and Tumor Institute for helpful discussions on developing the DNA-breakage assays. We also wish to express our gratitude to Ms. Susan Salinas for technical assistance.

This project has been supported by NIH Grant CA–10893-10, Pharmaceutical Manufactuers Association Foundation Research Starter Grant to J.E.S., and by Bristol Laboratories, Inc., Syracuse, New York.

References

1. C. W. Haidle, *Mol. Pharmacol.*, **7**, 645 (1971).
2. K. Nagai, H. Suzuki, N. Tonaka, and H. Umezawa, *J. Antibiot. (Tokyo)*, **22**, 569 (1969).
3. H. Umezawa, *Fed. Proc.*, **33**, 229 (1974).
4. L. F. Povirk, W. Wubker, W. Kohnlein, and F. Hutchinson, *Nucleic Acids Res.*, **4**, 3573 (1977).
5. H. Umezawa, H. Osakura, K. Oda, S. Hori, and M. Hori, *J. Antibiot. (Tokyo)*, **26**, 521 (1973).
6. S. L. Ross and R. E. Moses, *Biochemistry*, **17**, 581 (1978).
7. J. C. Bearden, R. S. Lloyd, and C. W. Haidle, *Biochem. Biophys. Res. Commun.*, **75**, 442 (1977).
8. C. Paoletti, J. B. Lepecq, and I. R. Lehman, *J. Mol. Biol.*, **55**, 75 (1971).
9. P. H. Johnson and L. I. Grossman, *Biochemistry*, **16**, 4217 (1977).
10. M. Salditt, S. N. Braunstein, R. D. Camerini-Otero, and R. M. Franklin, *Virology*, **48**, 259 (1972).
11. J. E. Strong and R. R. Hewitt, In *Isozymes III* (C. Markert, Ed.), pp. 473–483. Academic Press, New York (1975).
12. J. W. Lown and S. Sim, *Biochem. Biophys. Res. Commun.*, **77**, 1150 (1977).
13. E. A. Sausville, J. Peisach, and S. B. Horwitz, *Biochem. Biophys. Res. Commun.*, **73**, 814 (1976).
14. M. Konishi, K. Saito, K. Numata, T. Tsuno, K. Asama, H. Tsukuira, T. Naito, and H. Kawaguchi, *J. Antibiot. (Tokyo)*, **30**, 789 (1977).
15. W. T. Bradner, *The Bleomycins — Current Status and New Developments* (S. K. Carter, S. T. Crooke, and H. Umezawa, Eds.). Academic Press, New York, p. 343–355 (1978).
16. J. E. Strong and S. T. Crooke, *Cancer Res.*, **38**, 3322 (1978).
17. H. Asakura, M. Hori, and H. Umezawa, *J. Antibiot. (Tokyo)*, **28**, 537 (1975).

The Action of Bleomycin on DNA

FRANKLIN HUTCHINSON AND LAWRENCE F. POVIRK

First, this chapter summarizes our experimental results on the lesions produced by the antitumor drug, bleomycin, on DNA in solution. These results are compared critically with those reported in the literature in an attempt to assess the state of current knowledge in this area.

Second, there is a brief discussion of the binding of bleomycin to DNA, and the relation of this binding to the action of bleomycin. It is reasonable to conclude that the lack of activity of bleomycin on RNA is, at least in part, a failure of bleomycin to bind to RNA[1,2].

Third, there is a discussion correlating the effects of bleomycin on DNA with the biological effects produced by the action of bleomycin on living cells. There is strong support for the working hypothesis that many of the effects of bleomycin on cells can be explained by the known effects of bleomycin on DNA. There are still unanswered questions, however. Thus, it cannot be stated yet that the antitumor effects of bleomycin are caused by its damage to DNA, and not to some other process.

A word is in order here on the difficulty of getting reproducible results with bleomycin. From the literature, it is clear that the effect of bleomycin on DNA depends on the oxygen concentration[3,4], on the presence of suitable reducing agents, and on the presence of metal ions having about the correct oxidation-reduction potential, such as iron[4] [and perhaps copper,[5]]. These metal ions need be present only in micromolar amounts. Nucleic acids are well-known complexers of metal ions. Bleomycin is also a good chelating agent for metal ions, so that it is difficult to guarantee removal of trace metals by the use of chelating agents such as EDTA. Therefore, it is not surprising that experiments with bleomycin can show widely differing levels of activity at different times, and we have repeatedly found this to be the case.

In our experiments on the effect of bleomycin on two different DNAs, the DNAs are mixed together in the same solution and treated with bleomycin in the same tube. If two different strains of bacterial cells are being treated with bleomycin, the strains are mixed and treated in the same tube with bleomycin, and then the effects on the two different strains are determined by the use of suitable radioactive labels in the cell DNA, or by the use of suitable genetic markers in the cells. We found this to be the most reliable way to get repeatable data.

Lesions in DNA from Bleomycin

Base Release

Radioactively labeled DNA was isolated from *Escherichia coli* cells grown in the presence of either [*methyl*-^{14}C] thymine, [5-^{3}H] cytosine, or [2,8-^{3}H] adenosine (the latter labels both adenine and guanine through interconversion of the purines by metabolic processes). DNA isolation by the Marmur[6] procedure, followed by adsorption and elution from hydroxyapatite, reduced the radioactivity associated with RNA to less than 10%, as judged by solubility in 5% trichloroacetic acid after 30 min hydrolysis in 0.3 M NaOH at 37°C.

DNA with either [^{3}H] cystosine or with [^{3}H] adenine, -guanine was mixed with [^{14}C] thymine–DNA and the mixture was treated with bleomycin (2.5–250 μg/ml) after adding 16–160 μM Fe^{2+} and 4–20 mM 2-mercaptoethanol. Aliquots of the reaction mixtures were dried on Whatman #1 paper strips (1.9 X 50 cm) and then developed in various chromatographic systems. The paper strips were cut into 1-cm segments and counted. The yields of various products released was measured from the radioactivity, normalized to that of the [^{14}C] thymine released. The results are as follows[7].

1. The release of all four bases was identified by the comigration of radioactive label with authentic samples of thymine (in three chromatographic solvents), adenine (in two solvents), guanine (in two solvents), and cytosine (in one solvent system).

2. The yields were linear in bleomycin concentration for 1–10% thymine release, 1–4% cytosine release, and 0.2–2% purine release. The yields relative to thymine are given in Table 1.

3. Minor products containing the [*methyl*-^{14}C] thymine label, the [5-^{3}H] cytosine label, and the [2,8-^{3}H] purine label were detected, with yields as shown in Table 1. The thymine minor product definitely did not chromatograph with thymidine, and the other two minor products chromatographed somewhat differently from the corresponding deoxyribosides. No minor product was the corresponding nucleoside monophosphate.

The release of thymine has been reported before[8-10]. Uracil was released from PBS-1 phage DNA, a DNA which contains that base instead of thymine[11]. Müller *et al.*[8] could not detect the release of adenine from poly(dA–dT) and Ishida and Takahashi[12] did not detect the release of cytosine, adenine, or guanine from *Bacillus subtilis* DNA. The release of all four bases has been previously reported by Haidle *et al.*[11] and by Mueller *et al.*[13] but only at very high levels of treatment. Here, the three-dimensional DNA structure would be destroyed, and the specificity of action of bleomycin could be quite different.

Our results show unequivocally that all four bases are released by the action of bleomycin on DNA in the double-helical form. Several rather uninteresting reasons can be given for the differences between our results and previous ones. One intriguing possibility should be pursued, that is, that base selectivity may

Table 1: Relative Yields of Various Products for the Action of Bleomycin on DNA Which Has Roughly Equal Quantities of all Four Bases[a]

Product	Relative yield	Reference
Thymine	(1.0)	(7)
Thymine minor product	0.10 ± 0.03	(7)
Cytosine	0.63 ± 0.07	(7)
Cytosine minor product	0.09 ± 0.03	(7)
Adenine	0.15 ± 0.02	(7)
Guanine	0.08 ± 0.01	(7)
Purine minor product[b]	0.04 ± 0.02	(7)
Base release (total)	(1.0)	(7)
Unaltered bases	0.89	(7)
Minor products	0.11	(7)
Single-strand break (in alkali)	(1.0)	
True single-strand break	0.17 – 1.0	(15,16)
Alkali-labile bond	0 – 0.83	(15,16)
Double-strand break	0.05 – 0.2	(15,17)

[a]For example, E. coli DNA.
[b]As a fraction of total purine label.

depend on some unknown factor (metal ion, reducing agent?) which may also control the ratio of true single-strand breaks to alkali-labile bonds (see below).

True Single-Strand Breaks

Such breaks may be detected by treating supercoiled covalently closed DNA with bleomycin, and observing loss of supercoiling by neutral sedimentation or gel electrophoresis[14–17]. Breaks have also been observed by the reduced sedimentation rate in neutral gradients of ϕX174 single-strand DNA after bleomycin treatment[14]. The absence of 3'-phosphates (inorganic phosphate removable by exonuclease III) at the break, and the presence of comparable numbers of 5'-phosphates (removable by alkaline phosphatase) and 5'-OH groups (which can be charged by polynucleotide kinase), have been reported[18]. The presence of 5'-OH groups disagrees with other findings from the same group. No inorganic phosphate was released by bleomycin action; with DNA labeled with 5'-[^{3}H]deoxyribose, no product with the ^{3}H label was detected, such as a sugar fragment attached to a phosphate[9].

Alkali-Labile Bonds

When supercoiled covalently closed DNA is treated with bleomycin, more single-strand breaks are found on assaying the DNA under alkaline conditions than under neutral conditions[15,16]. The increase in number of breaks on treating with alkali is the number of alkali-labile bonds. Povirk et al.[15] report one to three alkali-labile bonds/true break for bleomycin treatment with and without

SH reagents. From the data of Ross and Moses[16], one can determine about five alkali-labile bonds/true break for bleomycin treatment without added -SH; they infer that there may be few alkali-labile bonds for treatment in the presence of -SH.

Schyns et al.[19] have demonstrated that the alkali-labile bond in bleomycin-treated DNA is an AP (apurinic or apyrimidinic) site, a structure well known to be opened by alkali[20,21]. First, Schyns et al. showed that there is a lesion in bleomycin-treated DNA which has the same rate of alkaline hydrolysis as do apurinic sites (formed by controlled heating of DNA), and also sites in alkylated DNA which had then been depurinated with gentle heating. Second, sites in bleomycin-treated DNA and in alkylated–depurinated DNA were opened at identical rates by a purified rat liver endonuclease specific for apurinic sites; furthermore, treatment with NaOH after incubation with the enzyme yielded the same number of breaks as treatment with alkali alone, showing that the same sites were opened by the enzyme and by base.

The presence of AP sites after bleomycin treatment could account for the presence of aldehyde function by iodine titration[8]; the aldehyde activity reacting with thiobarbituric acid[18] could also be a sugar fragment such as malonic dialdehyde.

The factors which control the ratio of alkali-labile bonds to true single-strand breaks are still unknown. It is possible that these factors may also control the specificity of base release, as discussed above.

Double-Strand Breaks

The formation of DNA double-strand breaks by bleomycin is well established[1,22]. Indeed, any agent which will form single-strand DNA breaks will also produce double-strand breaks by coincidences between breaks in the two complementary strands.

In addition to this source of double-strand breaks, it has been demonstrated that there is another nonrandom mechanism for making such breaks[15,17]. At low bleomycin exposures, where double-strand breaks from randomly produced single-strand breaks would be negligible, there was a linear production of double-strand breaks, one per nine true single-strand breaks, in supercoiled covalently closed ColE1 DNA (Fig. 1) for treatment with bleomycin in the presence of 25 mM 2-mercaptoethanol or 5×10^{-5} M Fe^{2+}[15]. Further, there was a linear production of double-strand breaks, one per ten single-strand breaks (measured in alkali) in a T2 bacteriophage DNA molecule, after treatment with bleomycin in the absence of added reducing agents[15]. Because of the presence of alkali-labile bonds, these yields, although they seem the same, are actually quite different.

When PM-2 DNA was first treated with restriction enzymes, then with bleomycin, gel electrophoresis of the digest showed specific bands appearing as the result of bleomycin activity[17]. This demonstrates the presence of specific sites at which double-strand breaks occur preferentially.

There are at least three possible mechanisms by which bleomycin could produce double-strand breaks.

1. Assume, as seems reasonable, that Fe^{2+} and oxygen cause some kind of free radical reaction which leads to the break[4]. Then, in a certain fraction of cases, one radical could lead to a break in one strand and a secondary radical to a break in the second strand. For this model to hold, the bleomycin molecule must play a more significant role than merely increasing the local concentration of Fe^{2+} near the DNA. Fe^{2+}, without bleomycin, produces no measurable DNA double-strand breaks (less than one per 70 true single-strand breaks, Fig. 1).

2. Fe^{2+} bound to a DNA–bleomycin complex could undergo successive oxidations and reductions, causing production of two or more radicals which attack both strands of the DNA near the adsorption site. However, this seems unlikely, since the ratio of single- to double-strand breaks is the same, adding Fe^{2+} or using 2-mercaptoethanol as a reducing agent (Fig. 1).

3. Two bleomycins could be bound close together to the DNA. If this happens because a bleomycin dimer is bound[17], the fraction of dimers, and hence the ratio of double- to single-strand breaks, would be expected to depend strongly on the bleomycin concentration, which is not observed (Fig. 1). If the two nearby bleomycins bind successively, the binding constant for the second molecule must be much higher than for the first (cooperative binding). This follows from the linear relation between double- and single-strand breaks (Fig. 1) and a linear dependence of single-strand breaks on bleomycin concentration under conditions in which the bleomycin reaction is stopped with certainty after a fixed time, by sedimenting the DNA in pH 5.6 gradients[23]. This model could give an explanation for the apparent site specificity for double-strand breaks found by Lloyd et al.[17].

Other Lesions

Ross and Moses[16] showed that heating supercoiled covalently closed DNA after treatment did not lead to any breaks in alkali, the way alkylating agents do.

Yields

Table 1 gives the available data on the relative yields of the various bleomycin-induced lesions. The following comments may be made.

1. The values given must be used with great caution, because the relative yields may vary with the experimental conditions used, the composition of the bleomycin mixture, and the base composition of the DNA.

2. The ratio of alkali-labile bonds to true breaks is particularly uncertain. In view of the early results suggesting that thymine is preferentially released, the ratio of the various bases released is also under some suspicion.

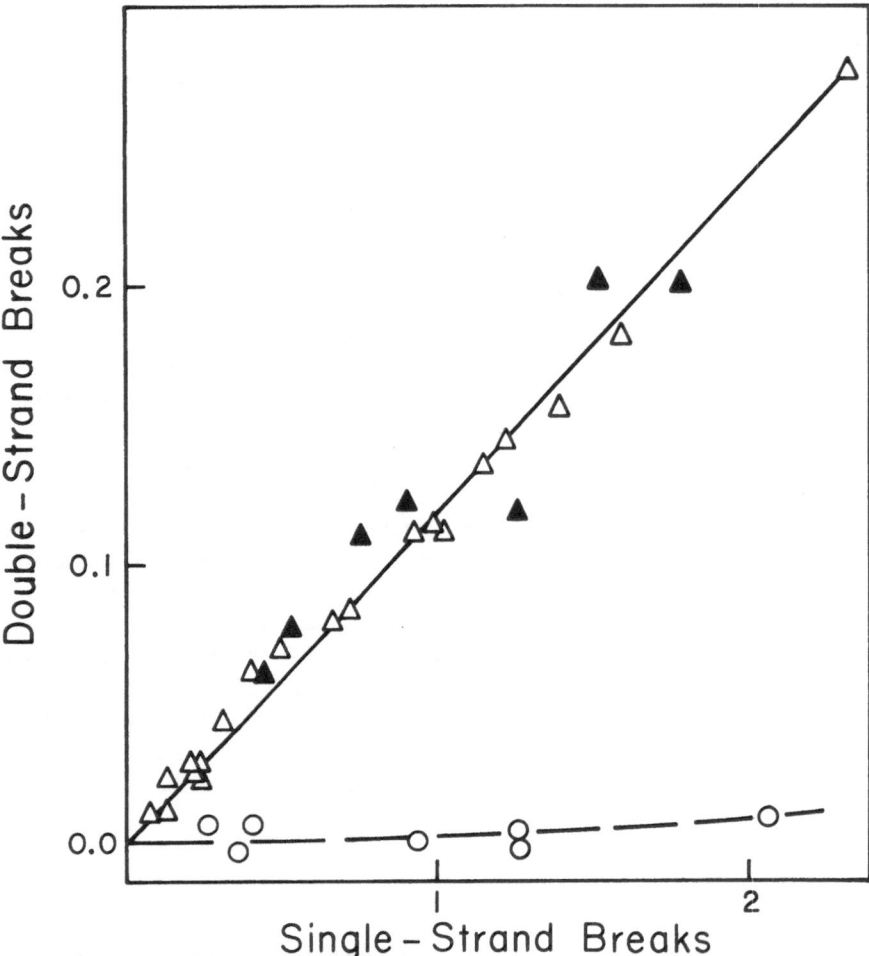

Fig. 1. Production of single- and double-strand breaks in ColE1 DNA by bleo-mycin plus 2-mercaptoethanol (△), bleomycin plus Fe^{2+} (▲), or Fe^{2+} alone (○) as determined from the proportion of supercoiled, nicked circular, and linear molecules seen after treatment. Treatment time was kept constant, and the bleomycin or Fe^{2+} concentration was varied. Reaction mixtures were: (△) 25 mM 2-mercaptoethanol, 100 μg/ml DNA, 0.05–1.0 μg/ml bleomycin, and 45 mM Tris, pH 8, incubated for 30 min at 37°C; (▲) 50 μM Fe^{2+}, 100 μg/ml DNA, 0.15–1.0 μg/ml bleomycin, and 45 mM Tris, pH 8, incubated for 30 min at 22°C; (○) 10–200 μM Fe^{2+}, 50 μg/ml DNA, 10 or 45 mM Tris, pH 8, incubated for 7 hr at 22°C. All reactions were stopped by adding EDTA to 20 mM, and samples were analyzed on agarose gels [see Ref. [15]].

3. It is reasonable to assume that, for low levels of bleomycin attack, the yield of bases released is within a factor of two the yield of single-strand breaks measured in alkali.

4. The numerical values of the yields are such that one minor base product could be released per double-strand break.

Intercalative Binding and Strand Breakage

It seems likely that bleomycin binding to DNA[24] is involved in DNA strand breakage. We have found that in 0.1 M NaCl–20 mM Tris, pH 8, supercoiled ColE1 DNA is broken 50% faster than relaxed closed circular ColE1 DNA,[23] in qualitative agreement with Lloyd *et al.*[17] Since supercoiling increases the affinity of DNA for intercalators, this result suggests that intercalative binding[25] may be involved in bleomycin-induced DNA breakage.

It is difficult to get enough bleomycin bound to DNA to observe structural changes usually associated with intercalation, without also degrading the DNA. However, at pH 5.5, we observed a relaxation of supercoiled ColE1 DNA, which appeared not to be due to strand breakage (Fig. 2A). At both pH 5.5 and 8, tripeptide S relaxed and then recoiled ColE1 DNA (Fig. 2B and C), eliminating any possibility that relaxation was due to strand breakage and implicating the bithiazole as the intercalating moiety. We determined an unwinding angle $\phi = 12°$ for tripeptide S by comparison with ethidium bromide, for which $\phi = 26°$[26]. This is smaller than the unwinding angle of most intercalators, but similar to that of daunomycin[27].

At pH 5.5, we have also observed length changes in DNA by measuring rotational correlation times of 150-base pair-long DNA molecules during orientation in an electric field [see Ref.[28]]. Each bound bleomycin or tripeptide S molecule lengthened DNA by 3.1 Å[23] which is within the range measured for other intercalators (N. Dattagupta, M. Hogan, and D. M. Crothers, in preparation). Thus, bleomycin binding probably involves intercalation of the bithiazole rings.

DNA Damage and the Biological Effects of Bleomycin

The types of DNA lesions induced by bleomycin are similar to those from agents (e.g., ionizing radiations) which produce major biological effects principally by creating lesions in the DNA of cells and viruses. It is well established that bleomycin produces a number of biological effects known to be caused by DNA lesions: loss of ability to form colonies for mammalian and bacterial cells[29], production of chromosomal aberrations[30,31], degradation of intracellular DNA to acid-soluble form[32], inhibition of DNA synthesis[33], release of DNA from a complex with membrane[34], induction of prophages in bacteria[35,36], and induction of synthesis of the *recA* protein[37]. The effects of bleomycin on the action of a variety of enzymes related to nucleic acids [review: Ref.[13]] can be interpreted as the result of lesions in DNA.

A possible way of linking a biological effect to bleomycin-induced lesions in DNA would be to alter the sensitivity of the DNA, for example, by incorporation of the thymine analog, bromouracil. This method has been useful in studying the effects of ionizing radiations and ultraviolet light. Unfortunately, bleomycin releases bromouracil and thymine from DNA at the same rate, and produces the same number of single-strand breaks in DNA with and without bro-mouracil[7].

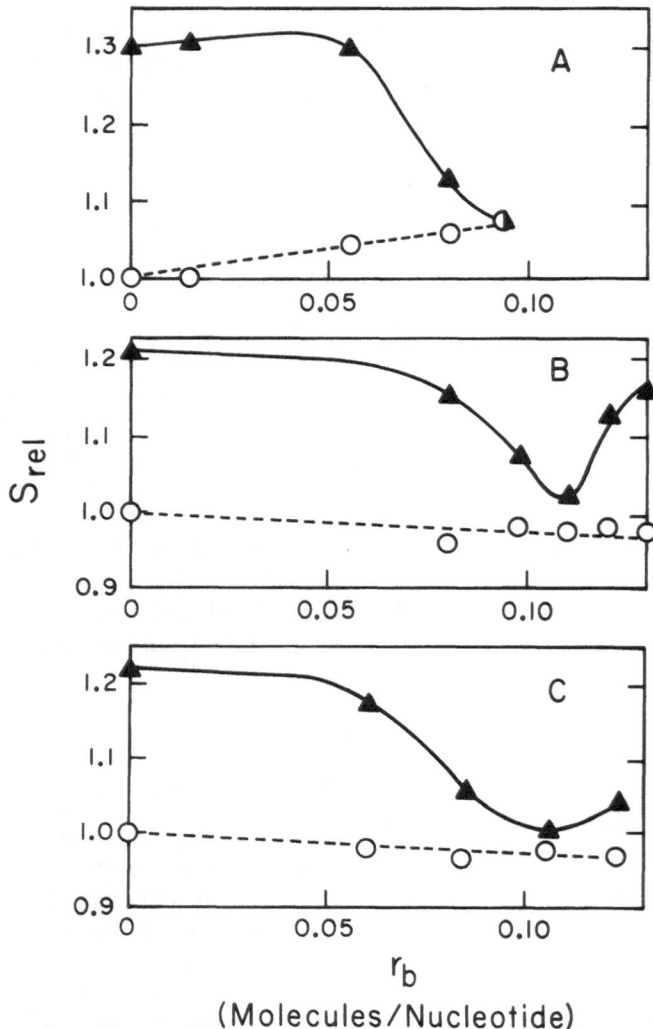

(Molecules/Nucleotide)

Fig. 2. Titration of superhelical turns in ColE1 DNA by bleomycin (A) or tripeptide S (B, C)[38]. Mixtures of supercoiled [^{14}C] DNA and nicked [^{3}H] DNA were sedimented in gradients containing various concentrations of bleomycin or tripeptide S. Median sedimentation coefficients of supercoiled (▲) and nicked (○) DNA were normalized to that of nicked DNA at zero drug concentration. The number of bound drug molecules per nucleotide, r_b, was calculated from the known free concentration of drug in the gradient, using binding curves obtained by equilibrium dialysis for bleomycin and by fluorescence quenching [using the method of Chien et al.[24]] for tripeptide S. Buffers were 0.8 mM sodium citrate, pH 5.5 (A, B), and 2 mM Tris–1 mM EDTA, pH 8.0 (C).

A good case can be made for the idea that bleomycin acts on cells principally by way of the lesions created in the DNA. However, results discussed next raise questions concerning this point of view.

Sensitivity of Bacterial Cells with Reduced DNA Repair Capacity

Table 2 summarizes, for *E. coli* and *B. subtilis* cells which carry *recA* or *lexA* or *uvr* genotypes, the reported sensitivities to bleomycin relative to that for the corresponding wild type cell. All reports support the idea that lack of the *uvr* excision system does not sensitize bacteria to bleomycin, as expected from the lesions detected to date in bleomycin-treated DNA.

From what is known of the sensitivity of *recA* and *lexA* mutants to physical and chemical agents, these mutations should sensitize cells to bleomycin. Table 2 shows that different authors get different results. While *polA* mutants are somewhat sensitized to X-rays, nothing like the 10-fold sensitization to bleomycin of a specific *polA* mutant reported by Moses and Ross[39] has been observed.

Dr. Kazuo Yamamoto at Yale has mixed *E. coli* AB2497K *rec⁺ his⁺ arg⁻* with AB2487K *recA⁻ his⁻ arg⁺*, treated with bleomycin, diluted by a factor of 10^4 to stop the reaction, and used selective plates to assay independently for colony-forming ability of the *rec⁺* and *recA⁻* cells. For cells grown in K medium and then starved for amino acids, *recA⁻* cells were fourfold more sensitive than *rec⁺*, as found by Yamagami *et al.*[40] and Moses and Ross[39]. For the two strains grown in rich nutrient broth, the *recA⁻* and the *rec⁺* strains had the same sensitivity, a result in agreement with Endo[41]. A *lexA* strain behaved as did wild type cells. For γ-rays, Dr. Yamamoto finds that the same *recA* and *lexA* strains are four- to sixfold more sensitive than the wild type for cell growth in either K medium or broth.

A clear-cut explanation of these results with these repair mutants should help in understanding the relation between bleomycin effects and DNA lesions.

Dependence of Sensitivity to Bleomycin on the State of the *E. coli* Cell

The sensitivity to bleomycin of colony-forming ability of *E. coli* cells is decreased by treatments such as starving for thymine[43] or inhibiting RNA synthesis[44]. Such treatments also decrease the number of DNA strand breaks measured after a short exposure to phleomycin[45]. The sensitivity to bleomycin of both colony-forming ability and strand breaks appears to depend much more strongly on the physiological state of the bacteria than has been seen in experiments with X-rays.

Simple explanations of this difference between bleomycin and X-rays are easy to imagine, but in fact there are few experimental facts to go on. An understanding of this feature in bacterial cells might help in understanding the remarkable difference in the response of normal mammalian and tumor cells to bleomycin.

Table 2: Sensitivity of *E. coli* Mutants to Bleomycin, Compared to That of Isogenic Wild Type[a]

				Reference		
	Endo (41)	Saunders and Schulz (42) (*B. subtilis*)	Onishi et al. (32)	Yamagami et al. (40) (in BLM-sensitive mutants)	Nakayama (43) (PLM)	Moses and Ross (39)
recA	1	More sensitive	—	4	More sensitive	3–4
lexA	1	—	—	—	—	3–4
uvr(A or B)	—	—	1	1	1	1
polA1	—	—	—	—	1	2–10
polA107	—	—	—	—	—	More sensitive

[a]Values represent increase (as a ratio) of bleomycin exposure needed to reduce colony-forming ability of the wild type to that of the mutant.

Acknowledgments

The authors are happy to thank Professor Wolfgang Köhnlein, with whom a number of experiments described here were done. They also thank Dr. Kazuo Yamamoto, whose work on the bleomycin sensitivity of *E. coli* cells inspired the last section of this paper. The original research reported here was supported by Grant CA 17938, awarded by the National Cancer Institute, United States Department of Health, Education and Welfare.

References

1. H. Suzuki, K. Nagai, E. Akutsu, H. Yamaki, N. Tanaka, and H. Umezawa, *J. Antibiot. (Tokyo)*, 23, 473 (1970).
2. H. Umezawa, H. Asakura, K. Oda, S. Hori, and M. Hori, *J. Antibiot. (Tokyo)*, 26, 521 (1973).
3. T. Onishi, H. Iwata, and Y. Takagi, *J. Biochem. (Japan)*, 77, 745 (1975).
4. E. A. Sausville, J. Peisach, and S. B. Horwitz, *Biochem. Biophys. Res. Commun.*, 73, 814 (1976).
5. M. J. Sleigh, *Nucleic Acid Res.*, 3, 891 (1976).
6. J. Marmur, *J. Mol. Biol.*, 3, 208 (1961).
7. L. F. Povirk, W. Köhnlein, and F. Hutchinson, *Biochim. Biophys. Acta*, 521, 126 (1978).
8. W. E. G. Müller, Z. Yamazaki, H.-J. Breter, and R. K. Zahn, *Eur. J. Biochem.*, 31, 518 (1972).
9. C. W. Haidle, K. K. Weiss, and M. T. Kuo, *Mol. Pharmacol.*, 8, 531 (1972).
10. M. Takeshita, A. P. Grollman, and S. B. Horwitz, *Virology*, 69, 453 (1976).
11. C. W. Haidle, M. T. Kuo, and K. K. Weiss, *Biochem. Pharmacol.*, 21, 3308 (1972).
12. R. Ishida and T. Takahashi, *Biochem. Biophys. Res. Commun.*, 66, 1432 (1975).
13. W. E. G. Müller, H. J. Rohde, and R. K. Zahn, *J. Mol. Med.*, 1, 173 (1976).
14. I. Shirakawa, M. Azengami, S.-I. Isihi, and H. Umezawa, *J. Antibiot. (Tokyo)*, 24, 761 (1971).
15. L. F. Povirk, W. Wübker, W. Köhnlein, and F. Hutchinson, *Nucleic Acids Res.*, 4, 3573 (1977).
16. S. L. Ross and R. E. Moses, *Biochemistry*, 17, 581 (1978).
17. R. S. Lloyd, C. W. Haidle, and D. L. Robberson, *Biochemistry*, 17, 1890 (1978).
18. M. T. Kuo and C. W. Haidle, *Biochim. Biophys. Acta*, 335, 109 (1974).
19. R. Schyns, M. Mulquet, and W. G. Verly, *FEBS Lett.*, 93, 47 (1978).
20. C. R. Bayley, K. W. Brammer, and A. S. Jones, *J. Chem. Soc.*, 1903 (1961).
21. T. Lindahl and A. Andersson, *Biochemistry*, 11, 3618 (1972).
22. C. W. Haidle, *Mol. Pharmacol.*, 7, 645 (1971).
23. L. F. Povirk, M. Hogan, and N. Dattagupta, *Biochemistry*, 18, 96 (1979).
24. M. Chien, A. P. Grollman, and S. B. Horwitz, *Biochemistry*, 16, 3641 (1977).
25. H. Murakami, H. Mori, and S. Taira, *J. Theoret. Biol.*, 42, 443 (1973).
26. J. C. Wang, *J. Mol. Biol.*, 89, 783 (1974).
27. M. Waring, *J. Mol. Biol.*, 54, 247 (1970).
28. M. Hogan, N. Dattagupta, and D. M. Crothers, *Proc. Nat. Acad. Sci. USA*, 75, 195 (1978).

29. H. Umezawa, M. Ishizuka, and T. Takeuchi, *Cancer*, **20**, 891 (1967).
30. K. Ohama and T. Kadotani, *Japan J. Human Genet.*, **14**, 293 (1970).
31. R. S. Bornstein, D. A. Hungerford, G. Haller, P. F. Engstrom, and J. W. Yarbro, *Cancer Res.*, **31**, 2004 (1971).
32. T. Onishi, K. Shimada, and Y. Takagi, *Biochim. Biophys. Acta*, **312**, 238 (1973).
33. H. Suzuki, K. Nagai, H. Yamaki, N. Tanaka, and H. Umezawa, *J. Antibiot. (Tokyo)*, **21**, 379 (1968).
34. M. Miyaki, T. Kitayama, and T. Ono, *J. Antibiot. (Tokyo)*, **27**, 647 (1974).
35. H. Aoki and H. Sakai, *J. Antibiot. (Tokyo)*, **22**, 87 (1967).
36. C. W. Haidle, K. K. Weiss, and M. L. Mace, *Biochem. Biophys. Res. Commun.*, **48**, 1179 (1972).
37. L. J. Gudas and A. B. Pardee, *J. Mol. Biol.*, **101**, 459 (1976).
38. H. Umezawa, *Fed. Proc.*, **33**, 2296 (1974).
39. R. E. Moses and S. L. Ross, *J. Supramolecular Structure, Supplement* 2, Abstract 49 (1978).
40. H. Yamagami, M. Ishizawa, and H. Endo, *Gann*, **65**, 61 (1974).
41. H. Endo, *J. Antibiot. (Tokyo)*, **23**, 508 (1970).
42. P. P. Saunders and G. A. Schultz, *Biochem. Pharmacol.*, **21**, 1657 (1972).
43. H. Nakayama, *Mutat. Res.*, **29**, 21 (1975).
44. S. S. Cohen and J. I, *Cancer Res.*, **36**, 2768 (1976).
45. M. J. Sleigh and G. W. Grigg, *Biochem. J.*, **155**, 87 (1976).

Polyamines and the Toxicity of the Bleomycins

SEYMOUR S. COHEN

This laboratory has been concerned for some years with the problem of the physiological roles of the polyamines[1,2] and has studied in some detail the binding of spermidine to double-stranded regions of the nucleic acids[3-5]. The discovery of various organic amines, including putrescine, agmatine, spermidine, and spermine, as terminal carboxamide substituents in the bleomycins[6] has led us to ask if these amines are not essential in the initial binding of the antibiotics to DNA, prior to the cleavage of the nucleic acid in various *in vivo*[7] and *in vitro*[8] systems. The early reports to the effect that bleomycinic acid was inactive tended to support this view. Bleomycinic acid is the carboxyl terminal glycopeptide of bleomycin from which these polycations were removed.

The bleomycins are known to possess minimal toxicity in some tissues commonly regarded as highly sensitive to many antitumor agents. Despite this favorable attribute, the chemotherapeutic efficacy of the bleomycins is limited by the cumulative toxicity of these agents to lung tissue. It seemed possible that if the cationic amide were involved in these effects one might interfere with the compounds locally, i.e., in lung tissue. In any case it appeared desirable to compare the toxicities of bleomycinic acid and various bleomycins in various cellular systems as a precondition to determining the nature of the various toxicities in order to minimize them *in vivo*.

As is well known, essentially all of the clinical work has been done with a mixture of bleomycins, comprised largely of bleomycins A_2 and B_2. The terminal carboxamines of A_2 and B_2 contain the salt of dimethyl sulfonium propylamine and of agmatine, respectively. These cationic moieties may be thought of as the respective decarboxylation products of methyl methionine and of arginine. It appeared desirable to compare these bleomycins, i.e., A_2 and B_2, with those containing a diamine (bleomycin A_2-b) and a triamine, bleomycin A_5. Bleomycin A_2-b contains 1,3-diaminopropane; bleomycin A_5 contains spermidine. All four of these individual bleomycins, as well as bleomycinic acid, have been tested in our laboratory systems. We have not been able to obtain other relatively pure bleomycins.

We have compared the antibiotics on several strains of *Escherichia coli*, studying the effects on the growth rate of the bacteria measured turbidimetrically, as well as effects on the viability of the cells[9]. The considerable differences in toxicity and apparent mode of action of the various bleomycins raise numerous questions concerning the present choice and use of these substances. In *E. coli* the lethality of bleomycin A_5 was found to require RNA synthesis during exposure to the drug.

The results with strains of *E. coli* have led us to test the same compounds on the growth and viability in culture of mouse fibroblasts thereby attempting to approach the mammalian system more closely. Our results were similar to those with the bacteria in many respects[10]. The bleomycin containing spermidine was most toxic. However, we were unable to find a requirement for RNA synthesis in demonstrating the lethality of bleomycin A_5 in fibroblasts. In addition, it was possible to show that a synthetic and natural derivative of spermidine, diamidinospermidine, or hirudonine, which has little toxicity by itself for fibroblasts, significantly and specifically protects animal cells from the toxicity of bleomycin A_5. The structures of the amines and guanido derivatives are presented below:

$$H_2NCH_2CH_2CH_2CH_2NH_2$$
Putrescine

$$\overset{\displaystyle NH}{\overset{\|}{H_2N-C-HNCH_2CH_2CH_2CH_2NH_2}}$$
Agmatine

$$\overset{\displaystyle NH}{\overset{\|}{H_2N-C-NHCH_2CH_2CH_2CH_2NH}}\overset{\displaystyle NH}{\overset{\|}{-C-NH_2}}$$
Arcaine

$$H_2NCH_2CH_2CH_2CH_2NHCH_2CH_2CH_2NH_2$$
Spermidine

$$\overset{\displaystyle NH}{\overset{\|}{H_2N-C-NHCH_2CH_2CH_2CH_2NHCH_2CH_2CH_2NH}}\overset{\displaystyle NH}{\overset{\|}{-C-NH_2}}$$
Hirudonine

Materials and Methods

Chemicals

Various species of copper-free bleomycins were obtained from Bristol Laboratories, Syracuse, New York, with the assistance of Dr. John Douros of the National Cancer Institute. Stock solutions of 1 mg/ml in mineral medium were stored at $-20°C$. The radioactive compounds $[2-^{14}C]$uracil, $[2-^{14}C]$thymine, and L-[^{14}C] arginine were obtained from New England Nuclear, Boston, Massachusetts. Chloramphenicol was obtained from Calbiochem, La Jolla, California. 6-Azauridine and 2'-deoxycytidine were obtained from P-L Biochemicals, Milwaukee, Wisconsin. 3-Deazacytidine was obtained from ICN Nucleic Acid Research Institute, Irvine, California. Hirudonine sulfate and arcaine sulfate were synthesized by the method of Robin and van Thoai[11].

Bacterial Growth

The growth and properties of the various strains of *E. coli* have been described. These included strain B[12] and two different stringent and relaxed pairs, strains 15 TAU rel A$^+$ and 15 TAU rel A[13] and strains CP78 and CP79[14]. The organisms were grown with aeration at 37°C in mineral medium[12], supplemented with 1 mg glucose per ml plus additional nutrients where required. Strain B had no additional requirements. The 15 TAU strains were grown in the presence of 2 μg thymine per ml, 20 μg L-arginine per ml, and 12 μg uracil per ml. Strains CP78 and CP79 were grown with 5 μg thiamine per ml and 40-μg quantities each of L-arginine, L-histidine, L-leucine, and L-threonine per ml. The cells were harvested in exponential growth at 2×10^8 cells/ml, washed once with mineral medium, and resuspended in the appropriate medium with or without the antibiotic. Increases of bacterial mass were followed turbidimetrically in a Klett colorimeter, equipped with a 420-nm filter. The mass doubling times of the organisms during exponential growth were 60, 58, 65, 60, and 58 min for strains B, 15 TAU rel A$^+$, 15 TAU rel A, CP78, and CP79, respectively. Viable counts were determined by plating on nutrient agar plates.

Animal Cells

The growth and multiplication of mouse fibroblasts (L cells) in suspension culture and the estimation of cell number and size in the Coulter counter have been described[15]. The plating technique for determination of cell viability has also been reported[16]. Cultures of cells in the exponential phase were diluted in fresh complete media[17] to 4 to 5×10^4 cells/ml and permitted to double to 10^5 by growth for 20 hr. The test compounds were then added to the exponential cultures. Studies on the incorporation of isotopic compounds into nucleic acids were carried out as described earlier[17].

Incorporation of [^{14}C] Thymine, [^{14}C] Arginine, and [^{14}C] Uracil

Aliquots (1 ml) of the test bacterial suspension were mixed at 4°C with an equal volume of cold 10% trichloroacetic acid. After 30 min, the mixture was filtered onto Millipore filter discs (pore size, 0.45 μm; diam, 24 mm) and was washed twice with 5% trichloroacetic acid, and then with 95% ethanol. The filter was dried and added to a vial containing 10 ml of Liquifluor (New England Nuclear) in toluene (42 ml/liter toluene), and the radioactivity was determined in a scintillation counter. The counts were proportional to the mass of bacteria from 2×10^8 to 8×10^8 cells. A multiplication factor of 1.34 was used to convert the incorporated radioactivity into micromoles; such a value was determined empirically by comparing counts of free [^{14}C] uracil or [^{14}C] thymine and of completely incorporated pyrimidine.

Lethality of Various Bleomycins

All *E. coli* strains tested were similarly sensitive to the spermidine derivative, bleomycin A_5 (1 μg/ml) in complete media, i.e., actively growing cultures at 1 to 2×10^8 colony-forming units/ml decreased exponentially to a survival of about 10^{-3} in 2 hr. At the same concentration bleomycinic acid is essentially inactive on strain 15 TAU rel A^+ (Fig. 1), scarcely impeding the increase of viable count or turbidity. On the other hand, the spermidine derivative, bleomycin A_5, is extremely lethal, killing 90% of the bacteria in 30 min. During this interval, the culture increased its turbidity at the same exponential rate as did the culture lacking the antibiotic. This increase in turbidity ceased in 1 hr, and a small amount of lysis was observed in the culture. Under the microscope, cells treated with antibiotics are seen to have increased significantly in volume. Under similar conditions cultures of some other strains, such as strain *B*, may increase in turbidity four to five times over a 90-min interval, while their viability falls precipitously.

As seen in Fig. 1, strain 15 TAU rel A^+ is not killed over a 90-min interval by bleomycins A_2, B_2, and A_2-b (1 μg/ml). Although the cells stop multiplying, they nevertheless continue to grow at almost a normal rate, as measured turbidimetrically. The cells become extremely elongated. The implication of these results is that the effects of the latter compounds are reversible and that inhibited cells can go on to divide to produce colonies on the agar plate. Since the only known metabolic effect of the bleomycins is that of cleavage of DNA, it may be asked if the reversibility of inhibition by A_2, B_2, and A_2'-b is due to DNA repair. This effect may be expected to result in high mutation rates. On the other hand it must then be asked if the immediate exponential decline of viability observed with A_5 masks such possible repair, or reflects toxic effects other than those which are repairable.

The Toxicity of Bleomycin A_5

As can be seen in Fig. 2, lower lethal concentrations of A_5 also begin to kill without lag or are bacteriostatic, as is A_2. A differential inhibition of multiplication without inhibition of growth (increase of turbidity) can be produced by low concentrations of bleomycin A_5. At 0.1 μg/ml (Fig. 2) division is inhibited whereas turbimetric increase is unaffected for 1 hr. At 0.5 μg/ml killing is exponential, although at a lower rate than that for 1 μg/ml. The latter concentration

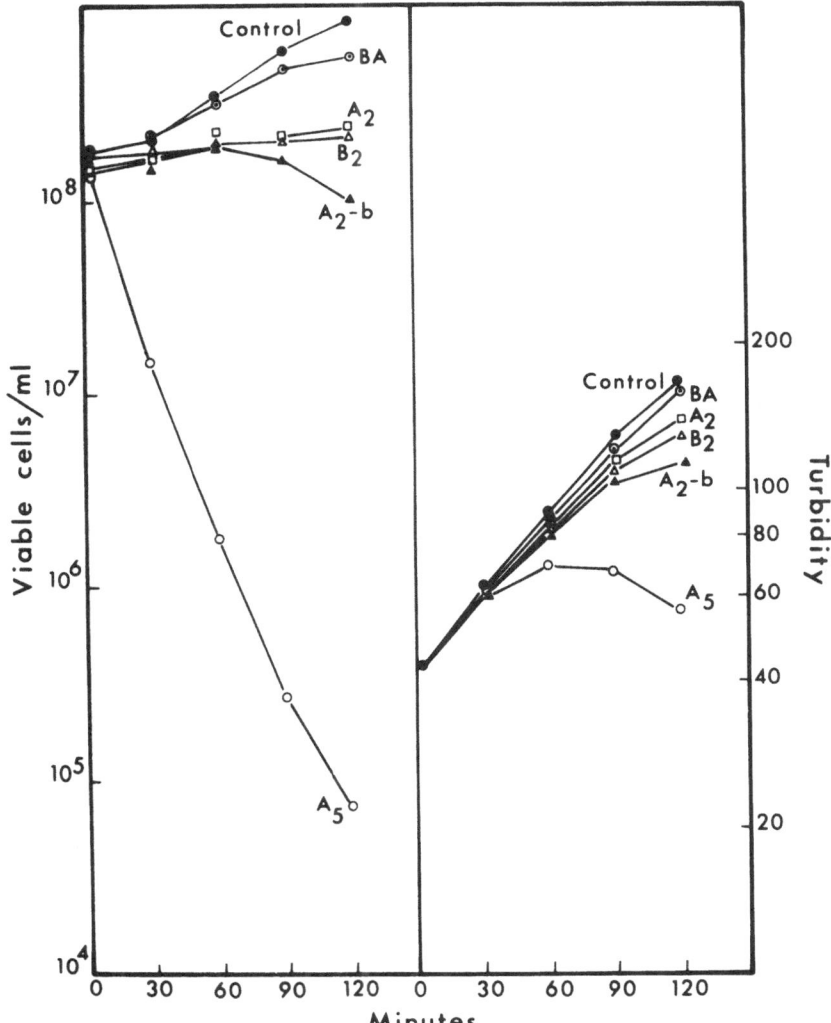

Fig. 1. The effects of bleomycins A_5, A_2-b, A_2, and B_2 and of bleomycinic acid (BA) (1 μg/ml) on the viability and turbidity of *E. coli* strain TAU rel A^+ in a complete growth medium.

is not maximally toxic, however, and the greatest rate of killing is obtained at 5 μg/ml. The very toxic levels prevent significant turbidmetric increase. It was convenient in subsequent tests to use the intermediate rate of killing obtained with 1 μg of bleomycin A_5 per ml. The great toxicity of A_5, at least 10-fold greater than A_2 or B_2, to Gram-negative bacteria raises the question of whether it would not be useful in the short-term treatment of urinary infections caused by some Gram-negative organisms.

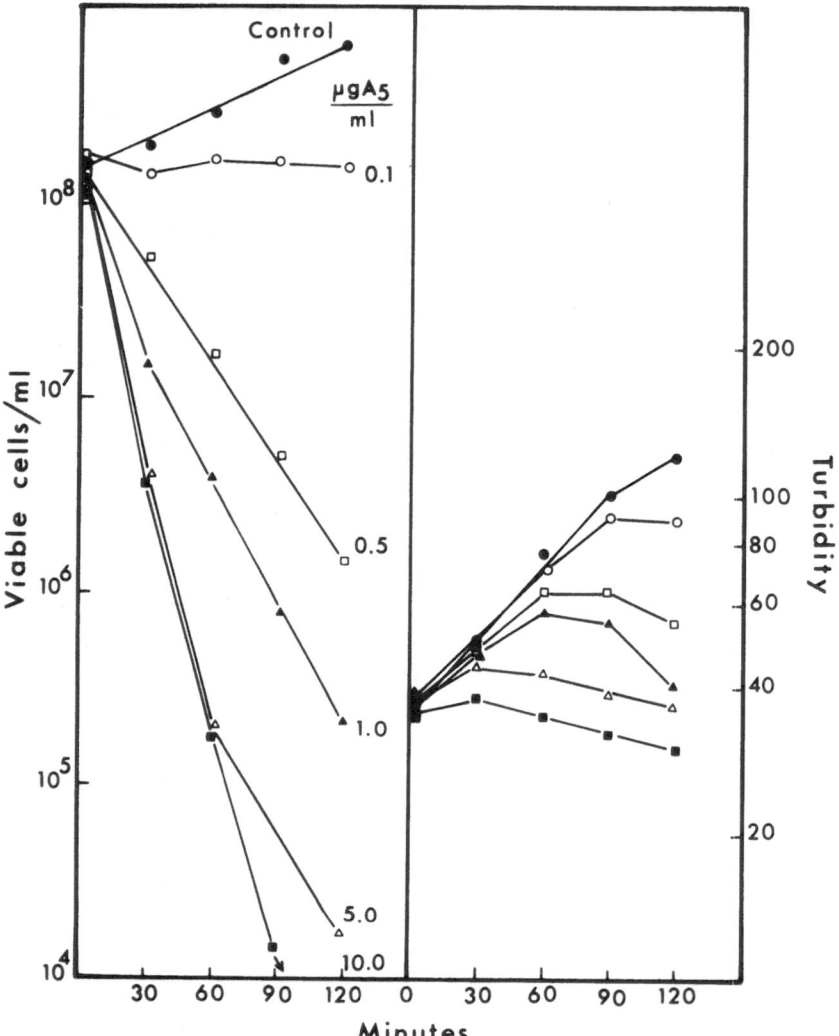

Fig. 2. The effect of various concentrations of A_5 on the viability and turbidity of *E. coli* strain 15 TAU rel A^+ in complete medium.

Attempted Protection by Spermidine

In Fig. 3 data are presented on the effects of various concentrations of spermidine on the toxicity of bleomycin–spermidine at 1 μg/ml, i.e., <1 μM. It can be seen that 0.1 and 1 mM spermidine, i.e., a >1000-fold excess in the latter instance, did not significantly affect the rate of killing by the antibiotic for about 1 hr, i.e., for a 2-log fall in viability. After this time, however, the killing was prevented. If spermidine is increased 10 mM, a concentration that did not

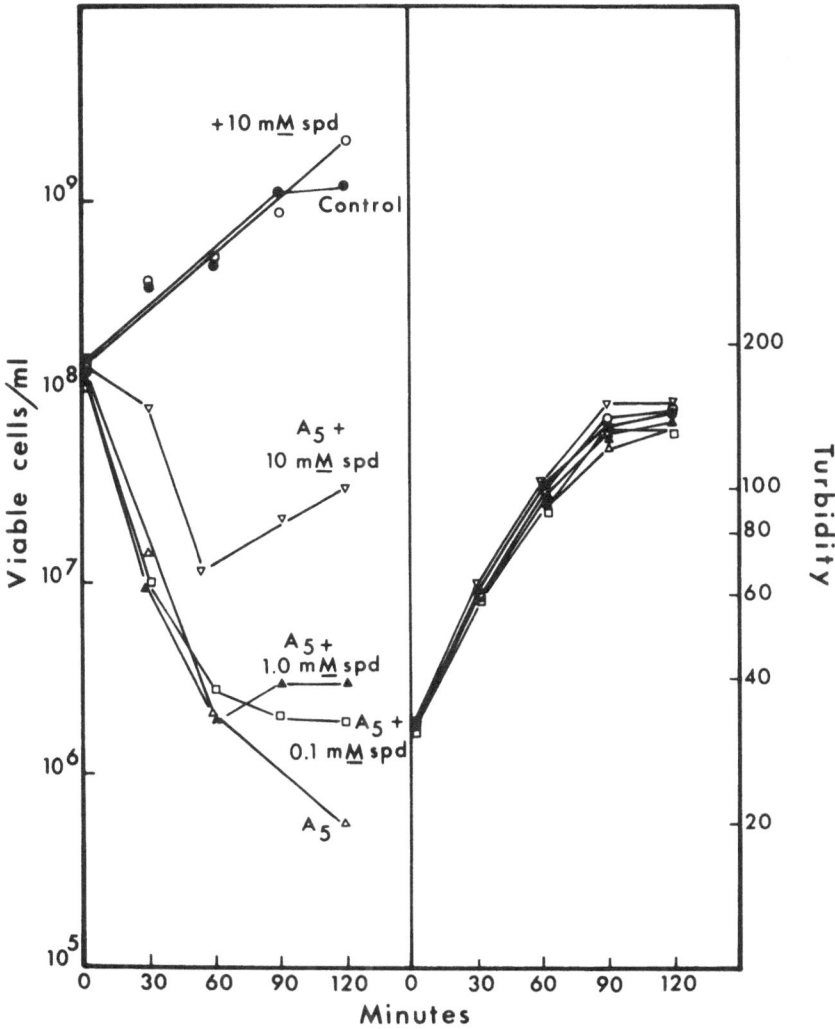

Fig. 3. The effect of spermidine on the lethality of bleomycin A_5 (1 µg/ml) to *E. coli* strain B. spd, spermidine.

inhibit growth of strain B, the inception of killing is significantly inhibited. After a fall of about 1 log, the surviving cells then went on to multiply at a rate only slightly slower than that of the control culture containing 10 mM spermidine alone. All of the cultures increased their turbidities similarly. We have concluded that in the bacterial systems tested relatively high concentrations of spermidine do not interfere with receptors for A_5 at the bacterial surface. Because 10 mM spermidine does not affect the growth of *E. coli* strain B, we think it likely that the triamine does penetrate the cell very slowly. Nevertheless some degree and

pattern of protection against killing by 10 mM spermidine suggests some interference with uptake by A_5, as well as some penetration of spermidine into the
cell and protection against DNA breakage.

Effects of Bleomycin A_5 on Polymer Synthesis

Killing by the compound occurs at similar rates in the stringent and relaxed
strains of *E. coli* in complete media. In Fig. 4 data are presented on the incorporation of [^{14}C] thymine, [^{14}C] arginine, and [^{14}C] uracil in complete media
in the presence or absence of the antibiotic. The stringent organism (15 TAU) is
slightly inhibited in DNA synthesis in the first hour, whereas the relaxed organism incorporates thymine during this interval at a rate characteristic of the
multiplying organism. The rates of arginine and uracil incorporation for at least
the first 30 min are those characteristic of the uninhibited organisms. Thus,
these rates of synthesis of DNA, RNA, and protein are virtually unchanged
despite extensive killing (>90%) of the cells. Obviously, in this system the ability
to form colonies on an agar plate is a far more sensitive parameter of the effect
of the antibiotic than that of the polymer synthesis tested.

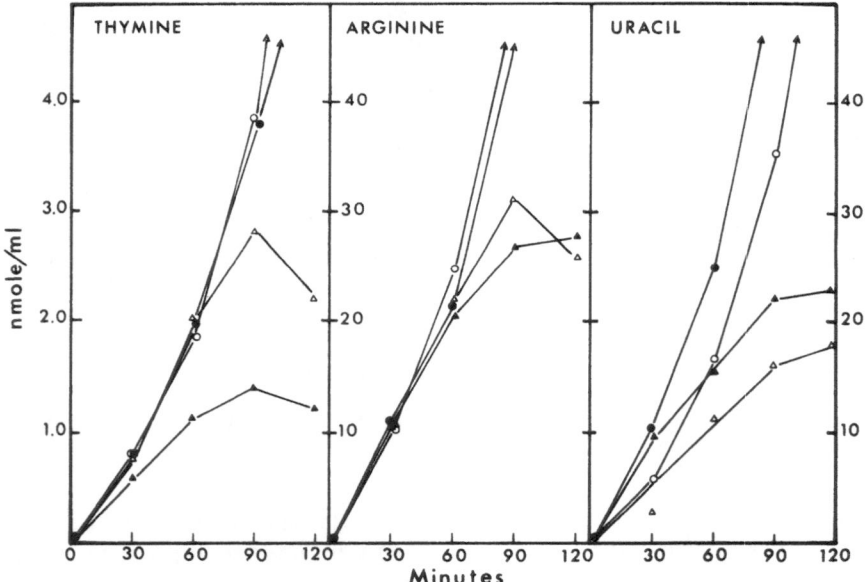

Fig. 4. Incorporation of thymine, arginine, and uracil into cultures of TAU rel
A^+ and TAU rel A in complete medium in the presence and absence of bleo-
mycin A_5 (1 μg/ml). [^{14}C] Thymine, 10^6 cpm/μmole; [^{14}C] arginine and [^{14}C]
uracil, 10^5 cpm/μmole. (●) TAU rel A^+; (▲) TAU rel A^+ + A_5; (○) TAU rel A; (△)
TAU rel A + A_5.

Dependence of Bleomycin Lethality on Polymer Synthesis

Although it appeared that the antibiotic did not have a gross quantitative effect on the synthesis of nucleic acids and proteins, it was of interest to determine whether the lethality of the antibiotic requires these syntheses. Hence, strain 15 TAU rel A^+ was exposed to bleomycin A_5 in the absence of thymine, arginine, or uracil. The lethality of thymine deficiency alone, i.e., thymineless death[18], is far slower than that provoked by the antibiotic. It was found that the antibiotic kills at a rate independent of the presence or absence of thymine, i.e., independent of the occurrence of DNA synthesis.

In the absence of uracil, however, it was observed that both the stringent and relaxed members of an isogenic pair were killed for a short time at a rate similar to that in the complete medium and that this lethality was then abruptly arrested. If cells were incubated for 30 min without exogenous uracil to reduce the pyrimidine content of the cellular pool and then given bleomycin, the antibiotic was completely ineffective (Fig. 5). Similar effects were obtained with bleomycin B_2 at lethal concentrations.

In the absence of arginine, a stringent organism was killed at a rate of one-third to one-half that in the complete medium (Figs. 6A and 7A). Since an amino acid deficiency in the stringent organism prevents the synthesis of rRNA and tRNA, relaxed organisms were also tested. Unlike the stringent organisms, TAU rel A and CP79 were killed by bleomycin in the absence of arginine at rates only slightly less than that in the presence of the amino acid (Figs. 6A and 7B). The differences in RNA synthesis in the absence of arginine in TAU rel A^+ and rel A are presented in Fig. 6B.

Effect of Chloramphenicol on Bleomycin Lethality

If RNA synthesis were the major metabolic determinant in the killing action of the drug, it could be supposed that the inhibited lethality expressed in the stringent organism, strain 15 TAU rel A^+, in the absence of protein synthesis could be markedly stimulated by chloramphenicol. This antibiotic inhibits protein synthesis and relaxes RNA synthesis organisms. In the stringent organism, chloramphenicol sharply stimulates RNA synthesis in the absence of arginine in the presence or absence of bleomycin. Nevertheless, chloramphenicol was markedly inhibitory to the lethal action of the drug in the presence or absence of arginine.

Similar results were obtained with the relaxed strain 15 TAU rel A. Chloramphenicol is much more inhibitory to lethality than is a mere arginine deficiency, although the RNA synthesis in the deprived organism with chloramphenicol is virtually identical to that in the absence of arginine alone. RNA synthesis in the presence of bleomycin, i.e., the killing condition, was observed to be identical with and without chloramphenicol in the absence of arginine, although the exponential rate of killing is decreased about 50% by chloramphenicol.

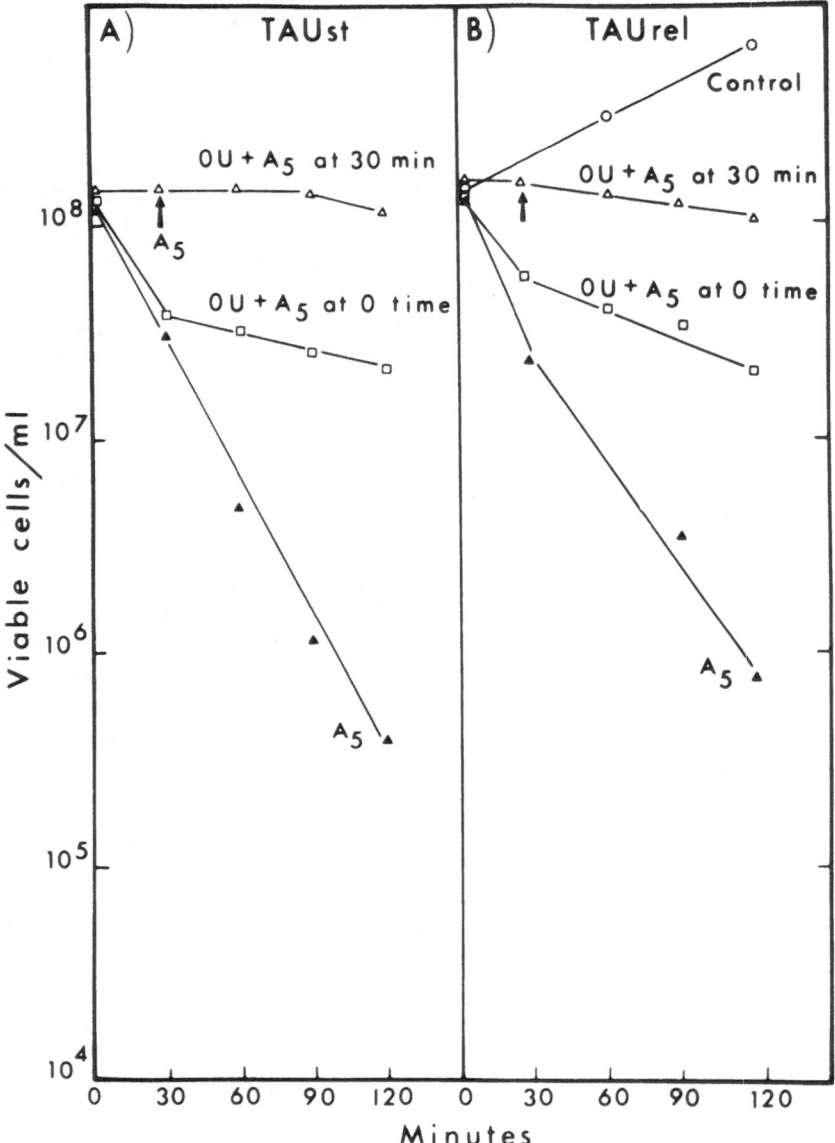

Fig. 5. The viability of the stringent and relaxed strains of *E. coli* strain 15 TAU on treatment with bleomycin in supplemented media (with thymine and arginine) in the presence or absence of exogenous uracil. One culture was incubated without uracil for 30 min before addition of bleomycin. Bleomycin A_5, 1 μg/ml. TAU st, TAU rel A^+; TAU rel, TAU rel A; OU, absence of exogenous uracil.

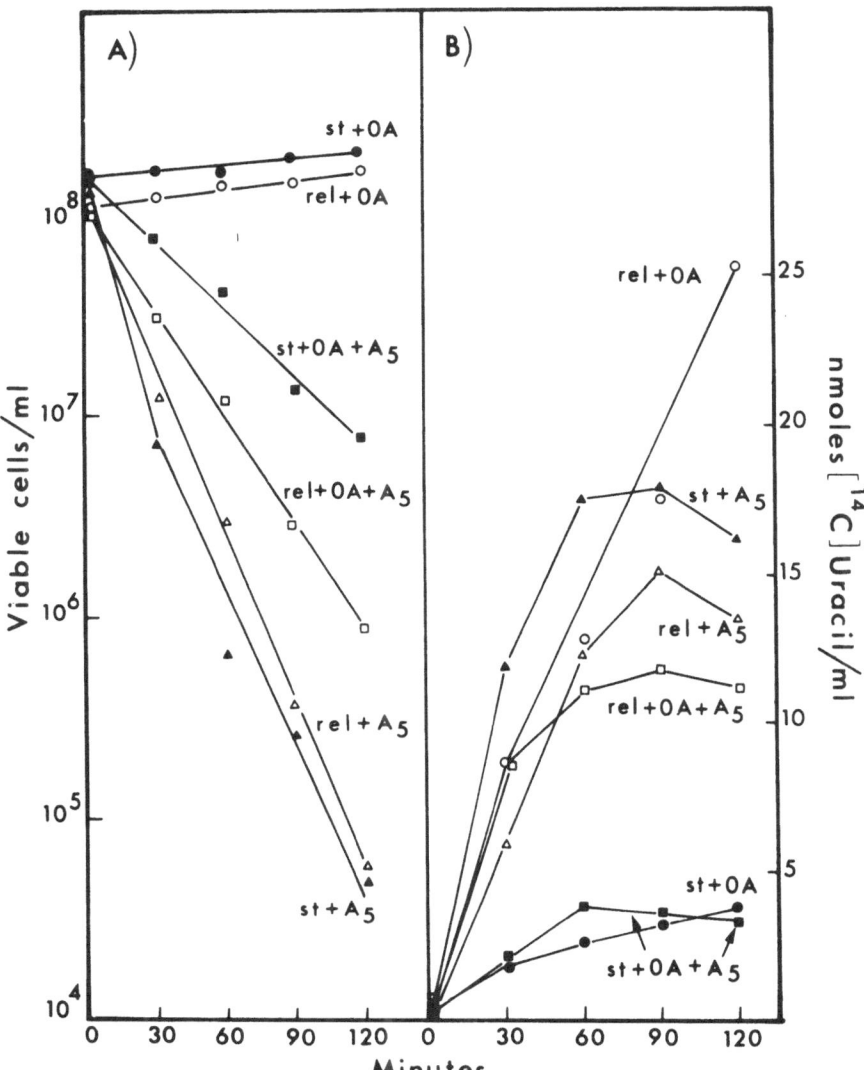

Fig. 6. The effect of arginine deprivation on viability (A) and RNA synthesis (B) in bleomycin-treated cultures of *E. coli* strains 15 TAU rel A$^+$ (st) and rel A (rel). The media contained thymine and [^{14}C]uracil (10^5 cpm/μmole). Bleomycin A$_5$, 1 μg/ml. OA, absence of exogenous arginine.

Lethality of A$_5$ to Mouse Fibroblasts

Figure 8 presents a comparison of the lethality to L cells of two concentrations of A$_5$. It can be seen that 20 μg/ml was required to produce an exponential decrease to a survival of less than 1% in viable cells in a 48-hr period. In a com-

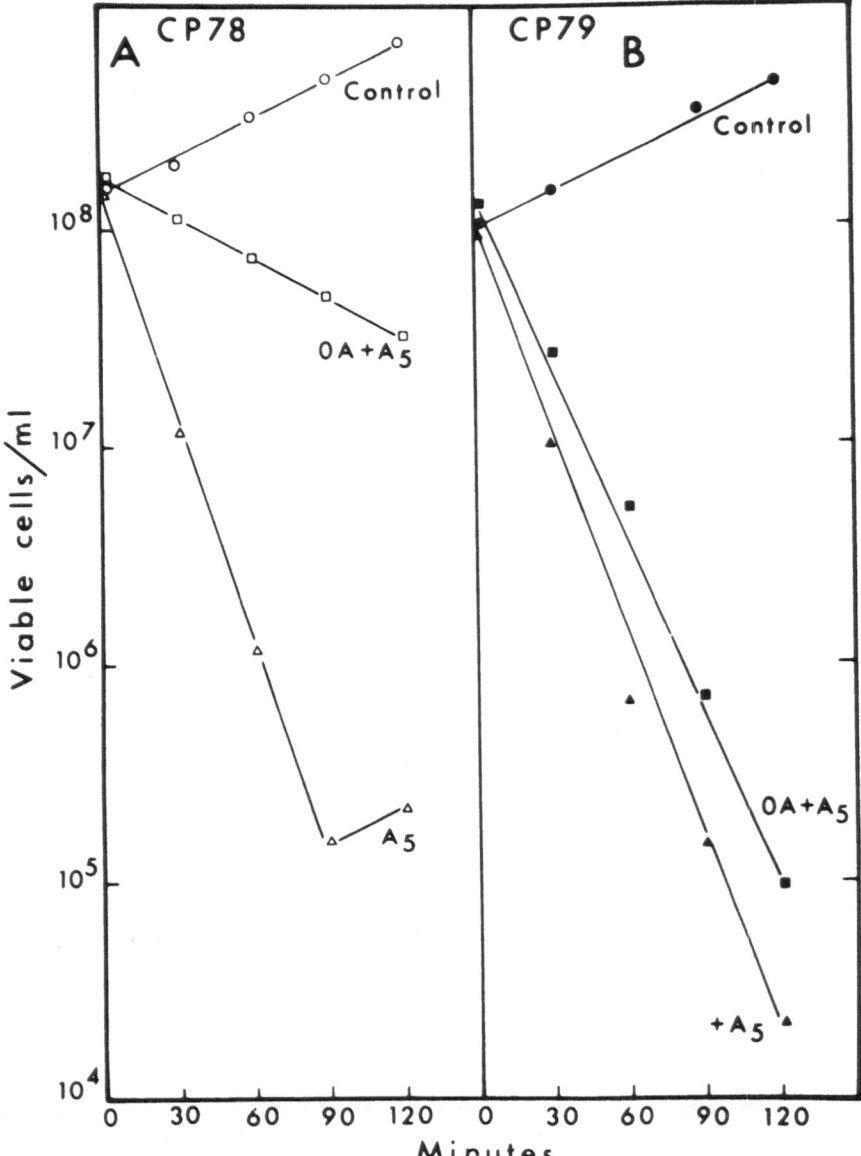

Fig. 7. The effect of arginine deprivation on the lethality of bleomycin A_5 to stringent CP78 and relaxed CP79. Bleomycin A_5, 1 μg/ml; OA, absence of exogenous arginine.

parable period of about two generation times, 1 μg A_5 per ml produced a greater than 10^{-3} drop in survival in *E. coli*. Figure 8 (inset) shows that, at 31 hr after addition of A_5, the cells were greatly enlarged, particularly in the culture containing 20 μg/ml. At 10 μg/ml there was virtually no lethality for 24 hr. At 31 hr most of the cells exposed to this concentration of drug were enlarged but to a lesser extent than those at 20 µg/ml.

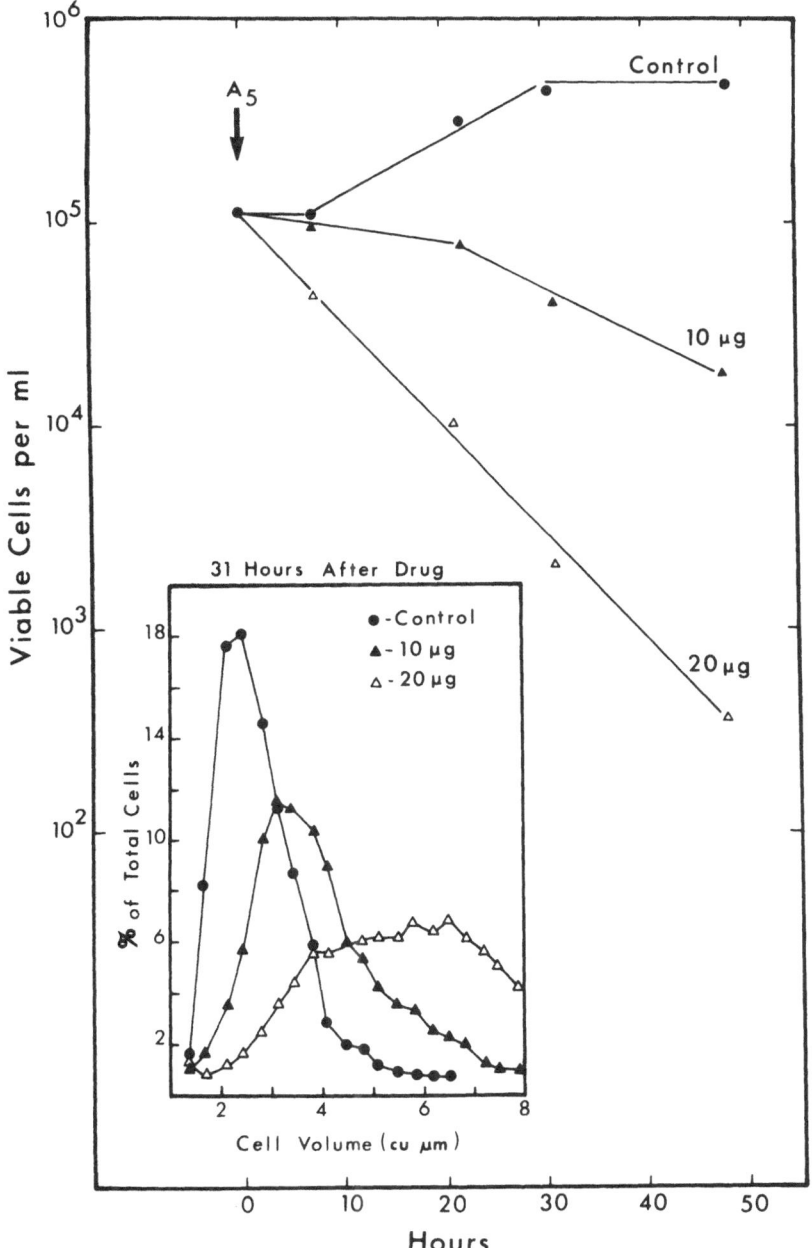

Fig. 8. The effects of A_5 on the viability of mouse fibroblasts (L cells). Inset, distribution of cell sizes of cultures 31 hr after addition of the antibiotic.

Lethality of Other Bleomycins to L Cells

Figure 9 presents a comparison of the lethality to L cells of various bleomycins at 20 μg/ml. It can be seen that at this concentration bleomycinic acid was only slightly inhibitory, whereas A_2-b was approximately as inhibitory as was A_5. As shown earlier in tests on *E. coli*, A_2-b was far less toxic than was A_5.

A_2 and B_2 were consistently less lethal than was A_5; the former antibiotics produced perhaps one-tenth as much cell kill as did A_5. These differences among effects of bleomycins on L cells, although significant, are far less so than in *E. coli*. B_2 was consistently more toxic in this system than was A_2; in the mixture used clinically, A_2 is frequently present in amounts 1.5 to 2 times that of B_2.

In Fig. 9 (inset) it can be seen that B_2 and A_5 affected cell size similarly, whereas A_2 was intermediate in its effect. Bleomycinic acid permitted an exponential increase in cell number for at least a full day. A_2 frequently permitted such an increase for 18 hr, followed by an abrupt arrest in multiplication and cell death. In contrast to those effects, A_5, A_2-b, and B_2 produced an immediate and very considerable inhibition of the increase in cell number. The intermediate sizes of cells treated with A_2 for 31 hr seemed to reflect a later inhibition of division by this drug. We think that it would be desirable to compare the effects of A_2 and B_2 more closely, to determine if the differences we have observed reflect significant differences in the action of these components of the mixture used clinically.

Attempted Prevention of Bleomycin Toxicity by Polyamines

In studies with A_2-b at 20 μg/ml, which caused a 2-log drop in viability in 48 hr, neither 10^{-2} M putrescine nor 10^{-2} M spermidine decreased the cell kill significantly in this time.

In studies with A_5 at 20 μg/ml, which also produced a 2-log drop in 48 hr, 10^{-2} M spermidine added to A_5 caused a lag of 24 hr in cell kill, followed by a precipitous killing to the level of A_5 alone. A similar result had been seen in a temporary protection by high concentrations of spermidine against A_5 with *E. coli*. In studies with the L cell system, the size distributions of cells treated with A_5 and cells treated simultaneously with A_5 plus spermidine were similarly heterodisperse at 24 and 48 hr. A brief protection against A_5 by 10^{-2} M spermidine (data not shown) had not prevented cell enlargement, i.e., synthesis and cell growth.

Inhibition of Nucleic Acid Synthesis and A₅ Lethality

In our studies on *E. coli*, we had observed that the lethality of A_5 was expressed without significantly affecting the rates of synthesis of DNA, RNA, and protein. At 20 μg/ml, A_5 has been reported to reduce DNA synthesis in mouse fibroblasts about 75% in 2 hr and to inhibit RNA and protein synthesis about 10 and 20%, respectively[19]. In our studies on the effect of A_5 (20 μg/ml) on [^{3}H] thy-

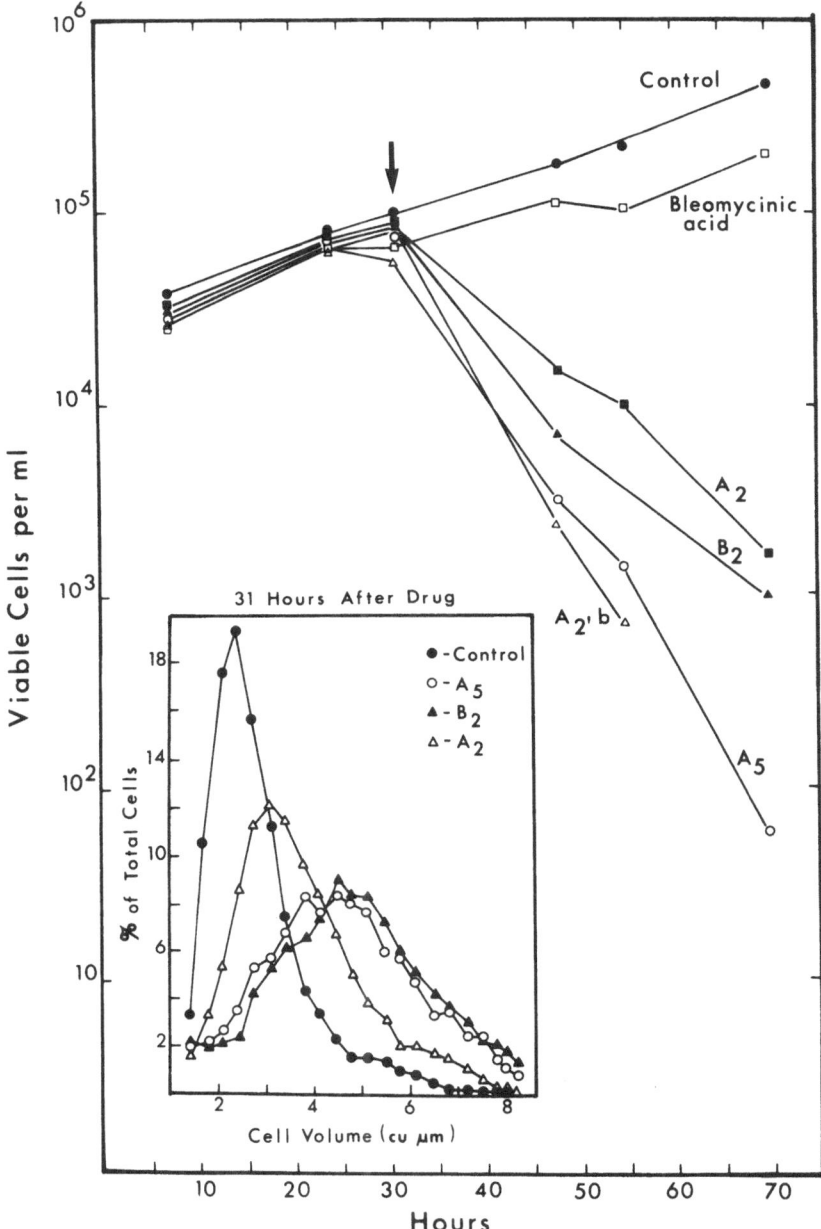

Fig. 9. The effects of various bleomycins on the viability of L cells. Various compounds were added at 20 μg/ml. Inset, distribution of cell sizes of cultures 31 hr after addition of the antibiotic.

midine and [^{14}C]uridine incorporation on L cells (3.7×10^5 cells/ml), no differences were found in these incorporations over a 4-hr period in control and treated cultures. After 24 hr, the A_5-treated culture was found to incorporate thymidine at a rate similar to the initial culture, unlike the control in which the rate of thymidine incorporation had doubled. After 24 hr, however, the A_5-treated culture had almost doubled its rate of uridine incorporation.

The lethal effects of A_5 and B_2 (J. I and S. S. Cohen, unpublished data) required RNA synthesis in *E. coli*. We have attempted to see whether this surprising result was also obtained with L cells. Despite the large amount of work on the inhibition of RNA synthesis in mammalian cells, few attempts have been made to determine whether cells inhibited in RNA synthesis were also killed. For example, actinomycin-D is an excellent inhibitor of RNA synthesis, but is also lethal under the conditions of prolonged incubation necessary to explore the lethality of bleomycin. Among the few compounds that appeared to be capable of inhibition of RNA synthesis without extensive killing were 3-deazauridine[20], 3-deazacytidine[21], and 6-azauridine; the effects of these agents on RNA synthesis are reversed by natural ribosyl nucleosides, such as uridine and cytidine. This reversal is particularly effective in the presence of deoxycytidine, which appears to prevent the lethal effects accompanying an inhibition of DNA synthesis.

At 100 µg of 3-deazacytidine per ml, cell number increased 20% in 24 hr and fell to 80% of the original cell number at 48 hr. The addition of 100 µg of deoxycytidine per ml to 3-deazacytidine at this concentration permitted cell multiplication of one-half the rate of the control. In any case neither of these inhibitor combinations inhibited the course of A_5 killing over a 33-hr interval.

This result led us to test 6-azauridine with or without deoxycytidine as an inhibitor of nucleic acid synthesis. It was found that, after a few hours of treatment, 10 µg of 6-azauridine per ml totally inhibited the increase of cell number and viability for 24 hr without killing. The cells began to die slowly 32 hr after addition of this agent. In the presence of 6-azauridine plus 40 µg of deoxycytidine per ml, the total number of cells as well as the number of viable cells increased slowly. Nevertheless, as in the experiments with 3-deazauridine, neither treatment markedly affected the lethality of A_5.

In short-term studies on uptake of [8–^{14}C]adenine into the nucleic acids of L cells, it was found that 6-azauridine produced its maximal inhibition (60%) after 3 hr. On the other hand, in the presence of deoxycytidine, the uptake of adenine into nucleic acids was normal for at least 7 hr.

Accordingly, these experiments on bleomycin killing in the presence of 6-azauridine were repeated with the addition of A_5 5 hr after inhibition with 6-azauridine. The addition of 6-azauridine to A_5 permitted the same rate of cell kill whether added before the antibiotic or at the same time. Preincubation with deoxycytidine and 6-azauridine did not significantly affect the rate of cell kill after addition of A_5. These results indicate that, contrary to results with *E. coli*, extensive inhibition of RNA synthesis did not hinder the lethal action of A_5.

Prevention of the Lethality of A_5 by Hirudonine

It was found that 10^{-4} M hirudonine was partially effective in blocking A_5 at 20 μg/ml. This was in marked contrast to the inefficacy of 10^{-2} M spermidine in blocking the toxicity of A_5. It can be seen in Fig. 10 that 10^{-4} M hirudonine was nontoxic, whereas 10^{-3} M hirudonine slowed the growth rate slightly. In the presence of A_5, 10^{-4} M hirudonine protected the cells considerably, whereas at 10^{-3} M the compound was reproducibly somewhat less protective. As presented in Fig. 10, hirudonine enabled cultures containing A_5 to increase in cell number at a significant rate.

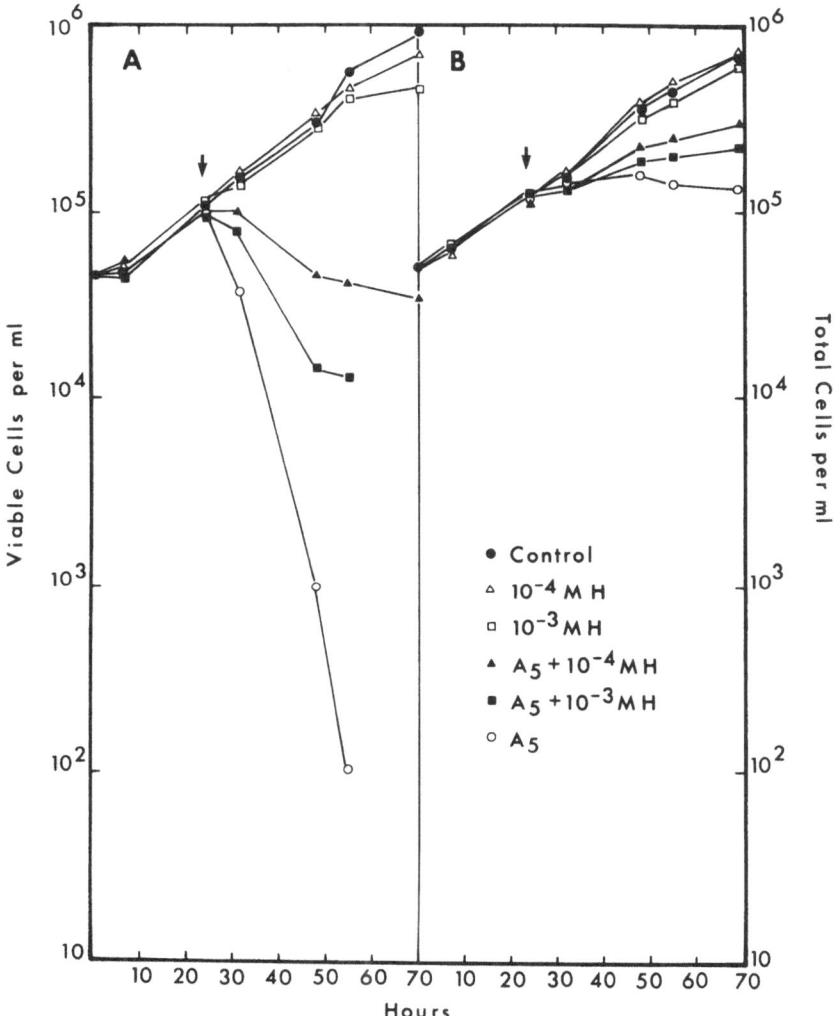

Fig. 10. The protection of L cells by hirudonine (H) against the lethal action of A_5 (20 μg/ml). (A) Cell survival; (B) increase of cell number.

These effects were also obtained in the distribution of cell sizes. Hirudonine $(10^{-3}\ M)$ alone affected cell size somewhat at 24 hr after addition of the compound, whereas A_5 alone changed the size distribution greatly. With $10^{-4}\ M$ hirudonine, a large number of cells remained in the normal size range in the presence of A_5.

Specificity of Hirudonine Protection

Aliquots of the same exponentially growing cultures were challenged by three different bleomycins, A_5, A_2, and B_2, at 20 μg/ml in the presence or absence of $10^{-4}\ M$ hirudonine or arcaine. It can be seen in Fig. 11 that hirudonine protected against A_5 but not against A_2 or B_2. Arcaine $(10^{-4}\ M)$ protected somewhat against A_5 (Fig. 11A) but not nearly as well as hirudonine. However, arcaine was ineffective against A_2 (Fig. 11B) or B_2 (Fig. 11C).

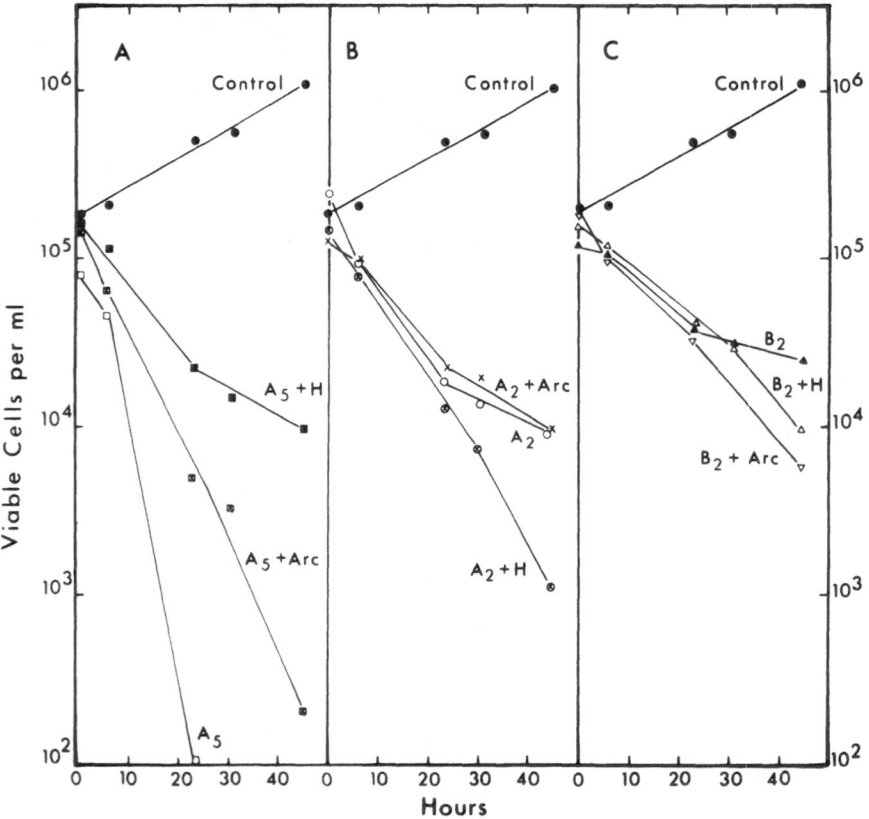

Fig. 11. The specificity of hirudonine in minimizing the lethality of A_5. H, hirudonine at $10^{-4}\ M$; Arc, arcaine at $10^{-4}\ M$; A_2, B_2, and A_5 bleomycins tested at 20 μg/ml. (A), (B), and (C), effects of A_5, A_2, and B_2, respectively, in the presence or absence of the diamidinopolyamines.

Discussion

The comparison of the five relatively pure species of bleomycins, including the bleomycinic acid, presented in this chapter clearly demonstrates the markedly greater toxicity of bleomycin containing spermidine, i.e., A_5, over the bleomycins containing the bicationic derivatives, i.e., A_2, B_2, and A_2-b. This result had been stated earlier[6], and the spermine derivative, bleomycin A_6, is said to be even more toxic. In my view it will be important to have a presentation of a direct comparison of the various pure species of bleomycins in bacteria and in animal cells, even as described in this chapter.

The carboxyl derivative appears to have very much less activity than any amide, but it is not certain that it is totally lacking in activity. It has been reported[22], albeit somewhat lacking in details, that bleomycinic acid amide, which lacks a terminal cation, is an active compound. This result suggests that the presence of numerous amino groups on the polyamine-free glycopeptide may be sufficient to permit some binding of this moiety to DNA. It will then be desirable to attempt a careful comparison of the effects of pure bleomycinic acid, bleomycinic acid amide, and bleomycin A_2-b on both a sensitive strain of *E. coli* and on DNA.

In any case, it appears that bleomycin A_5 is far more toxic to both *E. coli* and mouse fibroblasts than the commercial mixture comprised largely of A_2 and B_2. The fact that the material in clinical use is a mixture suggests that differences from batch to batch will be common, as has in fact been reported[23]. Furthermore, it will be difficult to relate data on the mode of action of the antibiotics in model systems to their clinical use. It may be suggested that an effort be made to employ single species of the antibiotic for tests on its mode of action, as well as for efficacy in particular clinical entities.

Which species of bleomycin should be used in the clinic? If bleomycin A_5 is many times more active than A_2 or B_2, A_5 seems a likely candidate. Our Japanese colleagues have reported[6] that it is quite easy to produce good yields of bleomycin A_5 by using bleomycinic acid and spermidine or spermine in the fermentation medium. The fact that a cheap synthetic inhibitor of the toxicity of bleomycin A_5, i.e., hirudonine, is readily available should make this suggestion more attractive. The new bleomycin isolated by Bristol Laboratories as a possible competitor to the Japanese mixture is also a spermidine derivative and it will be important to learn if the toxicity of this spermidine derivative can also be prevented by hirudonine.

As is well known, the utility of bleomycins is limited by their cumulative toxicity to lung tissue. It will be particularly important to know if a local administration of hirudonine to lung tissue can be used to prevent the toxicity of bleomycin A_5 or of the Bristol product.

Acknowledgments

These studies were supported by USPHS Grant A1-10424 and were done in collaboration with Mrs. Joseph I and Miss Lillie Lapi at the University of Colorado Medical Center, Denver, Colorado.

References

1. S. S. Cohen, *Introduction to the Polyamines*, Prentice-Hall, Inc., Englewood Cliffs, N.J. (1971).
2. S. S. Cohen, In *Advances in Polyamine Research*, Vol. I, p. 1. Raven Press, New York (1978).
3. S. S. Cohen, S. Morgan, and E. Streibel, *Proc. Nat. Acad. Sci. USA*, **64**, 669 (1969).
4. T. T. Sakai, R. Torget, J. I, C. E. Freda, and S. S. Cohen, *Nucleic Acids Res.*, **2**, 1005 (1975).
5. T. T. Sakai and S. S. Cohen, *Prog. Nucleic Acid Res. Mol. Biol.*, **17**, 15 (1976).
6. H. Umezawa, *Biomedicine*, **18**, 459 (1973).
7. T. Terasima, M. Yasukawa, and H. Umezawa, *Gann*, **61**, 513 (1970).
8. I. Shirakawa, M. Azegami, S. Ishii, and H. Umezawa, *J. Antibiot. (Tokyo)*, Ser. A., **24**, 761 (1971).
9. S. S. Cohen and J. I, *Cancer Res.*, **36**, 2768 (1976).
10. L. Lapi and S. S. Cohen, *Cancer Res.*, **37**, 1384 (1977).
11. V. Robin and N. van Thoai, *Compt. Rend.*, **252**, 1224 (1961).
12. S. S. Cohen and R. Arbogast, *J. Exp. Med.*, **91**, 619 (1950).
13. S. S. Cohen, N. Hoffner, M. Jansen, M. Moore, and A. Raina, *Proc. Nat. Acad. Sci. USA*, **57**, 721 (1967).
14. I. Fukuma and S. S. Cohen, *J. Virol.*, **12**, 1259 (1973).
15. A. Doering, J. Keller, and S. S. Cohen, *Cancer Res.*, **26**, 2444 (1966).
16. W. Plunkett and S. S. Cohen, *Cancer Res.*, **35**, 1547 (1975).
17. W. Plunkett and S. S. Cohen, *Cancer Res.*, **35**, 415 (1975).
18. D. Kanazir, H. D. Barner, J. G. Flaks, and S. S. Cohen, *Biochim. Biophys. Acta*, **34**, 341 (1959).
19. M. Watanabe, Y. Takabe, and T. Katsumata, *J. Antibiot. (Tokyo)*, Ser. A., **26**, 417 (1973).
20. W. M. Shannon, R. W. Brockman, L. Westbrook, S. Shaddix, and F. M. Schabel, *J. Nat. Cancer Inst.*, **52**, 199 (1974).
21. G. P. Khare, R. W. Sidwell, J. H. Huffman, R. L. Tolman, and R. K. Robins, *Proc. Soc. Exp. Biol. Med.*, **140**, 880 (1977).
22. H. Umezawa, *Lloydia*, **40**, 67 (1977).
23. T. Ohnuma, J. F. Holland, H. Masuda, J. A. Waligunda, and G. A. Goldberg, *Cancer*, **33**, 1234 (1974).

Cytotoxicity of Various Bleomycins to Cultured Mammalian Cells

ADRIAN D. NUNN AND JOHN LUNEC

The successful use of bleomycins as chemotherapeutic agents appears to be dependent upon their metal-binding capacities, but the area of influence, whether metal binding affects their *in vivo* distribution, stability, and/or activity, is as yet unclear. Much work has been done in the field of nuclear medicine to determine the utility of metal–bleomycin complexes, incorporating radioactive metals as tumor-visualizing agents. From this work it has been found that bleomycin can affect the *in vivo* distribution of a metal, that the excretion of the metal complexes appears to be similar to the metal-free bleomycin, and that the individual metals form complexes with bleomycin of widely differing stabilities. Of the metal complexes investigated the cobalt complex has been found to have the highest stability and the greatest activity for the tumor tissue (when measured as the concentration of radioactive cobalt). Unfortunately for nuclear medicine the physical characteristics of the best available cobalt radionuclide, cobalt-57, are not ideal for use as a tumor-imaging radionuclide.

In the chemotherapeutic field the involvement of metals has been recognized from the early stages but the nature of the involvement on the reported attack on DNA by bleomycin has until recently been confused. In initial *in vitro* experiments[1], DNA attack was observed only after the addition of sulfhydryl compounds to the reaction mixture. More sensitive assay procedures were subsequently used[2] to show that DNA attack occurred in the absence of sulfhydryl compounds indicating that such compounds merely enhanced DNA attack. It has been shown that under aerobic conditions Fe^{2+} enhances bleomycin attack on DNA in the absence of sulfhydryl compounds and we have heard of various mechanisms[3, 4].

Enhancement of attack has also been demonstrated for hydrogen peroxide[5] and the superoxide ion[6], but the addition of EDTA was reported to have no effect[7] or inhibited attack even in the presence of sulfhydryl compounds[8]. The enhancement of attack of naked DNA produced by sulfhydryl compounds has been found to be inhibited by Cu^{2+}, Co^{2+}, and Zn^{2+}[7].

It has been suggested that the role of the sulfhydryl compounds is to compete with the bleomycin for metals that would normally interact with it and inhibit the attack on DNA[7, 9].

In the clinical situation it was thought that copper would be the metal most likely to inhibit the effectiveness of bleomycin due to its concentration in the blood. Preliminary results with patients of a protocol designed to lower serum copper levels during bleomycin therapy are encouraging but assessment is incomplete[10].

As the oxidation of cobalt(II) amines to cobalt(III) amines frequently involves the formation of a superoxide-containing intermediate, and as it has been shown that cobalt(II)–bleomycin is rapidly oxidized to cobalt(III)–bleomycin, it was thought that it might be advantageous to study the effect of a possible superoxide-containing compound on cell viability and DNA damage.

In order to understand the effect of metal binding and to improve treatment with bleomycin, we decided to determine the effect of preformed metal–bleomycin complexes on the colony-forming ability of cultured mammalian cells. We also compared DNA damage with cell viability in the same groups of cells to see if there was any direct correlation between DNA damage and cell viability. As we were interested in trying to relate the results to the clinical situation, we used the mixture of bleomycin for injection rather than any of the pure, separated fractions.

Metal Complexes

The cobalt(III) and copper(II) complexes of bleomycin (BLM) were prepared by adding equimolar amounts of the bivalent metal salt in water to the bleomycin dissolved in water and were characterized by published methods[11–13]. The metal complexes were made just prior to use and were adjusted to pH 7.4 before adding to the cells. The cobalt(II) complex is known to be rapidly oxidized to a cobalt(III) complex; it is also known that it is the Co(III)–BLM complex that localizes in tumors (Fig. 1). Over a period of days the initially formed Co(III) complex undergoes hydrolysis to the final product unless the solution is adjusted to pH 7.4 when the reaction is rapid. The spectra of the aged and neutralized solutions are identical.

Cells

Chinese hamster V79 cells were chosen for the study and the line used has a doubling time of 12 hr in Eagle's minimum essential medium supplemented with Earle's salts and 15% fetal calf serum. The pH of the medium during subculture and colony growth was controlled by adding 2.2 g liter^{-1} of sodium bicarbonate and maintaining a 5% CO_2 atmosphere in the 37°C incubator.

Cells to be treated with the bleomycin were plated 2 days before exposure to yield approximately 5×10^5 cells per dish at the time of exposure. After 1 day, half of the cell samples were radioactively labeled with 0.02 μCi·ml^{-1} of [^{3}H] thymidine (40–60 Ci mmole^{-1}) by exposure for 15–16 hr. At the end of

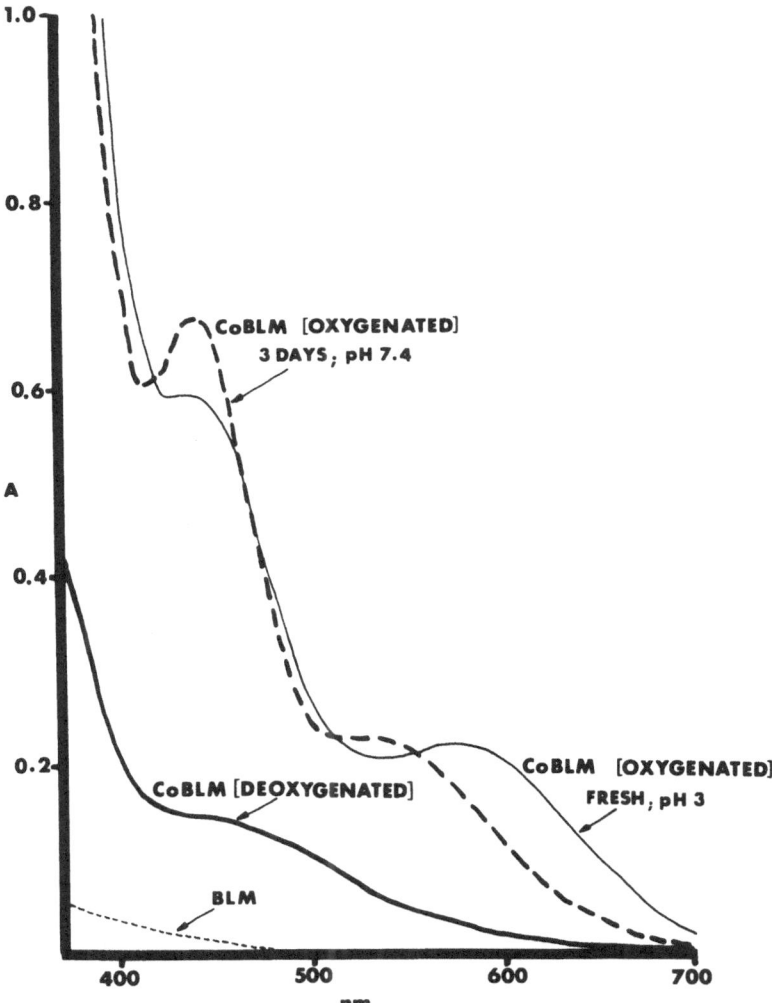

Fig. 1. Visible spectra of an equimolar solution of CO^{2+} and BLM in deoxygenated water, oxygenated water, pH 3, and the same solution 3 days later or neutralized and the spectrum taken immediately.

the labeling period the medium was aspirated out of the culture dishes and each cell sample was washed twice with 5 ml of label-free medium. The cells were then grown for 1 hr in medium free of [³H] thymidine prior to treatment with bleomycin. The same washing procedure was carried out on the unlabeled cells.

Small volumes of the bleomycin solution were added to the medium above the cells and mixed by gentle rocking. The cells were exposed for 1 hr while still attached and while in the incubator.

At the end of treatment the bleomycin-containing medium was removed from the culture dishes and each cell sample was washed once with 3 ml of fresh medium and once with 2.5 ml of 0.1% tryspin in 0.9% NaCl. The cells were then incubated for 10–20 min at 37°C with 0.3 ml of the trypsin solution (to detach the cells from the surface), and finally suspended in 5 ml of culture medium.

The labeled cells were used to assay for DNA damage and the unlabeled cells were diluted and replated to assess their clonogenic ability. A correction was made for the plating efficiency.

DNA Damage Assay

The hydroxyapatite chromatography system for determining the proportion of single- and double-stranded DNA (and hence the DNA damage) was used because it is suited to large numbers of samples. The method of treating the cells to unwind the DNA was similar to that described in detail by Rydberg[14] and has been published[15].

Initial experiments were performed on untreated cells to determine the optimum amount of [3H]thymidine required to label the DNA, and the time of lysis. In agreement with Rydberg[14] the rate of unwinding was found to be linear with time (Fig. 2). Thus the sensitivity of the system can be easily adjusted by altering the time of unwinding.

The degree of labeling with [3H]thymidine also has an effect on the sensitivity of the system as the DNA receives a radiation dose from the decay of the incorporated tritium calculated to be $1-2$ rads $\circ$ hr^{-1} nucleus^{-1} at the end of the labeling period with 0.02 μCi [3H]thymidine ml^{-1}.

It has been shown that the rate of unwinding and hence the fraction of double-stranded DNA observed is dependent on the cell cycle[14]. Since different labeling procedures may result in varying proportions of labeled cells in the different phases of the cell cycle, the labeling conditions were kept as constant as possible from run to run. The percentage of counts in the double-stranded fraction of the untreated cells was 93.6 ± 2.04 ($n=10$) and so no effect of the labeling conditions on the rate of unwinding was observed. The cell survival and DNA damage studies were performed in parallel; each study was performed on three or more occasions for each bleomycin formulation.

DNA Damage Results

The results are plotted as the percentage of counts in the double-stranded fraction related to the bleomycin concentration for each of the three formulations. For clarity the values for the untreated cells have been normalized to 95%. With the experimental condition employed significant deviation from the background DNA damage in untreated cells was not detectable until concentrations

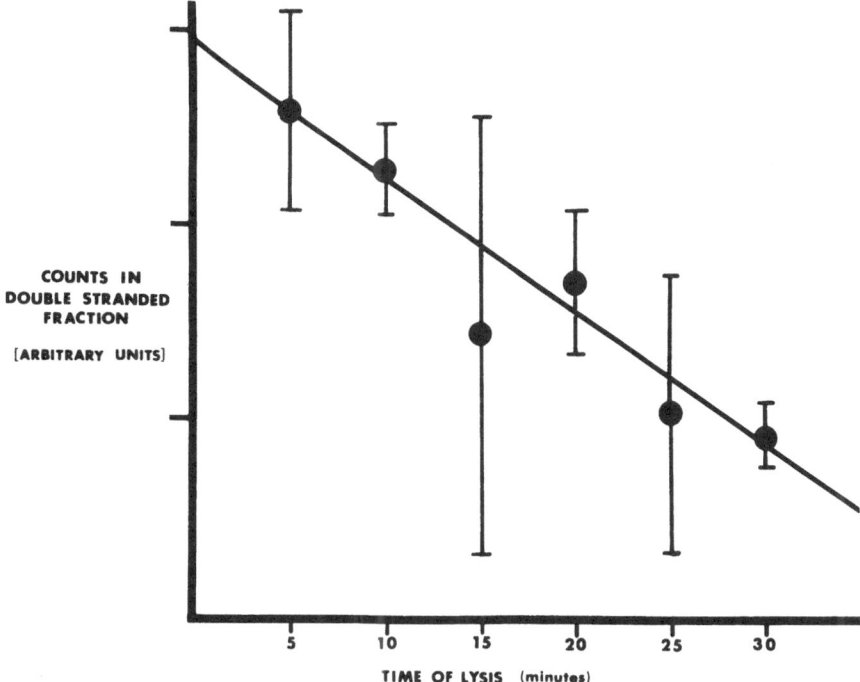

Fig. 2. Rate of unwinding of DNA during lysis: 2.0×10^{-2} μCi·ml^{-1} [^{3}H] thymidine, 10^5 cells. No bleomycin.

of approximately 0.1 unit bleomycin ml^{-1} were employed. This may be due to repair of DNA damage or may be due to a lack of sensitivity of the system. The latter could be increased at the expense of losing accuracy at higher doses by altering the lysing conditions[14].

Extrapolating from the *in vitro* work with naked DNA one would expect that there would be no attack of the DNA by the cobalt–bleomycin. This is indeed the case as seen in Fig. 3. Surprisingly there is a significant and reproducible increase in the double-stranded fraction (implying less DNA damage) compared with the control cells as the concentration of the cobalt–bleomycin is increased. It has been demonstrated on numerous occasions that ^{57}Co–bleomycin has a high affinity for DNA[9, 16, 17], higher than other metal–bleomycin complexes, and this is the basis for its use in nuclear medicine. The uptake of [^{14}C]BLM into tumor cells is increased when chelated with cobalt and parallels the uptake of cobalt. In addition the distribution within the cells is reported to be different (being associated with the DNA) than a variety of other metal complexes tested. The bond between cobalt–bleomycin and DNA is reported to be unstable under the alkaline conditions encountered in density gradient analysis[17], and it has been suggested that this is due to unwinding of the DNA. However, it appears

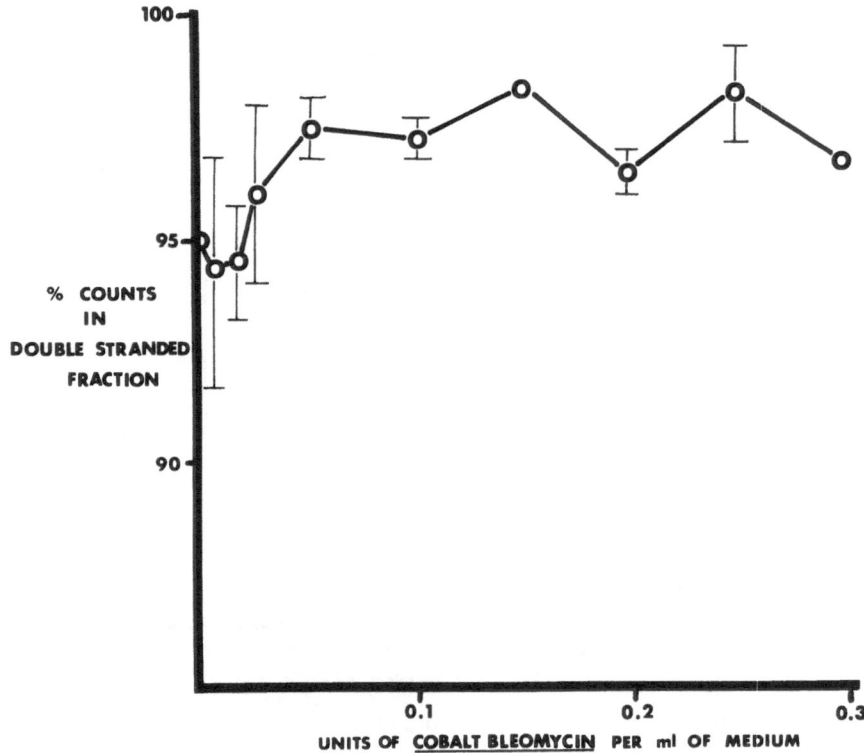

Fig. 3. Effect of various concentrations of cobalt(III) bleomycin on the proportion of single- and double-stranded DNA in Chinese hamster V79 cells.

that under conditions used here, less alkaline, a lower temperature, and shorter
time, the cobalt–bleomycin prevents unwinding of the DNA. Alternatively as
cobalt–bleomycin is known to bind to single-stranded DNA[17], it may be that
renaturation of the DNA during the experimental procedure which is negligible
normally[14] may be accelerated.

Turning to the metal-free and copper–bleomycin results (Fig. 4), it can be seen
that the curves obtained for each formulation are similar. This is not what would
be expected from extrapolating the *in vitro* naked DNA results where the cupric
ion inhibits DNA attack by bleomycin. It does, however, agree with the report
by Takahashi *et al.*[18] that copper is rapidly removed from the bleomycin complex *in vivo* and that DNA fragmentation by metal-free or copper–bleomycin is
similar.

Apart from the peculiar affinity for DNA that the cobalt–bleomycin complex
possesses, the main difference between the two metal complexes used here,
indeed between the cobalt complex and all the other metals examined, is the
high stability of the cobalt–bleomycin complex which is a result of the oxidation of the initially formed cobalt(II) bleomycin to cobalt(III)–bleomycin[11,13].

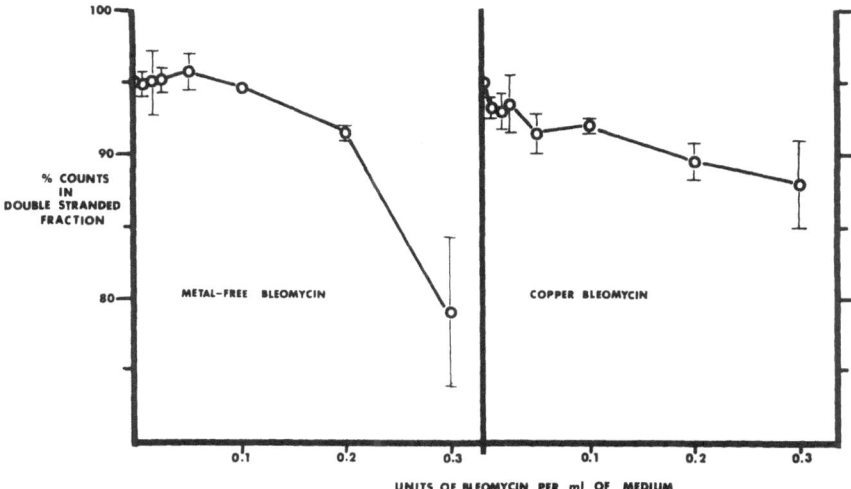

Fig. 4. DNA damage caused by metal-free and copper(III) bleomycin.

Of course due to the different coordination characteristics of the metals, there may be differences between the three-dimensional structures of the different metal–bleomycin complexes.

Cell Survival

Figure 5 shows the survival of the cells after treatment for 1 hr with the different bleomycin formulations expressed as percentage survival compared to the untreated controls. There was no correlation between the response to the bleomycin

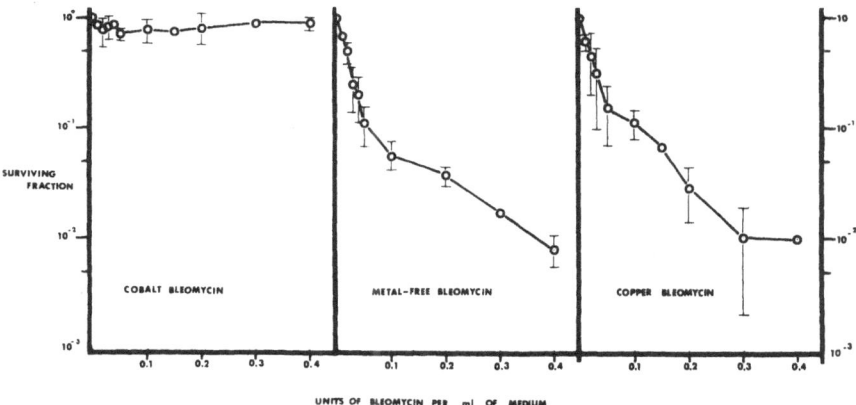

Fig. 5. Effect of Co(III) BLM, metal-free BLM, and Cu(II) BLM on the survival of Chinese hamster V79 cells.

treatment and the plating efficiency which varied between 50 and 75% for the untreated cells. The sensitivity of the cells to metal-free bleomycin is similar to that reported previously [19, 20], with the familiar rapid initial response up to approximately 0.05 units ml^{-1} followed by a reduced rate at higher concentrations.

As expected from the *in vitro* DNA damage results reported previously and the *in vivo* DNA damage reported here there is a negligible drop in cell survival produced by cobalt–bleomycin. In the light of recent experience the striking similarity between the metal-free and copper–bleomycin curves is not surprising.

Application to the Clinical Situation

These results indicate that the idea behind the protocol developed to reduce the serum copper levels in patients on bleomycin therapy was erroneous as it is apparent that copper does not alter the DNA damage or cell death produced by bleomycin. However, there were some rather interesting findings from the study. First, some of the patients (who had Hodgkin's disease and so had elevated serum copper levels) experienced a reduction in some of the symptoms during the week prior to the start of the bleomycin therapy when their serum copper levels were being diminished by a regime of increasing penicillamine therapy. In some cases this improvement increased when bleomycin therapy was started. Second, parallel experiments in animals to determine the toxicity of the pencillamine/bleomycin combination showed that the combination was extremely toxic to the bone marrow, more so than the sum of the constituents (Preece, personal communication). An almost complete lack of bone marrow toxicity is one thing that sets bleomycin apart from other cytotoxic agents. It thus appears that copper in some way protects the bone marrow from the toxic effects of the bleomycin. In relation to this it is interesting that of all the tissues in the body, the concentration of copper in the marrow is second highest only to that in the kidney, and the concentration in the skin and lungs, bleomycin-sensitive tissues, is less than half that in the marrow [21]. Furthermore, it has been shown in leukemias that there is a direct relationship between serum copper levels and the status of the bone marrow [22] and this has been proposed as a method of staging Hodgkin's disease. Bleomycin quantitatively increases the urinary excretion of copper when injected intravenously.

These observations suggest that in the clinical situation copper may be important even though it does not have any effect on DNA attack or cell viability in tissue culture. This may arise because the formation of the copper chelate in the blood effects the subsequent distribution of the bleomycin. Although a radioactive copper–bleomycin combination has proven disappointing as a tumor-localizing agent – probably because of exchange of the copper – copper injected as the citrate does have an affinity for tumor tissue notably in the lung [23]. Alternatively the bleomycin may be reacting *in vivo* with some other metals that affect its distribution or activity. As it has been shown that Fe^{2+} enhances the attack of DNA by bleomycin, this metal ion is an obvious candidate. Unfortunately the situation is more complex because bleomycin is known to affect the serum concentration of other metals besides copper, e.g., zinc[24] and the serum concentration of zinc and iron, and iron and copper are related. Furthermore, serum copper/iron ratios can become inverted in certain disease states; for instance in lung cancer or severe pulmonary TB[25], which may be pertinent to the problem of lung toxicity of bleomycin. One of the uses of penicillamine (which is used in Willson's disease to lower serum copper levels) is to treat chronic lung fibrosis. In this respect the ideas of Willson[26,27] on the disease- or therapy-related damage produced by the decompartmentalization of metal ions are very interesting. The problem is that bleomycin is a chelating agent and a very effective one at that. Is it the reaction of bleomycin with metals at the active site or is it the transport of metals to or from the active site which is important for the various effects produced by bleomycin? An example of the complexity of the problem is that simple metal complexes can either produce or destroy the free radicals that have now been implicated in the DNA attack by bleomycin[28-30]. Whereas iron EDTA or iron transferrin produces the superoxide ion, simple metal complexes such as copper salicylate have been shown to destroy it. It would be interesting to determine if copper–bleomycin acts as a model for superoxide dismutase. There is no reason to believe that different metal–bleomycin complexes cannot play such an ambivalent role *in vivo*.

Finally, looking at the survival curves for the cytotoxic bleomycins one is struck by the fact that doses required to produce a measurable loss in cell viability are many times greater than those that are effective in the clinical situation. There is no doubt that cells in culture are vastly different than cells *in vivo* as is naked DNA when compared to DNA in an intact cell (which has been a confusing factor in the past). Two questions come to mind. The first is: "Why are tumor cells *in vivo* so much more sensitive than cells *in vitro*?" If you calculate from the tissue culture studies the amount of BLM required to produce killing down to the 10^{-3} level (*in vivo*) it should be in the g/kg range. The obvious anoxia in the center of some tumors is not the answer as cells have been shown to be more sensitive to bleomycin under oxygenated conditions[31]. It also may be related to cell uptake and the existence of a cofactor in that for cells in tissue culture the cofactor titer is limited whereas in the clinical situation it may not be. The complexity of reactions involving metal complexes *in vivo* and *in vitro* coupled with decompartmentalization theories leads to the second question which is: "Is the DNA damage seen to be produced by bleomycin the initiating cause of tumor cell death in the clinical situation?" Metals can be considered after all as the smallest catalysts known.

References

1. H. Suzuki, K. Nagai, H. Yamaki, N. Tanaka, and H. Umezawa, *J. Antibiot. (Tokyo)*, **22**, 446 (1969).

2. I. Shirakawa, M. Azegami, S.-I. Ishii, and H. Umezawa, *J. Antibiot. (Tokyo)*, **24**, 761 (1971).
3. E. A Sausville, J. Peisach, and S. B. Horwitz, *Biochem. Biophys. Res. Commun.*, **73**, 814 (1976).
4. J. W. Lown and S.-K. Sim, *Biochem. Biophys. Res. Commun.*, **77**, 1151 (1977).
5. K. Nagai, H. Suzuki, N. Tanaka, and H. Umezawa, *J. Antibiot. (Tokyo)*, **22**, 624 (1969).
6. R. Ishida and T. Takahashi, *Biochem. Biophys. Res. Commun.*, **66**, 1432 (1975).
7. K. Nagai, H. Yamaki, H. Suzuki, N. Tanaka, and H. Umezawa, *Biochim. Biophys. Acta*, **179**, 165 (1969).
8. W. E. G. Müller, Z. I. Yamazaki, H.-J. Breter, and R. K. Zahn, *Eur. J. Biochem.*, **31**, 578 (1972).
9. A. D. Nunn, *J. Antibiot. (Tokyo)*, **29**, 1102 (1976).
10. A. W. Preece, P. A. Light, P. A. Evans, and A. D. Nunn, *Lancet*, 953 (1977).
11. A. D. Nunn, *Int. J. Nucl. Med. Biol.*, **4**, 204 (1977).
12. A. D. Nunn, *Eur. J. Nucl. Med.*, **2**, 53 (1977).
13. M. R. Zalutsky, A. M. Freidman, J. C. Sullivan, S. L. Ruby, and G. V. S. Rayudu, *Int. J. Nucl. Med. Biol.*, **4**, 216 (1977).
14. B. Rydberg, *Radiat. Res.*, **61**, 274 (1975).
15. A. D. Nunn and J. Lunec, *Eur. J. Cancer*, **14**, 857 (1978).
16. A. Kono, M. Kojima, and T. Maeda, *Jap. J. Clin. Radiol.*, **18**, 195 (1973).
17. A. Kono, *Chem. Pharm. Bull. (Tokyo)*, **25**, 2882 (1977).
18. K. Takahashi, O. Yoshioka, A. Matsuda, and H. Umezawa, *J. Antibiot. (Tokyo)*, **30**, 861 (1977).
19. T. Terasima. Y. Takabe, M. Watanabe, T. Katsumata, *Prog. Biochem. Pharmacol.*, **11**, 68 (1976).
20. S. C. Barranco and R. M. Humphrey, *Prog. Biochem. Pharmacol.*, **11**, 78 (1976).
21. C. A. Owen, Jr., *Am. J. Physiol.*, **207**, 446 (1964).
22. C. F. Tessemer, M. Hrgovcic, B. W. Brown, J. Wilbur, and F. B. Thomas, *Cancer*, **29**, 173 (1972).
23. C. Reynaud, D. Comar, M. Dutheil, P. Blanchiou, O. Monod, R. Parrot, and M. Rymer, *J. Nucl. Med.*, **14**, 947 (1973).
24. J. R. Baker, R. W. Fleishmann, G. R. Thompson, U. Schaeppi, V. Ilievska, D. A. Cooney, and R. D. Davis, *Toxicol. Appl. Pharmacol.*, **25**, 190 (1973).
25. J-Y. Le Tinier, J. Burnichou, J-M. Garnier, and J. Pre, *Nouv. Presse. Med.*, **5**, 649 (1976).
26. R. L. Willson, In *Iron Metabolism*, p. 331. CIBA Foundation, Symposium 51, Elsevier, Amsterdam (1977).
27. R. L. Willson, *Chem. Ind. (London)*, 183 (1977).
28. J. M. McCord and E. D. Day, Jr., *FEBS Lett.*, **86**, 139 (1978).
29. L. R. deAlvare, K. Goda, and T. Kimura, *Biochem. Biophys. Res. Commun.*, **69**, 687 (1976).
30. J. J. Van Hemmen and W. J. A. Meuling, *Arch. Biochem. Biophys.*, **182**, 743 (1977).
31. L. Roizin-Towle and E. J. Hall, *Br. J. Cancer*, **37**, 254 (1978).

Upward-Concave Dose–Response Relationship in Bleomycin Lethality of Mammalian Cells

T. Terasima, M. Watanabe, and Y. Takabe

Sterilization of tumor cells by exogenous agents is a major principle of tumor therapy. Studies of inactivation kinetics of tumor cells are, therefore, of great importance in providing a rational basis for the therapy. In some five years of studies of bleomycin action on mammalian cells, we have been concerned with various aspects of survival response. Among them we found the upward concavity of the dose–survival relationship to be particularly remarkable and characteristic for this antibiotic.

The upward concavity in bleomycin survival curves of cultured mammalian cells is shown in Fig. 1[1], which depicts the surviving fraction of four different cell lines as a function of dose, i.e., 1 hr incubation of cells at each drug concentration. The curves start with exponential inactivation, followed by another slope which is less steep. The overall sensitivity of each cell line to bleomycin seems to be roughly similar, but the extent of the initial portion appears to be different depending on the cell line.

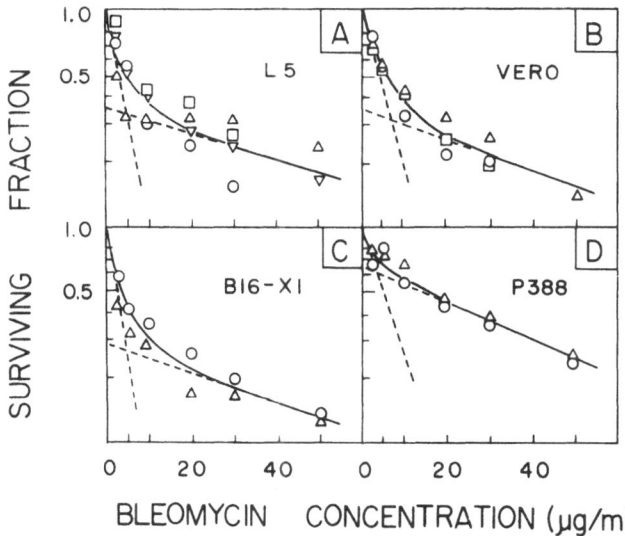

Fig. 1. Bleomycin dose–survival curves of four cultured mammalian cell lines.

Figure 2 illustrates the dose–survival relationship for three lines of transplantable mouse tumors: mammary carcinoma of C3H mouse, squamous cell carcinoma of WHT/Ht mouse, and Ehrlich ascites tumor of ICR/JCL mouse. These results were from experiments carried out separately by Urano et al.[2], Sakamoto and Sakka[3], and Takabe et al.[4]. The drug was injected at the graded doses shown into tumor-bearing mice; 1 hr later, the tumor was excised, minced, and trypsinized, and singly dispersed cells in each dilution step were either inoculated into recipient mice for TD_{50} assay or plated in vitro for colony survival assay. The upward-concave curvature is evident in Fig. 2, as is the finding that squamous carcinoma cells appear to be 100 times as sensitive to bleomycin as mammary carcinoma cells, when the inflection points of the curves are compared.

Essentially similar curves were found with human lymphocytes in response to phytohemagglutinin[5, 6]. The response of mouse bone marrow cells obtained by Twentyman and Bleehen[7] is interesting. Under normal rearing conditions, essentially all the bone marrow cells are nonproliferating. The response of such cells to bleomycin as determined by spleen colony assay (CFU-s) was rather less sensitive and might not be upward-concave. When bone marrow cells were stimulated to proliferation by injection of endotoxin into the donor mice, the response became obviously more sensitive than before and the shape was again upward-concave. This suggests that the effectiveness of bleomycin is somehow related to

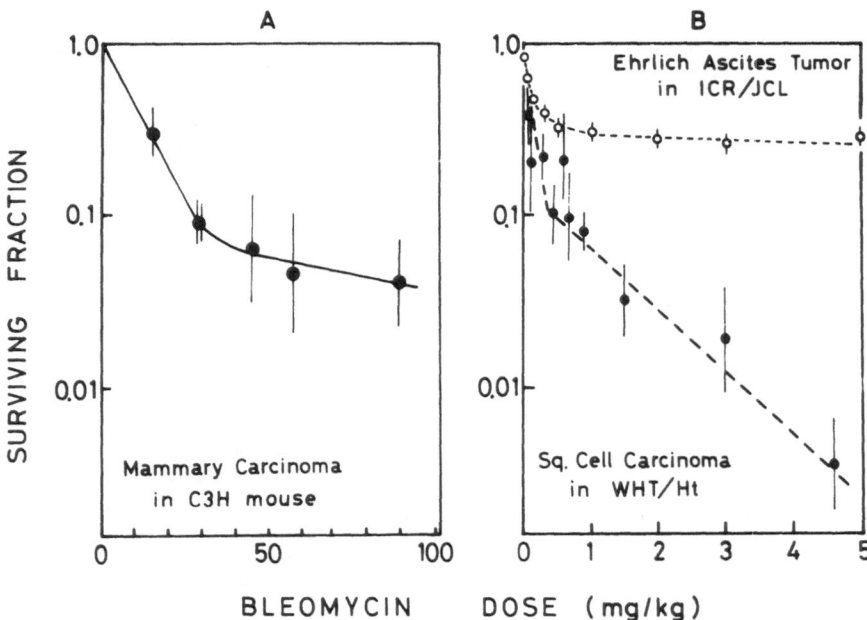

Fig. 2. Bleomycin dose–survival curves in three lines of transplantable mouse tumor.

proliferation. The survival data so far obtained reveal that every type of mammalian cell, irrespective of whether it was derived from normal or malignant tissue, from men or rodents, tested *in vitro* or *in vivo*, responds to bleomycin in essentially similar fashion, although the sensitivity varies depending on the type of cell [for details, see Table II of Ref.[8]].

What is the interpretation of upward concavity in the dose–survival curve? The hypothesis that immediately comes to mind is a mixture of cells of different sensitivities, which are genetically or physiologically determined. The test for survival response of cells obtained from the "tail portion" of the mouse L cell survival curve indicated that the "tail sensitivity" was not genetically determined. Based on the findings of Terasima and Tolmach, it has been established for increasing numbers of exogenous agents and cells that mammalian cell sensitivity changes during the cell cycle[9]. Accordingly, a randomly growing population is a mixture of cells of different sensitivity. The mixture can be readily resolved, if synchronous cells are tested. Figure 3 shows the response of synchronous L cells 3.5 hr after mitosis (G1). The survivals are practically the same as those of a randomly growing population (broken line), indicating that the upward-concave curve shape is not due to a mixture of cells of different sensitivities.

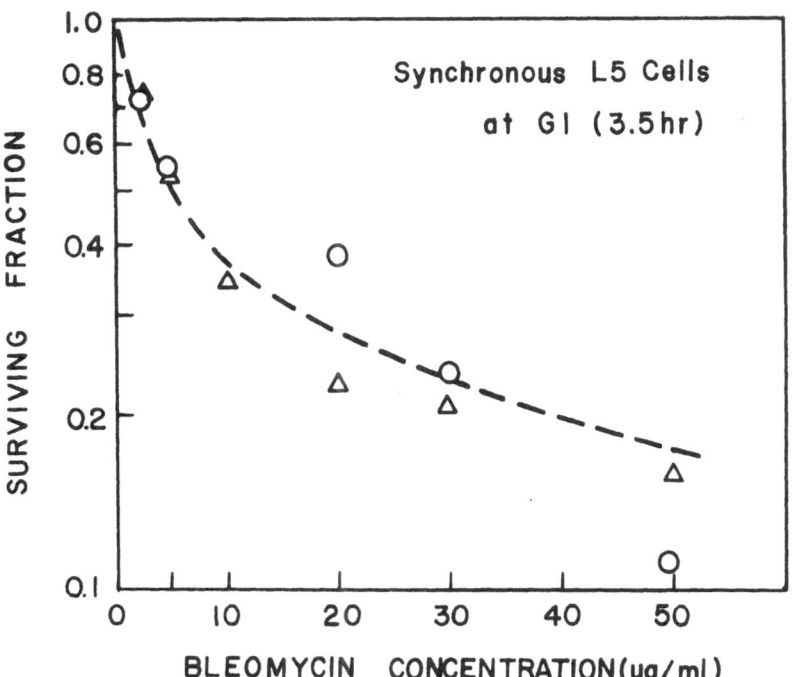

Fig. 3. Dose–survival curve of synchronous mouse L5 cells treated with bleomycin. The broken line represents the survival curve of a randomly growing L5 cell population.

In view of these results, it might be thought that the observed upward concavity resulted from a saturation of the lethal action of the antibiotic. Consequently, one may ask what a possible mechanism for the saturation phenomenon would be. In an effort to identify a mechanism, studies on responses as a function of time were carried out[1]. Figure 4 shows the time-inactivation curves of L cells with four different concentrations of bleomycin; all of these were, though unexpectedly, of biphasic curvature. Cells are inactivated exponentially, then the rate changes 15 to 30 min after the incubation with the drug, and the final slope appears.

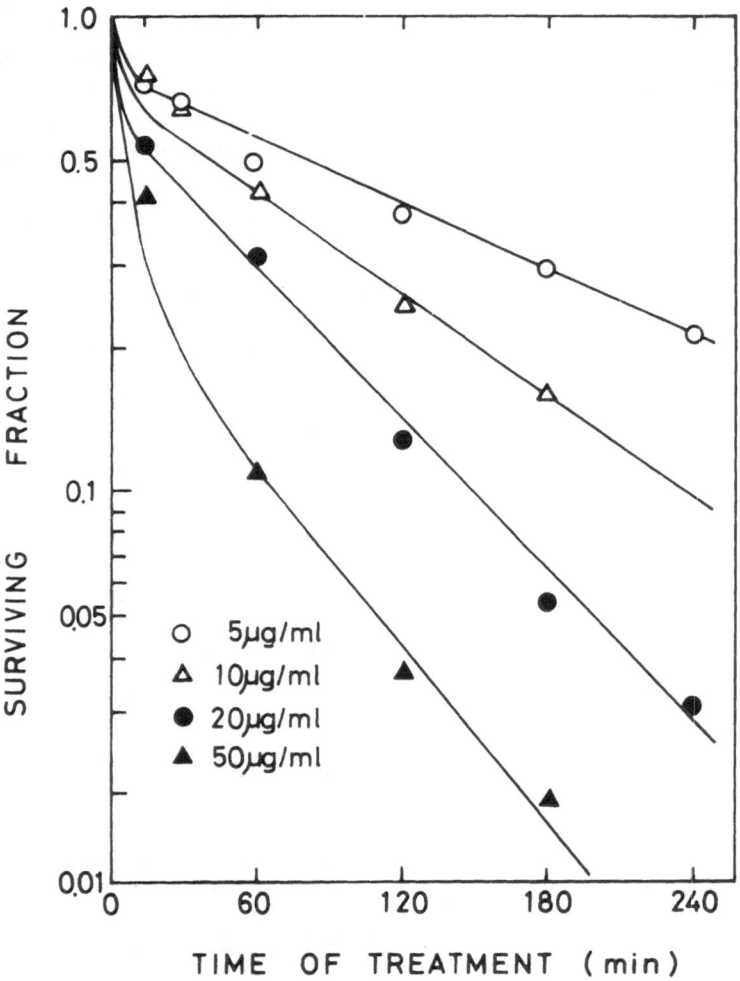

Fig. 4. Time-inactivation curves of mouse L5 cells treated with four different concentrations of bleomycin.

A possible interpretation of biphasic inactivation was provided by means of a two-dose fractionation experiment[1]. The time-inactivation curve of mouse L cells at 5 μg/ml bleomycin is represented by the closed circles in Fig. 5. Again, the "tail fraction" appeared 30 min after the initiation of bleomycin treatment. If a portion of dishes were washed to remove the drug at 120 min and then treated immediately with fresh durg-containing medium, cell survival was the same as that for uninterrupted treatment (not plotted in the figure). This clearly indicated that the biphasic phenomemon was not due to a thermal or metabolic breakdown of bleomycin in the medium, and suggested that the decreased sensitivity to belomycin ("tail") might be induced by the prolonged treatment with bleomycin. In fact, if the second treatment was given 1 hr later, the survival level was reduced as shown by the open circles in Fig. 5. This indicates that the induced resistance decays and that the original sensitivity appears again, at least in a portion of the population. From repeated experiments it was found that the observed reduction of survival level was greater as the interval between two

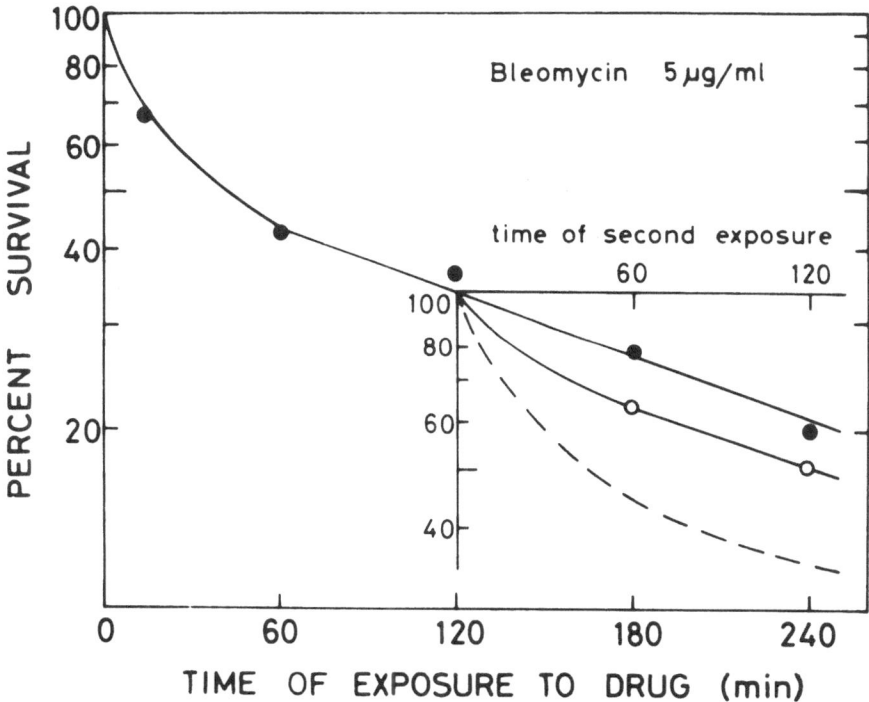

Fig. 5. Change in survival response of L5 cells after removal of bleomycin (two-dose fractionation experiment). The broken line is the survival level expected from complete reversion to the original sensitivity.

treatments increased and finally reached the level of original sensivity as shown by the broken line in the inset. Figure 6 shows the change in relative survival as a function of time interval between two 60-min treatments with 5 and 50 μg/ml bleomycin. The result indicates that the induced resistance decays rapidly at first, then slowly, after removal of the antibiotic and that the reversion to the original sensitivity is completed in 2 to 4 hr. It can also be noted that induction of resistance by bleomycin addition and subsequent decay of resistance after removal of the drug were repeatable in multifractionation experiments if the intervals between treatments were chosen appropriately[1,10].

The relationship between the dose–survival curve and the time–survival (time-inactivation) curve is shown schematically in Fig. 7. It indicates that the reduction of survival effected by 1 hr treatment is not directly proportional to the graded increase of drug concentration (left), and the observed disproportionality may result from induction of resistance which depends upon the concentration of the antibiotic (right).

The upward-concave response to bleomycin is not specifically confined to survival. Figure 8 shows the macromolecular synthetic rates of L cells as a function of drug concentration (Panel A) and time of incubation (Panel B)[11].

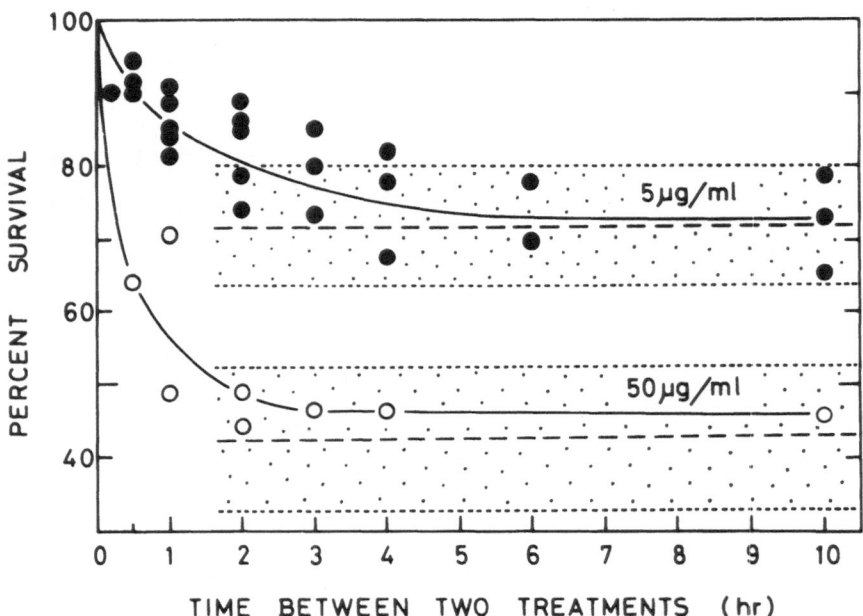

Fig. 6. Disappearance of induced resistance as a function of time after removal of bleomycin (two-dose fractionation experiments). The broken line with dotted band illustrates the average value and standard deviation of survival of the original population to each single exposure.

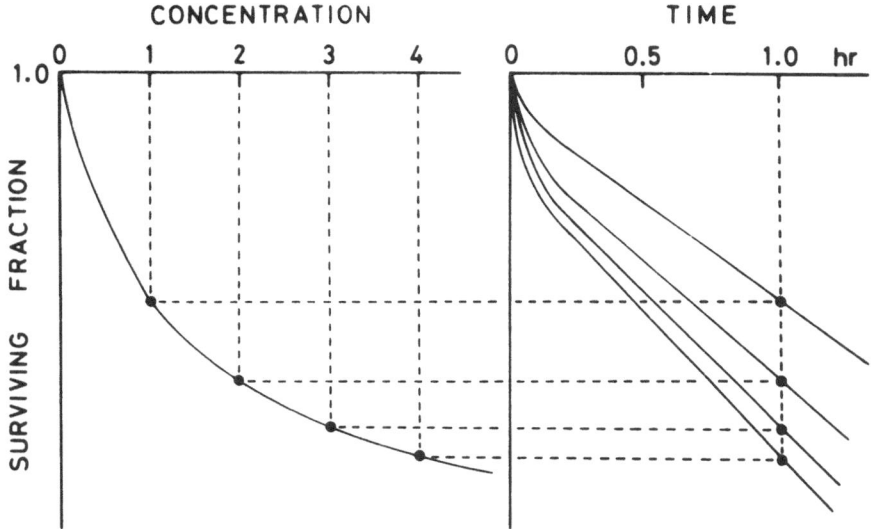

Fig. 7. Schematic illustration of the relationship between the dose–survival and time–survival curves.

Cultures were treated with a given concentration for 1 hr (A) or desired period (B), then pulse-labeled with [^{14}C] thymidine, [^{14}C] leucine, or [^{14}C] uridine for 30 min. The radioactivity in the acid-insoluble fraction was measured. These experiments indicated that, consistent with the survival experiments, each dose-response curve was of similar upward-concave nature and the time-inactivation curves were also biphasic with the inflection points at 15 or 30 min. Furthermore, two-dose fractionation experiments carried out in relation to DNA synthetic activity revealed that the induced resistance had broken down within 2 hr after removal of bleomycin from the medium. These findings strongly suggest that bleomycin action on three different cellular activities may be regulated or controlled in common.

In the search for the physical basis of the biphasic response manifested in the time-inactivation curve, several hypotheses may be considered, some of which are based on the notion that drug molecules may not be able to permeate a cell or reach a certain cellular target after the development of resistance. Alternatively, it might be the case that in the resistant cells bleomycin molecules are somehow converted to a less potent form(s). Specific mechanisms, then, might involve (i) a reduction in permeability or transport activity of resistant cells; (ii) rapid saturation of a target by drug molecules; or (iii) depletion of an intracellular or extracellular cofactor required for bleomycin action. In any case, the assumed change should be potentially reversible in nature.

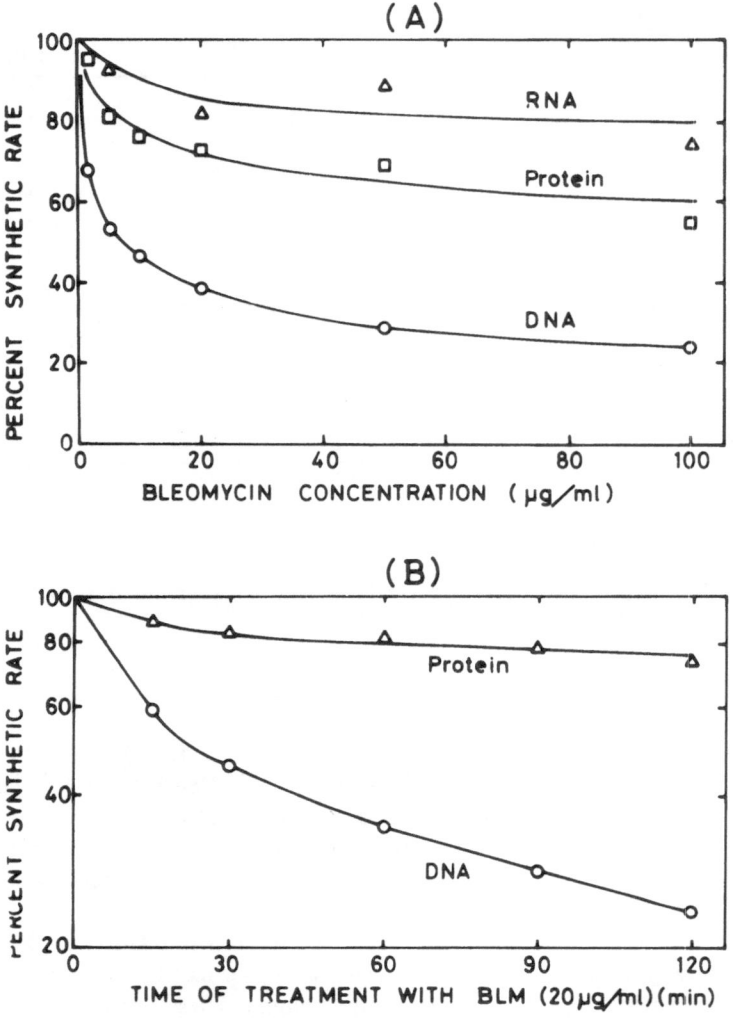

Fig. 8. Inhibition of macromolecular synthesis in L5 cells by bleomycin. Panel (A): dose (1 hr treatment)-inhibition relationship. Panel (B): time-inhibition relationship (bleomycin at 20 μg/ml).

In the experiments shown in Fig. 9, cultures were incubated with [^{14}C] bleo-mycin, saline-washed at specified times, and then used for measurement of uptake of radioactivity. The ^{14}C activity within the cells increased rapidly, nearly leveled off after 30 min, and then increased very slightly if at all. This observation seems consistent with biphasic time-inactivation. However, the location of radioactivity is not known. Miyaki et al.[12] have measured the distribution of [^{14}C] bleomycin among subcellular fractions in rat hepatoma cells. About 30–40% of whole cell radioactivity was present in the nuclei and 0.7% was found to be associated with

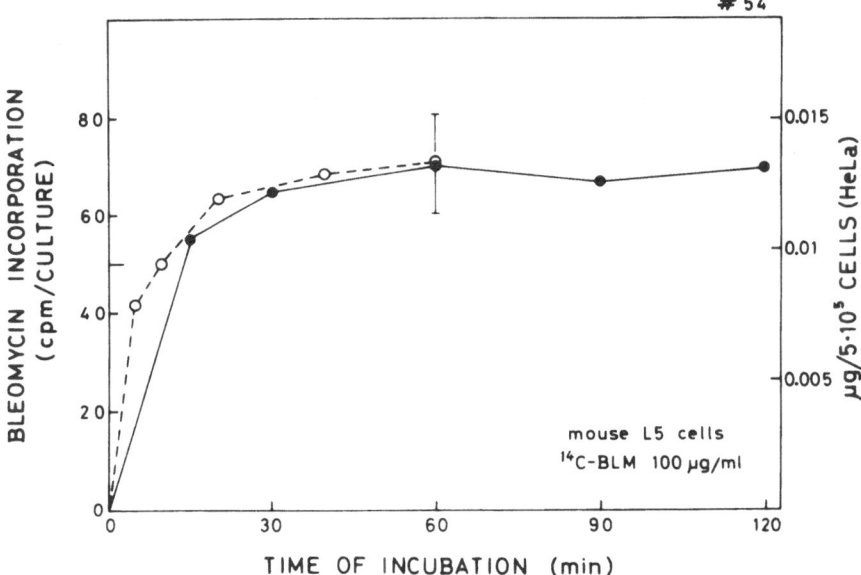

Fig. 9. Uptake of [^{14}C] bleomycin (100 μg/ml) by mammalian cells as a function of time. Solid circles (ordinate on left): mouse L5 cells. Open circles (ordinate on right): HeLa cells [14].

DNA after 1 hr of incubation. Fujimoto[13] has shown by means of autoradiography that [^{14}C] bleomycin was located preferentially at the cell surface of a certain mouse ascitic tumor during the first several hours and then accumulated in the vicinity of the nuclear membrane after 4 hr. However, quantitative kinetic data have not been available.

One of our experiments concerned with DNA breakage by bleomycin may have some bearing on the kinetics of the drug. If the number of breaks is related to the concentration of drug associated with DNA, the nuclear incorporation of bleomycin may be determined indirectly by measuring DNA breaks. Figure 10 shows a change in the molecular wieght of single-stranded DNA of mouse L cells during incubation in bleomycin (100 μg/ml)-containing medium. The molecular weight dropped within 5 min indicating the rapid transport of drug molecules into the nucleus. However, the DNA damage quickly induced repair reactions and, therefore, further damage produced by the dug could not be followed conveniently. In spite of rejoining reactions, the molecular weight did not return to the original level and remained constant after 40 min, suggesting that damage and repair processes were in equilibrium. This implies that the drug was being transported continually into the nucleus, but it is not known whether the amount of nuclear bleomycin underwent a biphasic change which could be responsible for the time-inactivation curve of the cells. To test hypothesis (i), it would obviously be useful to study the time-dependent uptake of the drug in subcellular fractions.

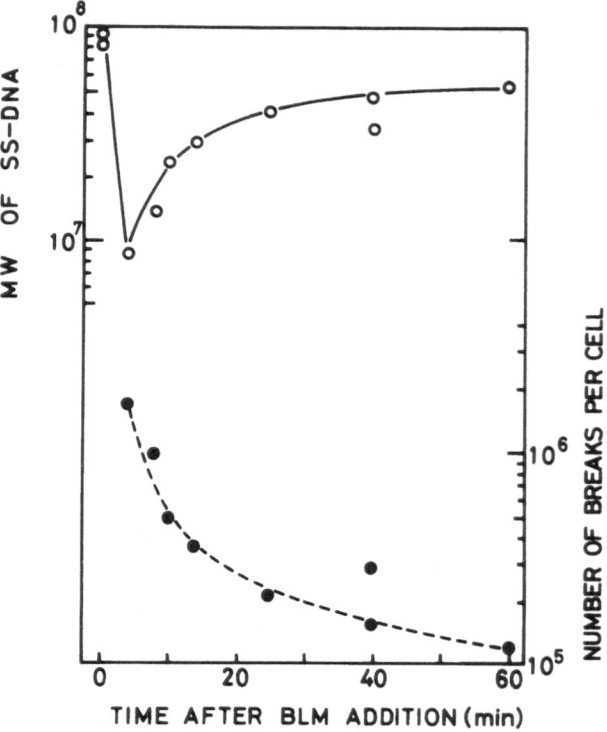

Fig. 10. Change in molecular weight of single-stranded DNA in L5 cells during incubation with bleomycin (100 μg/ml).

Cell surface interaction of the drug molecules needs further consideration. Early data on cell–drug interaction obtained by Yoshioka et al.[14] showed that cell:medium distribution ratio of [14C] bleomycin was extremely low, i.e., the amount of bleomycin taken up by Ehrlich tumor cells was 1/100 of the drug in the medium, whereas DNA-binding substances such as actinomycin-D and daunomycin showed ratios of 2 and 9. The extraordinarily low cell:medium ratio obtained with bleomycin suggests that adsorption of bleomycin molecules may alter a cell surface property required for further uptake of the drug and, in turn, this may be a basis for the induced resistance observed. Second, they have demonstrated that bleomycin molecules taken up by cells distributed rather homogeneously and approximately 60% of the molecules were released from cells while incubating them in drug-free medium for 30 min. These findings suggest that bleomycin has a low affinity for cellular components, which would be consistent with the decay of induced resistance noted after drug removal.

Hypothesis (ii) is strongly supported by the work of Urano et al.[2] who have reexamined the available dose–survival data in terms of the binding-saturation model. The model is based on the assumption that bleomycin is normally bound to a critical target for cell lethality (they assume it to be certain macromolecules); at higher concentrations, the target will be saturated with bleomycin molecules and simultaneously an equilibrium state will be reached. Let us consider the unit area of target surface and suppose that a fraction, σ, is occupied by adsorbed bleomycin molecules, $(1 - \sigma)$ being left free. A dynamic equilibrium between free and adsorbed molecules can be represented as in Eq. (1) (Langmuir's isotherm), where k_a and k_d are association and dissociation constants, respectively, and C is a concentration of bleomycin.

$$k_a \cdot C (1 - \sigma) = k_d \cdot \sigma \tag{1}$$

$$S = e^{-\alpha \cdot \sigma} \tag{2}$$

$$\frac{1}{\ln (1/S)} = \frac{1}{C} \cdot \frac{kd}{\alpha \cdot k_a} + \frac{1}{\alpha} \tag{3}$$

Equation (2) represents a simple exponential cell inactivation by bleomycin, where σ corresponds to a dose. Insertion of σ into Eq. (2) and subsequent conversion results in Eq. (3), which implies that the reciprocal of $\ln (1/S)$ is directly proportional to the reciprocal of concentration. Figure 11 illustrates the replotting of the data shown in Fig. 1 and suggests that their binding-saturation

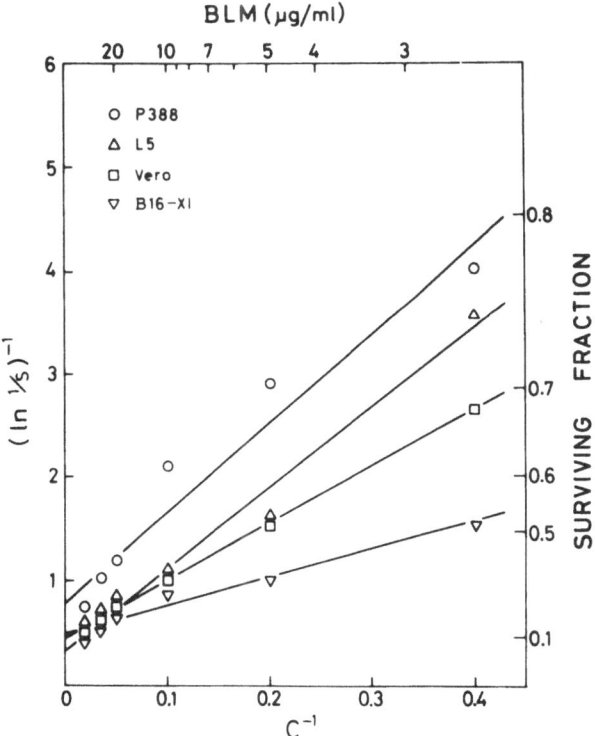

Fig. 11. Dose–response relationship, plotted on the basis of a binding-saturation model, of the four cultured mammalian cell lines shown in Fig. 1.

model is rather successful. This also seems to be the case for the relationship between the inhibition of DNA synthesis and bleomycin concentration (K. Takahashi, Nippon Kayaku Co., personal communication).

The last hypothesis is based on the remarkable role of ferrous ions on bleomycin action[15, 16]. If any cofactors were depleted, either intracellularly or extracellularly, with time after the addition of bleomycin, a biphasic action might possibly be expected. A plausible example in the present case might involve the depletion of ferrous ions by metal-free bleomycin. However, this particular hypothesis may not be tenable, since ferrous ion-bound bleomycin gave a biphasic dose–response curve similar to the one obtained using metal-free bleomycin (K. Takahashi, Nippon Kayaku Co., personal communication). The possible involvement of other factors cannot be dismissed.

Acknowledgments

The authors are grateful to Drs. Matsuda and Takahashi, Research Laboratory, Nippon Kayaku Co. (Tokyo) for their helpful discussions and kind permission to use unpublished data in the present chapter.

References

1. T. Terasima, Y. Takabe, T. Katsumata, M. Watanabe, and H. Umezawa, *J. Nat. Cancer Inst.*, **49**, 1093 (1972).
2. M. Urano, N. Fukuda, and S. Koike, *Cancer Res.*, **33**, 2849 (1973).
3. K. Sakamoto and M. Sakka, *Br. J. Cancer*, **30**, 463 (1974).
4. Y. Takabe, T. Miyamoto, M. Watanabe, and T. Terasima, *J. Nat. Cancer Inst.*, **59**, 1251 (1977).
5. R. Ohno, H. Nishiwaki, K. Kawashima, T. Uetani, M. Hirano, M. Miura, and K. Yamada, *Gann*, **62**, 267 (1971).
6. G. Tisman, V. Herbert, L. T. Go, and L. Brenner, *Blood*, **41**, 721 (1973).
7. P. R. Twentyman and N. M. Bleehen, *Br. J. Cancer*, **28**, 66 (1973).
8. T. Terasima, M. Watanabe, Y. Takabe, and T. Miyamoto, In *Gann Monograph* (S. K. Carter *et al.*, Ed.), Vol. 19, pp. 63–82. University of Tokyo Press, Tokyo (1976).
9. T. Terasima. H. Ohara, T. Miyamoto, M. Watanabe, Y. Takabe, and I. Watanabe, *J. Antibiot. (Tokyo)*, **30**, (Suppl.), 49 (1977).
10. Y. Takabe, T. Katsumata, M. Watanabe, and T. Terasima, *Gann*, **63**, 645 (1972).
11. M. Watanabe, Y. Takabe, T. Katsumata, T. Terasima, and H. Umezawa, *J. Antibiot. (Tokyo)*, **26**, 417 (1973).
12. M. Miyaki, T. Ono, S. Hori, and H. Umezawa, *Cancer Res.*, **35**, 2015 (1975).
13. J. Fujimoto, *Cancer Res.*, **34**, 2969 (1974).
14. O. Yoshioka, K. Takahashi, A. Matsuda, and H. Umezawa, *Proc. Japan Cancer Assoc.*, **31** (abstract), 122 (1972).
15. R. Ishida and T. Takahashi, *Biochem. Biophys. Res. Commun.*, **66**, 1432 (1975).
16. E. A. Sausville, J. Peisach, and S. B. Horwitz, *Biochem. Biophys. Res. Commun.*, **73**, 814 (1976).

Conjugation of Bifunctional Chelating Agents to Bleomycin for Use in Nuclear Medicine

CLAUDE F. MEARES, LESLIE H. DE RIEMER, AND DAVID A. GOODWIN

The affinity of bleomycin for tumorous tissue, coupled with its ability to chelate a variety of metal ions, has aroused considerable interest in its potential as a tumor-visualizing agent. The rationale behind this has been that the bleomycin chelate of a γ-ray emitting metal ion would localize in tumor tissue *in vivo*; the size and location of malignant tissue could then be determined for diagnostic purposes. ^{57}Co–bleomycin was the first bleomycin chelate to be used clinically as a tumor-locating agent[1]. Because of the 270-day radioactive half-life of ^{57}Co, several other bleomycin–metal chelates have since been investigated (see Table 1).

^{67}Ga, ^{59}Fe, and ^{62}Zn are all poorly chelated by bleomycin. Labeling of bleomycin with ^{59}Fe(III) has been attempted by Grove et al.[2] Fe(III) is readily hydrolyzed in neutral aqueous solution; it was not bound to bleomycin as determined by thin-layer chromatography. Ga(III)–bleomycin is unstable in neutral and alkaline solutions as determined by thin-layer chromatography[3]. *In vivo* studies in tumor-bearing mice have also demonstrated the instability of the ^{67}Ga(III)–bleomycin complex[2,3]. In a comparative study of the ^{57}Co, ^{62}Zn, and ^{111}In complexes of bleomycin, Taylor and Cottrall found that the tissue distribution of ^{62}Zn following injection of the ^{62}Zn–bleomycin complex mimicked that of an injection of ^{62}ZnCl$_2$; they concluded that the ^{62}Zn–bleomycin complex had dissociated *in vivo*[4].

Mixed reports have been given concerning the ability of bleomycin to chelate ^{99m}Tc and the clinical use of the ^{99m}Tc–bleomycin complex. Formation of the ^{99m}Tc–bleomycin complex is very dependent upon the pH and concentration of the reducing agent present, and different groups report yields ranging from 40 to 95%[5,6]. Mori et al. compared the ability of ^{99m}Tc–bleomycin and ^{67}Ga–citrate to visualize a variety of tumors in a total of 142 cases. ^{99m}Tc–bleomycin was found to be superior; 82% of malignant tumors were visualized with the ^{99m}Tc–bleomycin chelate as opposed to only 65% with ^{67}Ga–citrate[7]. In a more recent study, Lin et al. concluded that ^{99m}Tc could be used to locate only large or superficial tumors[8]. Despite the encouraging results of Mori et al., it is unlikely that the ^{99m}Tc–bleomycin chelate remains intact *in vivo*; it has been found that ^{99m}Tc is associated with human serum albumin 4 hr after injection of ^{99m}Tc–bleomycin[9].

Table 1: Native Bleomycin–Metal Complexes Investigated Clinically

	Radioactive metal ions			Native bleomycin complexes		
Isotope	$T_{1/2}$	γ Energy (keV)[a]	Stable in vitro (neutral pH)	Stable in vivo	Clinical results	References
^{57}Co	270 days	122 (87%), 136 (11%)	Yes	Yes	Very good	(18,3,20)
^{111}In[b]	2.81 days	173 (89%), 247 (94%)	Yes	No	Varied	(10–12,3)
^{64}Cu	12.8 hr	511 (38%)	Yes	No	Poor	(22,13–17)
^{67}Cu	2.4 days	92 (23%), 184 (40%)	Yes	No	—[c]	(13–17)
^{99m}Tc	6.0 hr	140 (90%)	—[d]	No	Limited use	(7–9)
^{62}Zn	9.1 hr	42 (20%), 510 (47%), 590 (22%)	Complex not formed	No	—[c]	(3,4)
^{67}Ga	3.2 days	93 (40%), 184 (24%), 296 (22%), 388 (7%)	No	No	Poor	(3,2)

[a] Percentages in parentheses refer to percentage of total decay events.
[b] Emits two γ-rays in decay.
[c] No information available on human studies.
[d] No information available on in vitro stability.

The bleomycin chelate of [111]In(III) has received a great deal of attention because of the ideal physical properties of [111]In (see Table 1). Lilien and co-workers were able to obtain images of 89% of a total of 357 tumors in human patients[10]. They achieved a 95% success ratio with adenocarcinomas of the gastrointestinal tract. Graham and co-workers reported an inability to visualize tumors smaller than 2.5 cm and mentioned problems in locating tumors over high background areas (i.e., kidney and bladder)[11]. In spite of these positive results, it has been shown that [111]In(III) does not remain bound to bleomycin *in vivo*. Within 4 hr after injection of ^{111}In–bleomycin, ^{111}In has been found bound to serum transferrin in human subjects[12]. Kono *et al.* have monitored the urine of mice injected with ^{111}In–bleomycin and found that most of the radioactivity was not bound to bleomycin[3].

Bleomycin occurs naturally as the copper chelate, so it was thought that a Cu(II)–bleomycin complex might be stable enough *in vivo* to use as a tumor-visualizing agent. Coates and co-workers clearly demonstrated the stability of the [67]Cu(II)–bleomycin complex *in vitro*; incubation of the complex for 30 min. with human plasma at 37°C followed by column chromatography resulted in all of the [67]Cu(II) remaining bound to bleomycin[13]. Results of animal studies with Cu(II)–bleomycin have varied. Hall *et al.* reported that [67]Cu–bleomycin is superior to ^{111}In–bleomycin in visualizing tumors in animals, while Eckelman and co-workers found ^{64}Cu–bleomycin to be vastly inferior to [57]Co–bleomycin in visualizing mammary carcinoma in rats[14,15]. Umezawa and co-workers have noted that Cu(II)–bleomycin and copper-free bleomycin are equally active in inhibiting growth of animal and bacterial cells, and have demonstrated that Cu(II) from Cu(II)–bleomycin is reduced and bound to a protein in mouse cells[16,17].

The most encouraging clinical results to date have been obtained with the [57]Co–bleomycin complex. Nouel, in an extensive study with 1000 patients, has had a success rate of 97.9% in visualizing pulmonary tumors, and was able to visualize tumors as small as 0.6 cm[18]. [57]Co–bleomycin is characterized *in vivo* by rapid blood clearance (80 to 90% excreted in urine within 24 hr), and a high affinity for the nuclei of tumor cells (more than twice that of non-cobalt-chelated bleomycin)[3,18].

The distinguishing feature of Co–bleomycin as opposed to the other metal ion-bleomycin complexes is the stability of the complex. Although the complex is prepared by combining $CoCl_2$ with bleomycin, chelated Co(II) may be air-oxidized to give Co(III)[19,20]. Complexes of the latter are inert to ligand exchange. Co-bleomycin has been shown to be inert; incubating it with a threefold excess of EDTA for 30 days resulted in no transfer of cobalt[3]. Incubation of the Co-bleomycin complex for 18 hr with a 10–fold excess of Cu(II) caused less than 5% of Co to be displaced, while incubation of In–bleomycin with the same amount of Cu(II) led to a greater than 50% displacement of In(III) after 1 hr[20].

The *in vivo* stability of Co–bleomycin has been determined by monitoring the radioactivity excreted in the urine of humans and mice injected with [57]Co–

bleomycin, with the finding that the ^{57}Co was still bound to bleomycin[3,18]. Kono et al. have also compared the distribution of ^{57}Co-bleomycin to Co-[^{14}C] bleomycin in tumor-bearing mice and found them to be very similar[3]. Thus of all the metal ion–bleomycin complexes investigated to date, only that of ^{57}Co has been found to remain intact in vivo; apparently as a result of this stability, ^{57}Co–bleomycin gives the best tumor images.

While its chemical inertness and in vivo behavior would appear to make ^{57}Co-bleomycin an ideal compound for tumor imaging, ^{57}Co has the undesirable physical property of a 270-day radioactive half-life. Ideally, a radionuclide used for diagnostic purposes should have a half-life of several hours to a few days and should emit only γ-radiation, with energy between 100 and 400 keV[18]. The longer lived radionuclides, such as ^{57}Co, pose serious contamination and health problems. Unfortunately, none of the other isotopes of cobalt has the desired physical properties. If bleomycin is to be used widely as a diagnostic tool, an alternative to current methods for binding radionuclides other than ^{57}Co to bleomycin must be explored.

Bifunctional Chelating Agents

The introduction of a powerful metal-chelating group into a biological molecule such as bleomycin leads to an alternative means of labeling with useful metal ions. With such applications in mind, we have prepared "bifunctional" chelating agents: analogs of ethylenediaminetetraacetic acid (EDTA) which may be attached to other molecules while retaining their ability to form highly stable metal chelates[21]. Most of our work has involved para-substituted derivatives of 1-phenyl-EDTA, whose general structure is given in Fig. 1.

From 1-(p-aminophenyl)-EDTA (Fig. 1, R = NH$_2$), a number of compounds may be prepared. The diazonium salt (Fig. 1, R = N$_2^+$) is a versatile protein-labeling reagent which couples to the side chains of several different amino

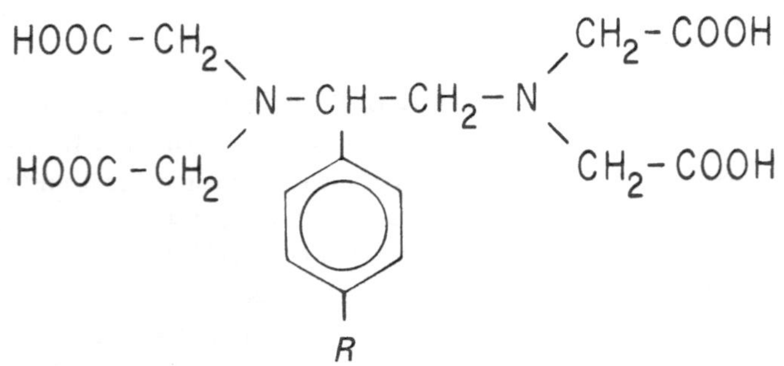

Fig. 1. General structure of the bifunctional chelating agents described in the text. R = NO$_2$, NH$_2$, N$_2^+$, BrCH$_2$CONH, etc.

acids, including lysine, histidine, and tyrosine[22]. The bromoacetyl derivative (Fig. 1, R = BrCH$_2$CONH—) is an alkylating reagent which reacts with several different nucleophiles including mercaptans, thioethers, imidazoles, and amines[24]. Long-chain acyl derivatives (e.g., Fig. 1, R = CH$_3$(CH$_2$)$_{14}$—CONH—) bind noncovalently to hydrophobic entities such as cell membranes and phospholipid vesicles[25]. A variety of other useful groups such as amino acid residues may be attached to the amino group of 1-(p-aminophenyl)EDTA.

The thermodynamic stability constants of chelates formed from 1-phenyl-EDTA and several metal ions have been found to be very similar to those of the corresponding EDTA chelates[26]. It is quite probable that the *para*-substituted analogs of 1-phenylEDTA shown in Fig. 1 also form metal chelates with similar stability constants. However, the work of Margerum and others clearly indicates that steric constraints due to substituents on the ethylene carbons of EDTA can markedly reduce the *rates* of chelate formation and decomposition[27-29]. Reduction of the rate of decomposition of a metal chelate is particularly important if the chelate is to be used to carry radioactive metal ions *in vivo*. The injection of radioactive metal chelate into the circulation places it suddenly in a condition of almost infinite dilution in the presence of significant concentrations of other metal-binding species. If such a metal chelate is to be a valid radiotracer *in vivo*, the rate of loss of radioactive metal ion from the chelate must be slow in comparison to the rate of tissue localization and the rate of excretion of the metal chelate. We have shown that *para*-substituted phenyl-EDTA chelates of indium(III) have bound lifetimes much longer than two weeks in human serum *in vitro*; the behavior of these chelates *in vivo* also indicates that all the indium remains chelated[30].

The kinetic inertness of a metal chelate can be due not only to the properties of the chelating agent, but also to the properties of the metal ion [viz. Co(III)]. It is noteworthy that the ligand-exchange reactions of indium(III) are relatively rapid[31]. Therefore, other metal ions whose chelates with the compounds in Fig. 1 are likely to be stable under physiological conditions may be chosen either on the basis of high thermodynamic stability of their EDTA chelates or the kinetic inertness of the metal ions with respect to ligand substitution. Figure 2 illustrates those metal ions which we consider likely to form bifunctional chelates which will be useful in this context, chosen by qualitative consideration of their kinetic and/or thermodynamic properties. Experiments in our laboratories indicate that technetium also forms bifunctional chelates which are stable under physiological conditions.

Alkylation of Bleomycin-Demethyl A$_2$

The clinical bleomycin mixture Blenoxane was obtained from Dr. William T. Bradner of Bristol Laboratories. This was mixed with a slight molar excess of aqueous copper(II) chloride and fractionated on a Sephadex C-25 cation-

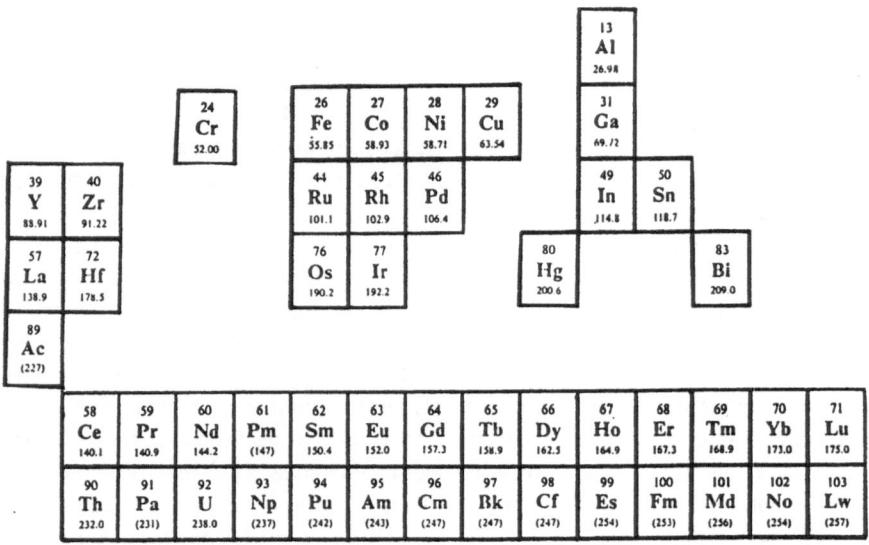

Fig. 2. Elements whose ions exhibit kinetic inertness and/or very stable EDTA chelates, promising properties for use with bifunctional chelating agents under physiological conditions.

exchange column by the method of Fujii *et al.*[32] The A_2 fraction was lyophilized and then pyrolyzed at $100°C$ and 2 Torr for 2 hr, converting it to bleomycin-demethyl A_2. This product was purified using a Sephadex C-25 column and a linear (0.01 to 0.05 M) gradient of ammonium formate. A 65% overall yield of bleomycin–demethyl A_2 (copper chelate) was obtained; the chromatographic and spectroscopic properties of this material were in agreement with literature values[33, 34].

1-(*p*-Aminophenyl)EDTA was prepared as described previously[35], except that the reduction of nitrophenyl-EDTA was performed at pH 11.5–11.8. The amino compound (0.24 mmole) was dissolved in 0.5 ml water, neutralized, and bromoacetylated by addition of a small excess of bromoacetyl bromide (0.34 mmole) in 5-μl aliquots with vigorous vortex mixing. The progress of the reaction may be monitored with fluorescamine[36]. The product (Fig. 1, $R = BrCH_2CONH-$) may be isolated by precipitation at $0°C$, pH 2.3, but yields are below 50%. In order to avoid losing any 1-(*p*-bromoacetamidophenyl)EDTA, the acidic aqueous reaction mixture was extracted eight times with equal volumes of ethyl ether to remove bromoacetic acid and bromoacetyl bromide; progress of the extraction was monitored by testing the organic layer with nitrobenzyl pyridine[37].

The acidic aqueous solution of 1-(*p*-bromoacetamidophenyl)EDTA was adjusted to pH 4.8 by addition of 1 M sodium citrate, and a 5% molar excess of 2 M $CuCl_2$ was added (resulting pH = 3.7). To this solution was added 0.04 mmole of copper-chelated bleomycin-demethyl A_2 (final pH = 4.0); the resulting deep blue solution was stirred for 6 hr at 37°C. The progress of this reaction was monitored by thin-layer chromatography on silica gel F-254 using a solvent containing equal volumes of methanol and 10% aqueous ammonium acetate. Three fluorescence-quenching spots were observed: bleomycin (R_f = 0.8), 1-(p-bromoacetamido-phenyl)EDTA (R_f = 0.9), and product (R_f = 0.5).

The reaction mixture was fractionated on a 1.5 × 40-cm Sephadex A-25 anion-exchange column using a (0.01 to 1.0 M) gradient of ammonium formate. Fractions with significant absorbance at 280 nm were examined by thin-layer chromatography. Two components were found with R_f ≈0.5; the major component (I) eluted from the A-25 column immediately after unmodified bleomycin-demethyl A_2 ([HCOO⁻] ≈0.02 M), while the minor component (II) eluted much later ([HCOO⁻] ≈0.4 M).

The ultraviolet absorption spectrum of component I is given in Fig. 3; the spectrum of component II appears to be identical. The peak at 247 nm and the

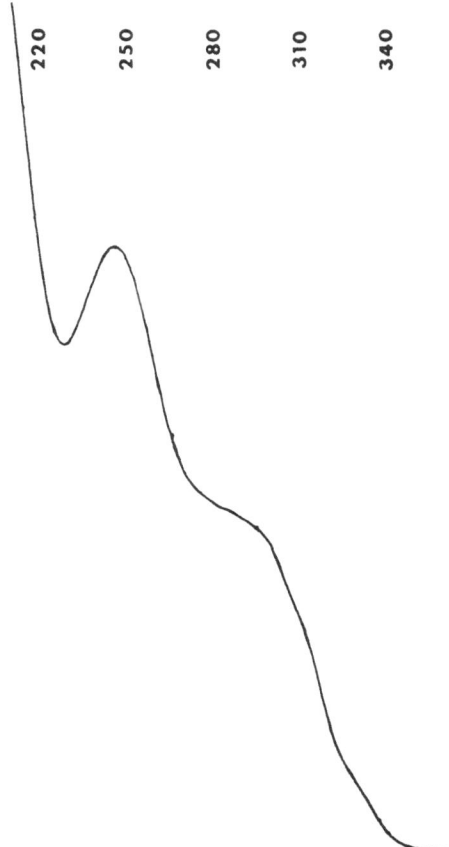

Fig. 3. Ultraviolet absorption spectrum of a 2 × 10⁻⁵ M solution of bleomycin–EDTA compound I. The ordinate is 0-1 absorbance; wavelength in nanometers is given at the top of the figure.

shoulder at 292 nm are consistent with the presence of the acetamidophenyl chromophore and the bithiazole chromophore, respectively. As shown in Fig. 4, the aromatic region of the proton NMR spectrum of copper-free component II is consistent with that expected for the addition of a *para*-disubstituted benzene to bleomycin. Since an attempt to alkylate bleomycin B_2 with 1-(*p*-bromoacetamidophenyl)-EDTA under similar conditions led to no reaction after 88 hr, it is reasonable to assume that alkylation of bleomycin-demethyl A_2 takes place at the thioether sulfur atom of the terminal amine group. Because component I is slightly but distinctly retarded during passage through an anion-exchange column, while component II is more firmly bound to the column, we tentatively propose that the structures of these compounds are as shown in Fig. 5. More work is needed in order to reliably establish the structures.

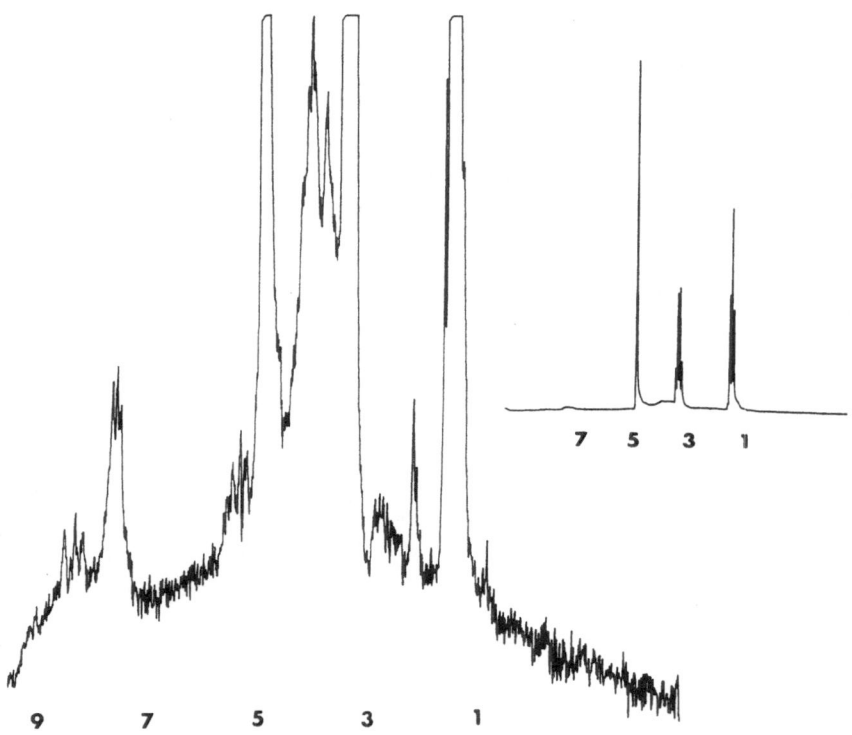

Fig. 4. The partially relaxed Fourier transform proton NMR spectrum of bleomycin–EDTA compound II in D_2O at a pH meter reading of 5.2. Chemical shift is given in ppm downfield from DSS. Region between 7 and 9 ppm contains imidazole, bithiazole, and *para*-disubstituted benzene resonances. The spectrum was taken in the presence of a large excess of triethylammonium chloride; lower amplitude spectrum in inset shows HDO peak at 4.8 ppm and ethyl resonances to the right.

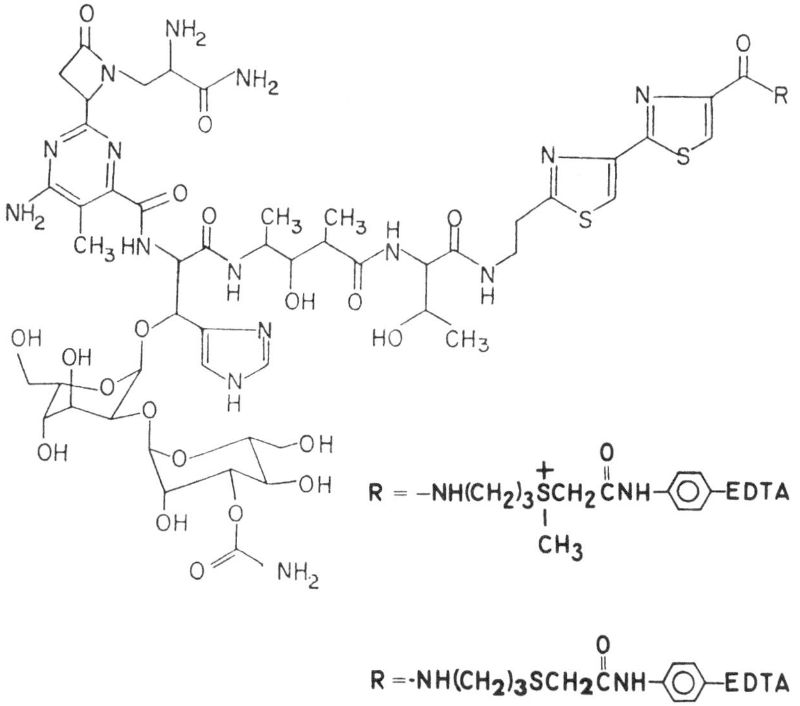

Fig. 5. Tentative structures for bleomycin–EDTA compounds I and II.

Complex Equilibria

After removal of copper by extraction with dithizone or oxine, the bleomycin-EDTA conjugates I and II are ready for labeling with other metal ions. In order to assure that addition of a simple metal salt leads to rapid, quantitative binding of the metal ion to the EDTA group (rather than binding directly to the metal-binding region of the bleomycin, or precipitating as the metal hydroxide), the metal-complexation reaction is carried out in a buffer solution with weak metal-chelating properties.

The principal concern is that binding of the desired metal ion to the EDTA group should greatly predominate over all other possible reactions. The choice and concentration of buffer, and the pH, may be varied in order to achieve this. Here we consider the addition of trivalent indium to the chelating group; we have found that 0.1 M citrate provides a useful buffer medium for such reactions[30, 35]. The degree of completeness of this reaction may be determined most simply by the use of conditional stability constants[38, 39].

For the addition of a metal ion (M) to an EDTA group (H_4Y) in a citrate-buffered solution (H_3Cit) at somewhat acid pH, the metal ion will enter into a side reaction with citrate, and the EDTA group will enter into side reactions with H^+. Beginning with the thermodynamic stability constant

$$K_S = \frac{A_{MY}}{A_M A_Y} = \frac{\gamma_{MY}}{\gamma_M \gamma_Y} \frac{[MY]}{[M][Y]}$$

where A_i denotes the thermodynamic activity of species i, $[i]$ the molar concentration of i, and γ_i the activity coefficient of i, it is possible to arrive at the more practically useful conditional stability constant

$$K_{cond} = \frac{[MY]_t}{[M]_t [Y]_t} = K_S \frac{\gamma_M \gamma_Y}{\gamma_{MY}} \frac{\alpha_{MY}}{\alpha_M \alpha_Y}$$

which relates the total concentration of metal–EDTA chelate, $[MY]_t$, to the total concentration of other metal-containing species.

$$\begin{aligned}
[M]_t &= [M] + [M\text{-Cit}] + [M\text{-BLM}] + \cdots \\
&= [M] \{1 + K_{cit} [Cit] + K_{BLM} [BLM] + \cdots \} \\
&= \alpha_M [M].
\end{aligned}$$

and the total concentration of metal-free EDTA species

$$\begin{aligned}
[Y]_t &= [Y] + [HY] + [H_2 Y] + [H_3 Y] + [H_4 Y] \\
&= [Y] \{1 + \beta_1 [H^+] + \beta_2 [H^+]^2 + \beta_3 [H^+]^3 \beta_4 [H^+]^4 \} \\
&= \alpha_Y [Y].
\end{aligned}$$

As indicated, the functions α_M and α_Y may be calculated if the appropriate equilibrium constants and concentrations are known; if mixed chelates are not formed in significant concentrations, $\alpha_{MY} = 1$.

In Fig. 6, the conditional stability constant for aqueous EDTA–indium in the presence and absence of citrate is shown as a function of pH. It may be seen that the thermodynamic stability constant is related only distantly to the ratio of total concentrations in dilute aqueous solution. Further, Fig. 6 indicates that in the presence of 0.1 M citrate, $K_{cond} \approx 10^{10}$ between pH 2 and pH 6. Thus the addition of carrier-free $^{111}In^{3+}$ to a 10^{-4} M solution of an EDTA analog in 0.1 M citrate at $2 < pH < 6$ should lead to quantitative formation of the EDTA–indium chelate.

The stability constant of the chelate between indium and native bleomycin is not available, but our experiments have shown that bleomycin does not compete appreciably with EDTA for binding indium in 0.1 M citrate between pH 2 and pH 4.

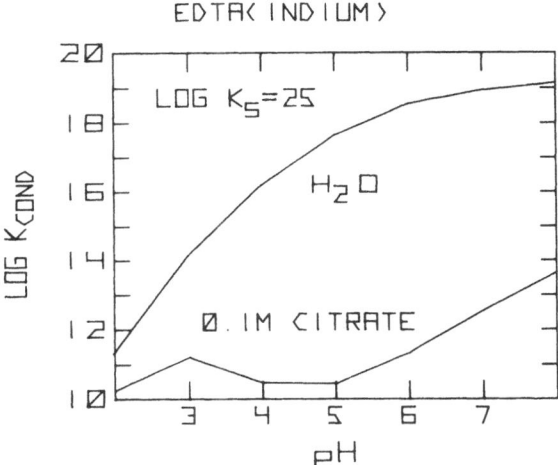

Fig. 6. Logarithmic plot of the conditional stability constant for the EDTA–indium chelate in aqueous solution at zero ionic strength (upper curve) and in 0.1 M citrate (lower curve). The tabulated thermodynamic stability constant is $K_S = 10^{25}$ [38].

Organ Distribution in Tumor-Bearing Mice

The bleomycin–EDTA conjugates were rapidly and specifically labeled with $^{111}In^{3+}$ by adding a small aliquot of a dilute solution ($\approx 10^{-4}$ M) of either compound I or II in 0.1 M citrate (pH 2.8) to a container of dry $^{111}InCl_3$, and incubating for 5 min at room temperature. The solutions were then diluted with normal saline solution for injection.

Following the injection of ^{111}In-labeled compound II into the tail veins of specially prepared BALB/c mice, the organ distribution and tumor uptake of radioactivity were determined. A tumor line, "KHJJ," derived from a primary mammary carcinoma arising in a mouse and maintained for over 100 transplant generations was used for the assay[40]. Transplantation was by subcutaneous implantation of tumor fragments about 1 mm in diameter into the flank. The studies were carried out after 14 days of growth, when the tumor had reached a size of about 1 cm^3. On histological examination, the tumor has a "carcinoma-like" pattern with a predominance of islands of round or polygonal malignant cells with little stroma and a generally undifferentiated appearance. After transplantation, the tumor takes in almost all animals and grows without metastasizing or killing the mice within 14 days. For the distribution assay, a volume of 0.2 ml containing approximately 0.1 μCi of labeled compound II was injected into the tail veins of five mice. After 18 hr, each mouse was anesthetized with ether, and blood was collected from the jugular vein into two preweighed capillary tubes. The mouse then was killed instantly by cervical dislocation and the major organs were excised. Samples of muscle, skin, bone (left femur plus marrow), tail, and tumor also were taken. All tissue samples were weighed immediately after excision and counted in a well-type scintillation counter.

As shown in Table 2, the uptake of radioactivity in liver and bone is quite low, indicating no loss of indium from the chelate. Favorable tumor/organ radioactivity concentration ratios are seen for all organs examined except the kidneys.

Scintillation Scanning of Tumor-Bearing Rabbits

Solutions of the indium-111 chelates of compound I or compound II were injected into the ear veins of white rabbits bearing type VX2 adenocarcinomas implanted under the right foreleg and the right hind leg. Two hours after injection, and again 18 to 24 hr after injection, the rabbits were scanned with a gamma camera, which provides a two-dimensional display of the distribution of radioactivity in the body. Figure 7 shows scans obtained 24 hr after injection of 700 μCi of ^{111}In-labeled compound I. A small tumor 1 cm in diameter under the right foreleg is clearly visualized (Fig. 7A), while a tumor 2 to 3 cm in diameter under the right hind leg is strikingly evident (Fig. 7B). Our experience with the characteristics of this rabbit tumor line is very limited at present, but the results shown in Fig. 7 appear to be highly favorable. Similar results were obtained with compound II. Scans taken 24 hr after injection were superior to those taken 2 hr after injection because at 2 hr, high concentrations of radioactivity in the kidneys and bladder interfered with tumor visualization.

Discussion

The results of these experiments suggest that bifunctional chelating agents may be attached to bleomycin without greatly modifying its biological properties *in vivo*. This is further supported by a positive result from the scan of a human

Table 2.   Distribution and Uptake in BALB/c Mice with KHJJ Tumor.

| Organ | % of injected radioactivity per gram[a] | | |
	^{111}In-bleomycin	^{111}In-1-phenyl-EDTA	Compound II
Blood	0.48 ± 0.20	0.19 ± 0.29	0.205 ± 0.066
Lungs	1.22 ± 0.58	0.029 ± 0.045	0.254 ± 0.067
Liver	2.77 ± 1.03	0.66 ± 0.43	0.263 ± 0.068
Spleen	1.93 ± 0.40	0.83 ± 0.65	0.327 ± 0.088
Kidneys	8.98 ± 2.24	0.22 ± 0.18	1.023 ± 0.333
Tumor	1.98 ± 0.38	0.098 ± 0.088	0.393 ± 0.015
Muscle	0.31 ± 0.026	0.006 ± 0.010	0.063 ± 0.015
Bone	1.26 ± 0.087	0.015 ± 0.026	0.188 ± 0.080
Skin	1.18 ± 1.02	0.037 ± 0.036	0.166 ± 0.052

[a]Mean ± SD for 3 mice.

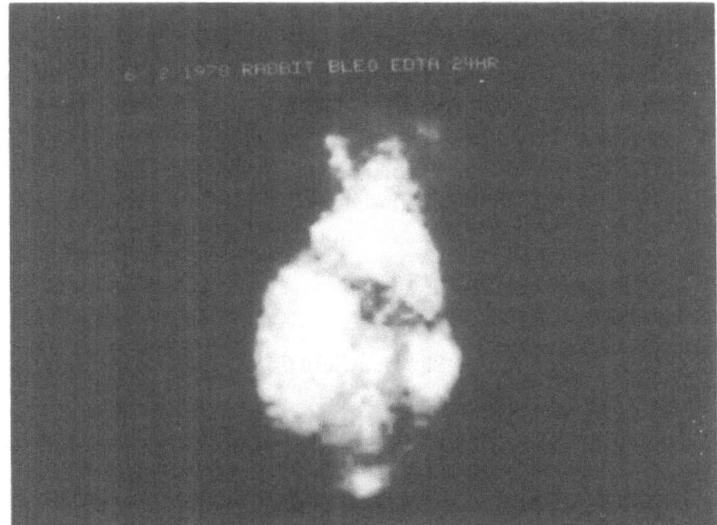

(A)

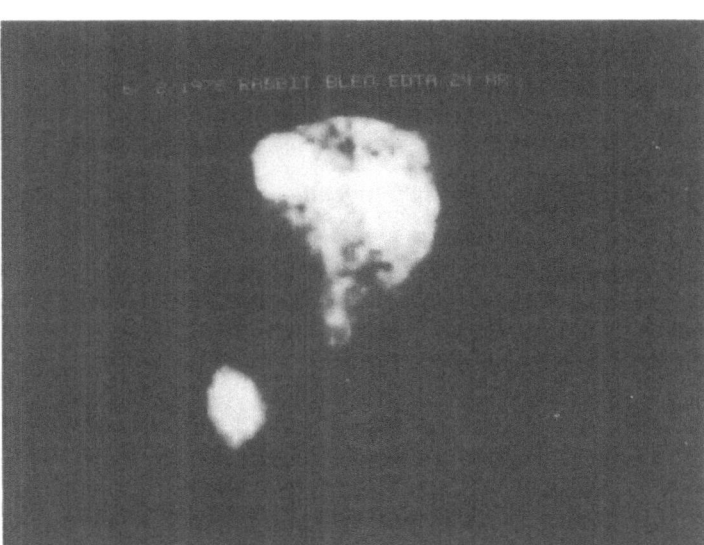

(B)

Fig. 7. Gamma camera scans (front view) of a tumor-bearing rabbit 24 hr after injection with bleomycin–EDTA (^{111}In) compound I. Radioactivity remaining in the heart, liver, and kidneys is shown by large bright regions in (A); a tumor 1 cm in diameter under the right foreleg is visible as a bright, narrow projection just left of center and slightly above the large bright regions. (B) shows a tumor 2–3 cm in diameter under the right hind leg as a distinct, very bright region in the lower left; radioactivity remaining in kidneys is seen as the large bright region in the upper part of (B).

patient. In future work, conjugation of nonradioactive cobalt bleomycin-demethyl A_2 to bifunctional chelating agents may lead to enhanced uptake in tumors[3,15].

Because they form stable chelates with a wide variety of metals (see Fig. 2), the bifunctional chelating agents are unusually versatile spectroscopic probes which may be used in fluorescence measurements, energy transfer, electron paramagnetic resonance, γ-ray angular correlations, radiation scattering, and other physical experiments. Thus the bleomycin–EDTA conjugates may also prove useful in studies of the mechanism of action of bleomycin.

Acknowledgments

We thank Charles Leung, David Sherman, and Carol Diamanti for their technical assistance. This work was supported by Grant No. CA16861 from the National Cancer Institute, DHEW, by Grant No. 3204 from the Veterans Administration, and by Special Grant No. 685 from the American Cancer Society, California Division.

References

1. J. P. Nouel, M. Renault, J. Robert, C. Jeanne, and L. Wicart, *Nouv. Presse Med.*, 2, 95 (1972).
2. R. B. Grove, W. C. Eckelman, and R. C. Reba, *J. Nucl. Med.*, 14, 917 (1973).
3. A. Kono, Y. Matsushima, M. Kojima, and T. Maeda, *Chem. Pharm. Bull. (Tokyo)*, 25, 1725 (1977).
4. D. M. Taylor and M. F. Cottrall, In *Radiopharmaceuticals* (G. Subramanian, B. A. Rhodes, J. F. Cooper, and V. J. Sodd, Eds.), p. 458. Society of Nuclear Medicine, New York (1975).
5. M. L. Thakur, *Int. J. Appl. Radiat. Isot.*, 24, 357 (1973).
6. P. J. Bartels, B. G. Dekker, C. L. DeLigny, and S. J. Oldenburg, *Int. J. Appl. Radiat. Isot.*, 29, 15 (1978).
7. T. Mori, K. Hamamoto, and K. Torizuka, *J. Nucl. Med.*, 14, 431 (1973).
8. M. S. Lin, D. A. Goodwin, and S. L. Kruse, *J. Nucl. Med.*, 15, 338 (1974).
9. H. Orii and H. Oyamada, *Proc. 1st World Congress Nucl. Med.*, Tokyo, 1974, p. 931.
10. D. L. Lilien, S. E. Jones, R. E. O'Mara, S. E. Salmon, and B. G. M. Durie, *Cancer*, 35, 1036 (1975).
11. L. S. Graham, R. C. Verma, J. J. Touya, M. J. Silverstein, and L. R. Bennett, In *Radiopharmaceuticals* (G. Subramanian, B. A. Rhodes, J. F. Cooper and V. J. Sodd, Eds.). Society of Nuclear Medicine, New York (1975).
12. M. L. Thakur, M. V. Merrick, and S. W. Gunasekera, In *New Developments in Radiopharmaceuticals and Labeled Compounds*. IAEA, Vienna (1974).
13. G. Coates, N. Aspin, P. Y. Wang, and D. E. Wood, *J. Nucl. Med.*, 15, 484 (1974).
14. J. N. Hall, R. E. O'Mara, and P. Cruz, *J. Nucl. Med.*, 15, 498 (1974).
15. W. C. Eckelman, R. C. Reba, H. Kubota, and J. Stevenson, *J. Nucl. Med.*, 15, 489 (1974).
16. H. Umezawa, *Gann Monograph Cancer Res.*, 19, 3 (1976).

17. K. Takahashi, O. Yoshioka, A. Matsuda, and H. Umezawa, *J. Antibiot. (Tokyo)*, **30**, 861 (1977).
18. J. P. Nouel, *Gann Monograph Cancer Res.*, **19**, 301 (1976).
19. F. A. Cotton and G. Wilkinson, *Advanced Inorganic Chemistry*, p. 874ff. Interscience Publishers, New York (1972).
20. A. D. Nunn, *Eur. J. Nucl. Med.*, 2, 53 (1977).
21. *(London)*, **250**, 587 (1974).
    A. D. Nunn, *Eur. J. Nucl. Med.*, 2, 53 (1977).
22. H. Renault, J. Rapin, and L. Wicart, *C. R. Acad. Sci. Paris*, **273**, 2013 (1971).
23. E. W. Gelewitz, W. L. Riedeman, and I. M. Klotz, *Arch. Biochem. Biophys.*, **53**, 411 (1954).
24. G. E. Means and R. E. Feeney, *Chemical Modification of Proteins*, p. 12. Holden-Day, Inc., San Francisco (1971).
25. C. F. Meares and S. M. Yeh, *Experientia*, in press.
26. N. Okaku, K. Toyoda, Y. Moriguchi, and K. Ueno, *Bull. Chem. Soc. Japan*, **40**, 2326 (1967).
27. D. W. Margerum, D. L. Janes, and H. M. Rosen, *J. Am. Chem. Soc.*, **87**, 4463 (1965).
28. J. D. Carr, R. A. Libby, and D. W. Margerum, *Inorg. Chem.*, **6**, 1083 (1967).
29. J. D. Carr, K. Torrance, C. J. Cruz, and C. N. Reilley, *Anal. Chem.*, **39**, 1358 (1967).
30. C. F. Meares, D. A. Goodwin, C. S-H. Leung, A. Y. Girgis, D. J. Silvester, A. D. Nunn, and P. J. Lavender, *Proc. Nat. Acad. Sci. USA*, **73**, 3803 (1976).
31. M. Eigen, *Pure Appl. Chem.*, **6**, 97 (1963).
32. A. Fujii, T. Takita, K. Maeda, and H. Umezawa, *J. Antibiot. (Tokyo)*, **26**, 396 (1973).
33. T. Takita, A. Fujii, T. Fukuoka, and H. Umezawa, *J. Antibiot. (Tokyo)*, **26**, 252 (1973).
34. H. Umezawa, *Pure Appl. Chem.*, **28**, 665 (1971).
35. M. W. Sundberg, C. F. Meares, D. A. Goodwin, and C. I. Diamanti, *J. Med. Chem.*, **17**, 1304 (1974).
36. S. Udenfriend, S. Stein, P. Böhlen, W. Dairman, W. Leimgruber, and M. Weigele, *Science*, **178**, 871 (1972).
37. O. M. Friedman and E. Boger, *Anal. Chem.*, **33**, 906 (1961).
38. A. Ringbom, *Complexation in Analytical Chemistry*, p. 35 ff. Interscience Publishers, New York (1963).
39. J. N. Butler, *Ionic Equilibrium: A Mathematical Approach*, p. 378 ff. Addison-Wesley, Palo Alto (1964).
40. S. C. Rockwell, R. F. Kallman, and L. F. Fajardo, *J. Nat. Cancer Inst.*, **49**, 735 (1972).

Bleomycin-Induced Pulmonary Toxicity:
A Model for the Study of Pulmonary Fibrosis

ILENE H. RAISFELD

The use of bleomycin in the chemotherapy of tumors of lung and skin is limited by the development of interstitial pulmonary fibrosis[1]. Older individuals, particularly those with pulmonary disease, are predisposed to this potentially fatal and irreversible toxic effect of the drug.

The development of animal models in which lung damage is induced by systemic administration of bleomycin has facilitated studies of the pathogenesis of interstitial pulmonary fibrosis. These models have been less useful in elucidating the relation between drug structure and pulmonary fibrosis because lung lesions do not develop with regularity. The present chapter describes conditions under which a single intratracheal injection of bleomycin in mice consistently produces pulmonary fibrosis. The pulmonary toxicity of several bleomycins is compared and a relationship is established between bleomycin terminal substituents and pulmonary fibrosis.

Methods

Animals

Male DBA, C57B1, and CD-1 mice, 28 weeks of age or older, were purchased from the Charles River Laboratories, Wilmington, Massachusetts. Groups of six animals were housed in polycarbonate cages containing corn cob bedding, and fed Purina rat chow (Ralston Purina Co., St. Louis, Missouri) and water *ad libitum*. Animals were caged individually after receiving bleomycin and died or were sacrificed 2 to 4 months later.

Intratracheal Administration of Drug

Surgery was performed under sodium pentabarbital anesthesia, administered intraperitoneally in a dose of 60 mg/kg. Bleomycin (Blenoxane, Bristol Laboratories, Syracuse, New York) was dissolved in sterile 0.14 M NaCl at concentrations sufficient to yield doses of 0.064, 0.128, or 0.256 U/0.05 ml.[1]

---

[1] As different batches of Blenoxane vary in potency, doses are expressed in terms of units. One unit is approximately 0.60 mg bleomycin. Blenoxane is a mixture of several bleomycins, the major components being bleomycin A_2 (50%) and bleomycin A_1 and B_2.

A 0.5– to 1.0-cm midline incision was made in the neck using a #15 scalpel blade and opened to expose two pockets of fat overlying the tracheal membrane. Using #5 forceps, the tracheal membrane was cut and 0.05 ml of the bleomycin solution was introduced through a 30- or 32-gauge needle. Air, 0.2 ml, was quickly injected to clear the airway, followed by 0.05 ml of 0.14 M NaCl and 0.2 ml air to force the solution into the lungs.

The wound was closed with a #7601 steel wound clip. Following surgery, animals were given 80 mg/kg doxapram HCl (A. H. Robins, Richmond, Virginia) to counteract the respiratory depressant effects of the anesthetic. For every three experimental animals treated with bleomycin, one animal received 0.05 ml of 0.14 M NaCl as a control. Less than 2% of the mice died as a result of the procedure.

Preparation of Tissue

Lungs were fixed *in situ* by intratracheal injection of 1 ml of 10% formalin buffered with phosphate, pH 7.1. Lungs were removed from the chest cavity *in toto* and immersed in formalin for at least 24 hr. Uncut lungs were embedded in paraffin and sections (5 μm) were cut tangential to pleural surfaces. Sections of whole lungs were mounted on slides, stained with hematoxylin and eosin (H & E) and Mallory's trichrome, and evaluated by light microscopy.

Definition and Evaluation of Pathology

Three lesions were observed, confluent fibrosis, interstitial fibrosis, and epithelial metaplasia. Each lesion was evaluated separately. Confluent fibrosis obliterated normal lung architecture. Ratings of 1, 2, 3, and 4 corresponded, respectively, to involvement of 10, 10–25, 25–50, and > 50% of lung area by this lesion.

In contrast to confluent fibrosis, the integrity of alveolar spaces and pulmonary vasculature was preserved in the presence of interstitial fibrosis. Interstitial fibrosis tended to involve contiguous areas of lung, but, occasionally, was patchy. Lesions of interstitial fibrosis were graded on a scale of 1 to 4 according to the stage of the lesion and the severity of fibrosis. Lesions graded as 1, 2, and 3 consisted of proliferating fibroblasts and mild, moderate, and pronounced collagen deposition. Grade 4 lesions were well-organized scars. Patchy or contiguous interstitial fibrosis that involved more than 30% of the area of both lungs was considered "extensive," and was graded as present (+) or absent (–).

Although prominence of single epithelial cells was a common finding in this study, the term epithelial metaplasia was used to denote proliferation of epithelial cells in sheets that effaced normal architecture. Epithelial metaplasia was graded on a scale of 1–4. A grade of 4 denoted the most prominent lesions.

Measurement of pO_2 and pCO_2

The opthalmic venous plexus was ruptured with the tip of 60-μl heparinized capillary blood sampling tube (Radiometer Corporation, Copenhagen). Blood from sealed, iced tubes was introduced within 15 to 30 min into an Instrumentation Laboratories Ultra Micro pH/Blood Gas System equipped with a Clark electrode for pO_2 measurement and a Severinhgaus electrode for pCO_2 measurement. pO_2 and pCO_2 values were read directly.

Results

Pathology

Lungs from mice with fibrotic disease were heavy and firm. Confluent and extensive fibrotic lesions were evident by visual inspection of stained slides, appearing as dense areas with the lung (Fig. 1).

Interstitial fibrosis was found as early as 4 days after treatment (Fig. 2). Confluent fibrosis that completely obliterated lung architecture (Fig. 3) was observed during the first week of treatment. Interstitial pneumonia or inflammation was rarely observed.

(a)   **(b)**

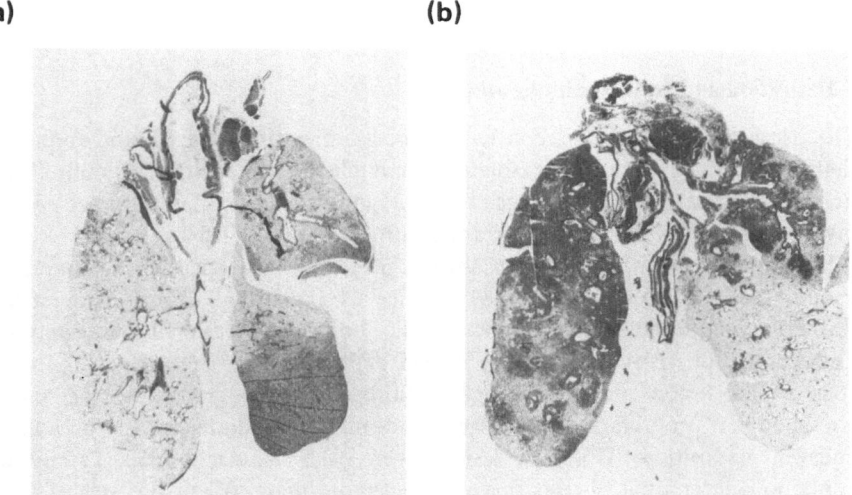

Fig. 1. Photograph of lung sections stained with Mallory's trichrome. Dense areas consist of interstitial and confluent fibrosis, and pale areas are histologically normal. Magnification X5. (a) Lungs from DBA mouse 10D (Table 1). Left lung is normal. Grade 4 interstitial fibrosis is present in right upper lung field; the right lower lung field displays grade 4 confluent fibrosis. (b) Lungs from DBA mouse sacrificed 83 days after treatment with 0.064 U bleomycin. Dense lesions are grade 4 confluent fibrosis, grade 2 interstitial fibrosis, and grade 3 epithelial metaplasia. In addition to bleomycin-induced lesions, bronchopneumonia is present in left lower lobe. Right lower lobe is normal.

(a) **(b)**

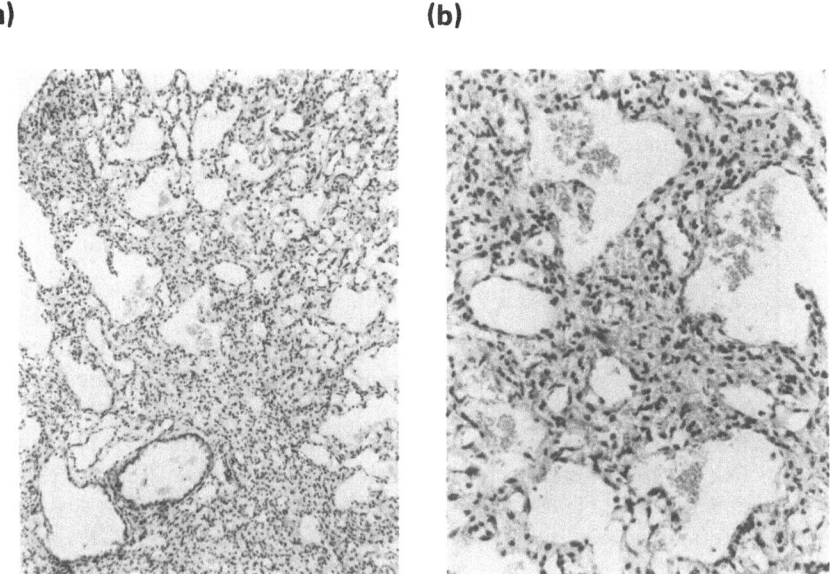

Fig. 2. Interstitial fibrosis from lung of mouse 2D (Table 1). Animal succumbed to this lesion 13 days after 0.128 U bleomycin. (a) Alveolar spaces are visible in upper portion of figure. Magnification ×70. (b) Fibroblasts and collagen fill interstitial spaces. Inflammatory response is minimal. Magnification ×160.

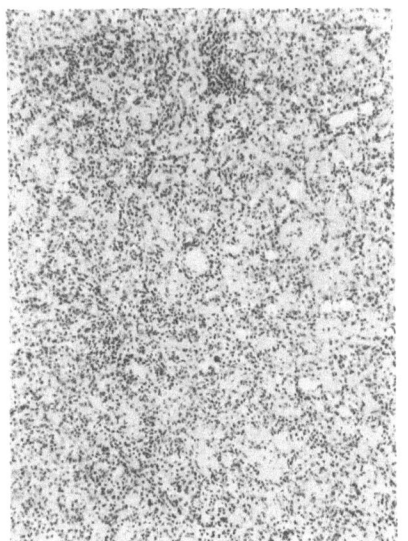

Fig. 3. Confluent fibrosis, grade 4, obliterates normal structure in lung from animal 10D (Table 1). Most of these cells are fibroblasts. Magnification ×70.

Epithelial metaplasia was a common finding. Morphology of metaplastic cells was varied, as illustrated in Fig. 4, but metaplastic changes in an individual animal tended to be of one morphologic type. These cells stained purple with Mallory's trichrome and contrasted with blue areas of fibrosis. Epithelial cells were often prominent in an area undergoing fibrotic change, but epithelial metaplasia was a process independent from fibrosis and occasionally occurred in the absence of fibrosis (Fig. 4d).

Incidence of Pulmonary Toxicity in Different Strains of Mice

All three strains of mice developed fibrosis and epithelial metaplasia. Differences between strains, with respect to the incidence of fibrosis, were apparent when lower doses of bleomycin were employed.

DBA mice were most sensitive to the toxic pulmonary effects of bleomycin (Fig. 5). Five of seven animals treated with a dose of 0.064 U bleomycin developed moderate to extensive interstitial fibrosis; a similar fraction developed epithelial metaplasia and two animals displayed confluent fibrosis. One animal with extensive confluent fibrosis survived indefinitely (Fig. 1b).

sive disease. Of 13 animals affected, 12 (92%) manifested extensive interstitial fibrosis. In one animal (11D), epithelial metaplasia was present but interstitial fibrosis was minimal. Eleven animals (85%) displayed confluent fibrosis and nine (70%) developed epithelial metaplasia. All animals affected at this dose died between 12 and 32 days after administration of drug (Table 1).

Pathologic findings in C57B1 and CD-1 mice were similar to those obtained in DBA mice. At doses of 0.064 and 0.128 U, respectively, 40 and 80% of C57B1 mice and 57 and 85% of CD-1 mice developed lesions (Fig. 5). At a dose of 0.256 U all animals were affected.

In contrast to DBA mice, 45 and 20% of affected CD-1 and 33 and 43% of C57B1 animals survived indefinitely after 0.128 and 0.256 U of bleomycin. Pathology was less severe in survivors than in animals which died. Focal fibrosis was more extensive and epithelial metaplasia occurred more frequently in animals which died, and confluent fibrosis and extensive interstitial fibrosis were found in only one survivor (animal 14B) (Table 2).

Pulmonary Function

Pathologic findings were associated with impaired pulmonary function as measured by pO_2 or pCO_2 in venous blood. Since bleomycin-treated animals frequently died during cardiac puncture, blood gas measurements were performed on blood obtained from the orbital sinus. A comparison of values obtained from bleomycin-treated animals and saline-treated controls indicated that interstitial fibrosis was associated with a venous pO_2 of less than 30 mm Hg and/or a venous pCO_2 of over 60 mm Hg (Fig. 6).

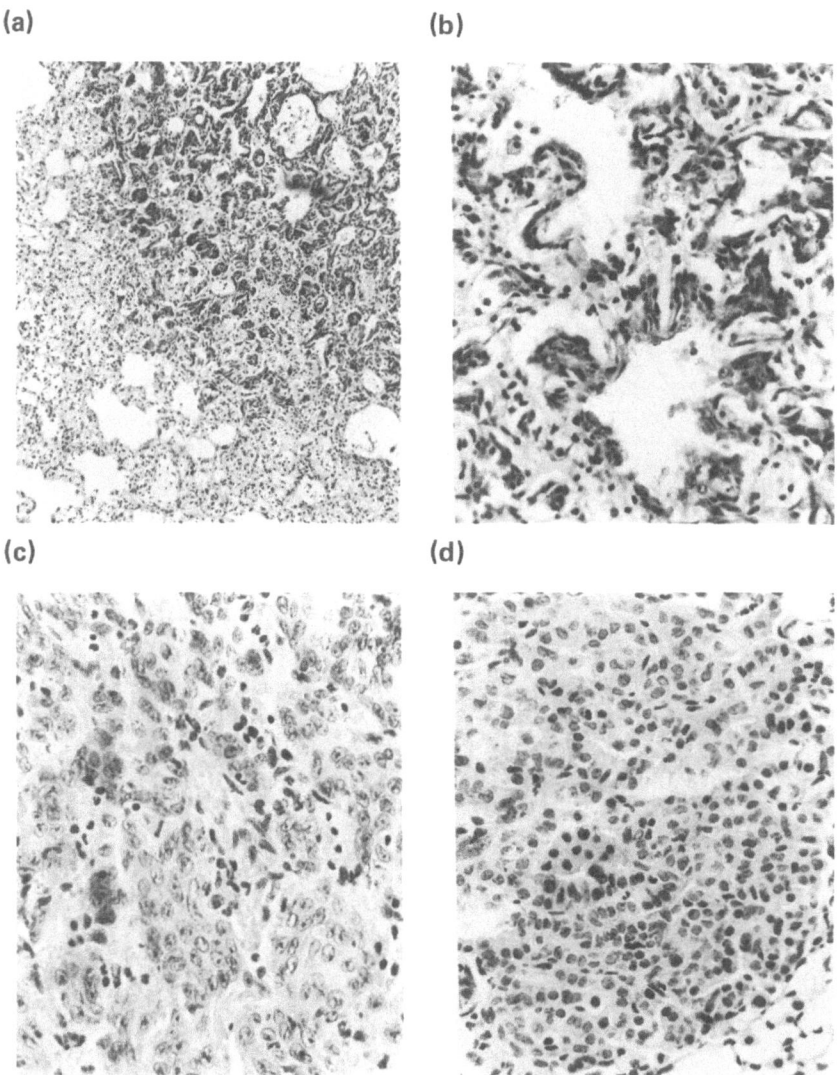

Fig. 4. Metaplastic epithelial cells display varied morphology. (a) In the upper right, metaplastic cells arise within an area of interstitial fibrosis. Magnification ×70. (b) Elongated squamoid cells with flattened nuclei surround alveolar lumina in photomicrograph taken from upper right field of section shown in (a). Magnification ×275. (c) Cells with large epithelial pale nuclei and prominent nucleoli form solid clusters that obliterate alveolar spaces. Fibroblasts are present but are not prominent. Magnification ×275. (d) Cuboidal cells with dense cuboidal nuclei proliferate and fill alveolar spaces. The probable origin of these cells, from the alveolar lining, can be seen in the lower right portion of this figure. Magnification ×275.

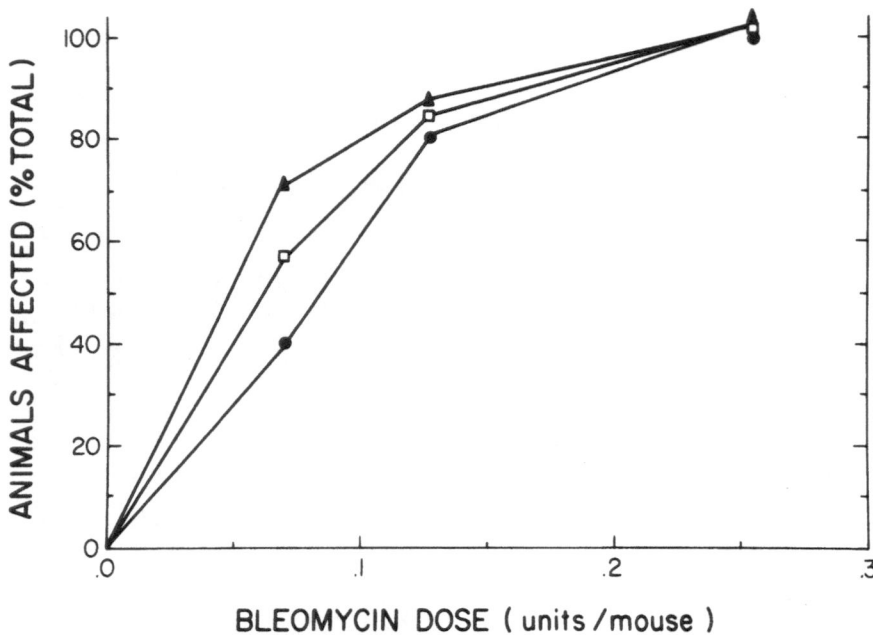

Fig. 5. Incidence of bleomycin pulmonary toxicity in three mouse strains (▲, DBA; □, CD-1; ●, C57B1). In each strain, 7 animals received 0.064 U, 15 animals received 0.128 U, and 5 animals received 0.256 U bleomycin.

Table 1: Lung Pathology in DBA Mice after Single Dose of 0.128 U Bleomycin[a]

Animal	Survival (days)	Fibrosis			Epithelial metaplasia
		Confluent	Interstitial	Extensive	
1D	12	−	−	−	−
2D	13	2	3	+	−
3D	14	2	4	+	1
4D	14	−	2	+	4
5D	14	3	2	+	4
6D	15	2	4	+	−
7D	16	1	3	+	3
8D	16	3	4	+	−
9D	18	3	2	+	−
10D	20	4	4	+	1
11D	21	−	±	−	4
12D	21	3	3	+	2
13D	28	2	2	+	3
14D	32	4	4	+	1
15D	57[b]	−	−	−	−

[a]Data compiled here are the sum of two experiments. All treated animals are listed, including two on which no pathologic changes were found. Six control animals were sacrificed: two on day 15, one on day 16, one on day 28, one on day 32, and one on day 57. One control animal sacrificed on day 15 displayed patchy bronchopenumonia; other lungs were normal.
[b]Designates sacrifice of animal.

Table 2:   Relation between Lung Pathology and Survival in C57Bl Mice[a]

Animal	Survival (days)	Fibrosis			Epithelial metaplasia
		Confluent	Interstitial	Extensive	
Bleomycin (0.064 U)					
1B	40	–	4	+	–
2B*	104	–	2	–	3
Bleomycin (0.128 U)					
3B	5	–	2	–	2
4B	7	–	4	+	1
5B	11	4	4	+	1
6B	12	–	3	+	1
7B	13	–	3	+	1
8B	15	4	4	+	2
9B	17	–	2	–	2
10B	34	4	4	+	–
11B*	58	–	2	–	1
12B*	78	–	1	–	3
13B*	80	–	1	–	1
14B*	105	1	4	+	3
Bleomycin (0.256 U)					
16B	8	2	1	–	–
17B	11	4	4	+	–
18B	12	2	4	+	1
19B	13	3	4	+	3
20B*	64	–	1	–	1
21B*	64	–	1	–	–
22B*	68	–	2	–	–

[a]Only animals with pathologic changes are tabulated. Animals not marked with an asterisk died on the day indicated. Animals marked with an asterisk were sacrificed on the day indicated after receiving bleomycin and are the long-term survivors cited in the text. A total of eight control animals were sacrificed: one at 8 days, two at 13 days, one at 40 days, two at 80 days, and two at 105 days after saline treatment. One control animal sacrificed at 13 days had pathologic changes of bronchitis; other control lungs were normal.

Structure Activity Relationships

The structures of bleomycins A_1, A_2, and B_2 are shown in Table 3. Differences in the incidence and severity of fibrosis were apparent at a dose of 1.6 μg. Fibrosis occurred more frequently following bleomycin A_2 and A_1 than bleomycin B_2 and Blenoxane (Table 4). Blenoxane, and bleomycin A_2, but not bleomycin A_1 or B_2, produced extensive fibrosis at this dose. After treatment with 16.0 μg, the incidence and severity of fibrosis was similar in all groups (Table 4).

Terminal groups of bleomycins A_2'-a, A_2'-b, A_2'-c, A_5, A_6, and B_2, listed in Table 3, were administered to DBA mice without bleomycin. All terminal groups produced fibrosis. The incidence of interstitial fibrosis is shown in Table 5. At doses of 20 μg, the incidence of fibrosis was highest after 1,4-diaminobutane and spermidine, followed by agmatine. After doses of 40 μg, more than 80% of animals treated with 1,4-diminobutane, 1,3-diaminopropane, and spermidine developed fibrosis.

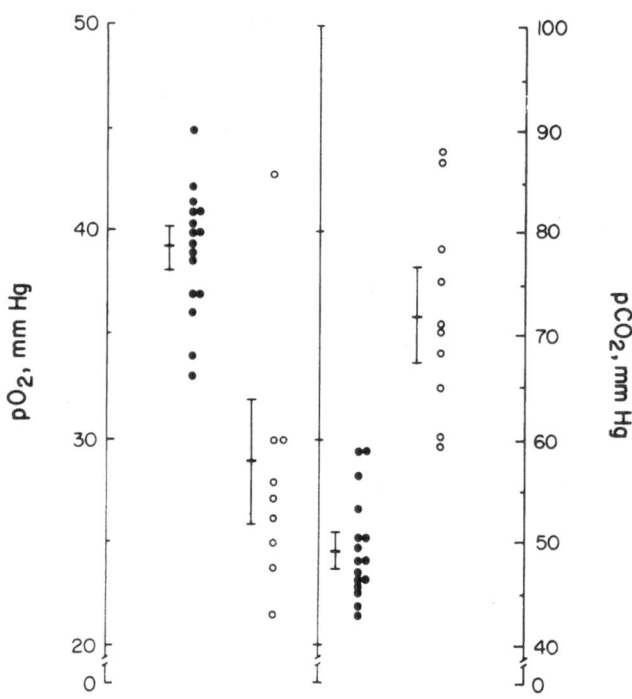

Fig. 6. Venous pO_2 and pCO_2 in mice treated with 0.128 U bleomycin. (●) NaCl-treated controls; (○) mice treated with 0.128 U bleomycin. Mean pO_2 for 15 control mice is 39.5 ± 2.8 SD, for 10 treated mice 29.3 ± 6.6 SD ($p<0.001$, t test). Mean pCO_2 for 15 control mice is 48.8 ± 4.3 SD; for 10 treated mice 72.2 ± 10.1 SD ($p<0.001$, t test).

Table 3: Terminal Structures of Bleomycins

Bleomycin	Terminal structure[a]	Name
A_1	$R-NH-(CH_2)_3-SO-CH_3$	3-Methylsulfinyl-propylamine
A_2	$R-NH-(CH_2)_3-S^+-(CH_3)_2X^-$	3-Aminopropyldimethylsulfonium salt
A_2'-a	$R-NH-(CH_2)_4-NH_2$	1,4-Diaminobutane
A_2'-b	$R-NH-(CH_2)_3-NH_2$	1,3-Diaminopropane
A_2'-c	$R-NH-(CH_2)_2-\overset{N}{\underset{\underset{H}{N}}{\diamond}}$	Histamine
A_5	$R-NH-(CH_2)_3-NH-(CH_2)_4-NH_2$	Spermidine
A_6	$R-NH-(CH_2)_3-NH-(CH_2)_4-NH-(CH_2)_3-NH_2$	Spermine
B_2	$R-NH-(CH_2)_4-NH-\overset{NH}{\overset{\|}{C}}-NH_2$	Agmatine

[a] R is bleomycinic acid.

Table 4: Incidence and Severity of Interstitial Fibrosis Produced by Bleomycins

| Dose | Fibrosis (% animals affected) | |
	Interstitial	Extensive
1.6 μg		
Blenoxane	33	22
A_1	50	0
A_2	55	22
B_2	40	0
16.0 μg		
Blenoxane	96	91
A_1	90	70
A_2	89	89
B_2	90	80

Discussion

Pulmonary fibrosis has been induced in various animal species, including mice, by systemic administration of bleomycin. Administration of a cumulative dose of 130–400 mg bleomycin to dogs over a period of 10 weeks produced subpleural fibrosis that involved up to 22% of lung parenchyma[2]. Intensive treatment for periods as long as 14 months increased the incidence and extent of interstitial pneumonia but fibrosis remained subpleural; in addition, animals suffered disabling renal and cutaneous toxicity[3]. Sixty-seven percent of baboons cumulatively dosed with 60–440 mg bleomycin developed interstitial fibrosis 6 months after treatment[4]. Pheasants develop epithelial metaplasia and interstitial penumonia after receiving 70–120 mg of bleomycin, but the incidence of pulmonary fibrosis was low[5].

A useful model for studies of bleomycin fibrosis has utilized mice in which either a single intravenous injection of 3 mg per animal was given or cumulative

Table 5: Incidence of Interstitial Fibrosis Produced by Terminal Groups[a]

| Terminal group | Dose | |
	20 μg	40 μg
1,4-Diaminobutane	80	80
1,3-Diaminopropane	20	86
Histamine	40	40
Spermidine	80	80
Spermine	20	60
Agmatine	60	40

[a]Percentage of animals affected.

intraperitoneal injections totaling 4.0 and 8.0 mg per animal[6-8]. By these regimens, 30% of treated mice developed interstitial fibrosis in 12–16 weeks.

The limitations of these experimental animal models are the low prevalence of pulmonary fibrosis, patchy distribution of lesions, lengthy time required for evolution of disease, and the need for large quantities of a drug that is expensive and limited in supply. Our experiments demonstrate that, after one intratracheal dose of 1.6 to 80 μg of bleomycin, approximately 90% of DBA mice developed extensive pulmonary lesions in a matter of days (rather than months or years) and the toxic effects are limited to the lungs.

The amounts of bleomycin initially used for intratracheal injection in this study was based on several factors. Adamson and Bowden produced interstitial fibrosis in mice after cumulative systemic administration of 4.0 mg per mouse[6]. Umezawa et al. recovered 1% of a dose of radioactive bleomycin from lung tissue 1 hr after systemic administration[9]. Bleomycin extracted from lung tissue was pharmacologically active because bleomycin hydrolase, the enzyme that inactivates bleomycin by hydrolysis of the α-aminocarboxamide moiety, is deficient in lung tissue, especially in elderly animals[9]. The dose originally selected for intratracheal administration in this study is 1% of 4.0 mg or 0.064 U and assumes full absorption of drug by the lung.

Pathologic changes observed after intratracheal and systemic administration of bleomycin are similar and consist of epithelial metaplasia, and alveolar, interstitial, and diffuse fibrosis. Interstitial pneumonia and vasculitis, common findings when bleomycin is administered systemically[2, 5, 10], are uncommon after intratracheal instillation of the drug. In comparison to lesions produced by systemic administration, fibrosis is more extensive after intratracheal instillation and epithelial metaplasia is often pronounced. Reproduction of pathological changes indistinguishable from lesions observed in patients[10] by direct instillation of bleomycin into animal lungs suggests that the parent drug or possibly a pulmonary metabolite, but not a systemic metabolite, is the agent responsible for pulmonary toxicity.

In addition to the small amounts of drug required and the high prevalence of pathology observed, intratracheal instillation of bleomycin has other advantages. Cutaneous and systemic toxicity, which have complicated other studies[2-4], are avoided. Furthermore, C57B1 and CD-1 mice may be used for long-term studies requiring extended survival of animals, while the DBA strain of mice is useful for short-term studies. The development of fibrosis in more than 90% of DBA mice simplifies selection of animals for biochemical studies.

If pulmonary function studies are required, this technique can be applied to larger animals. The drug can be instilled by bronchoscopy into both lungs or into one main bronchus so that one lung becomes affected while the contralateral lung serves as a control. Such models could be useful for investigations concerned with biochemical aspects of chronic pulmonary fibrosis.

In the present study, the incidence of fibrosis produced by Blenoxane and the major components of Blenoxane, bleomycins A_1, A_2, and B_2, is similar. The incidence and severity of fibrosis produced by these bleomycins differed only at a dose of 1.6 μg, the lowest dose employed in this study. At this dose, bleomycin A_2 and A_1 caused the highest incidence of fibrosis and Blenoxane and bleomycin A_2 produced the most extensive lesions. However, at 16 μg, all bleomycins tested were highly toxic.

Bleomycins differ from each other by terminal substituents linked to bleomycinic acid[11]. The pulmonary toxicity of different bleomycins has been reported to vary, suggesting a relation between terminal structure and pulmonary fibrosis[11].

In this study, intratracheal administration of terminal groups of six different bleomycins, rather than the bleomycins themselves, produced pulmonary fibrosis. This suggests an important role of bleomycin terminal groups in the genesis of pulmonary fibrosis. In contrast to the pathologic changes produced by intact bleomycins tested in this study, the fibrogenic proclivities of terminal groups differ from each other.

The technique described in this report will be useful in evaluating the ability of a drug to produce fibrosis. This may also be a useful way to screen compounds as potential terminal groups of new bleomycins.

Acknowledgments

The author is grateful to Gary Blank, Michael Grollman, Mark Reyer, and Jeffrey Zauderer for their contribution to this project. The author thanks Dr. Philip Kane and Dr. Marvin Kuschner for reviewing the pathology found in this study.

This work was supported in part by grants from the Stonywold Foundation and NIH BMRG 5-S07RR0573605.

References

1. A. Yagoda, B. Mukherji, C. Young, E. Etcubanas, C. Lamonte, J. R. Smith, C. T. C. Tan, and I. H. Krakoff, *Ann. Int. Med.*, **77**, 861 (1972).
2. R. W. Fleischman, J. R. Baker, G. R. Thompson, U. H. Schaeppi, V. R. Illievski, D. A. Cooney, and R. D. Davis, *Thorax*, **26**, 675 (1971).
3. U. Schaeppi, R. Phelan, S. W. Stadnicki, R. W. Fleischman, I. A. Heyman, V. Illievski, and R. A. Redding, *Cancer Chemother. Rep.*, **58**, 301 (1974).
4. B. McCullough, J. F. Collins, W. G. Johanson, and F. L. Grover, *J. Clin. Invest.*, **61**, 79 (1978).
5. C. W. Bedrossian, S. D. Greenberg, D. H. Yawn, and R. M. O'Neal, *Arch. Pathol. Lab. Med.*, **101**, 248 (1977).
6. I. Y. R. Adamson and D. H. Bowden, *Am. J. Pathol.*, **87**, 569 (1977).
7. I. Y. R. Adamson and D. H. Bowden, *Am. J. Pathol.*, **77**, 185 (1974).
8. I. Y. R. Adamson, *Environ. Health Perspect.*, **16**, 119 (1976).
9. H. Umezawa, T. Takeuchi, S. Hori, T. Sawa, and M. Ishizuka, *J. Antibiot. (Tokyo)*, **25**, 409 (1972).
10. C. W. M. Bedrossian, M. A. Luna, B. Mackay, and B. Lichtiger, *Cancer*, **32**, 44 (1973).
11. H. Umezawa, *Lloydia*, **40**, 67 (1977).

Concluding Remarks at the Symposium on Bleomycin

D. MIZUNO

"Bleomycin is a minimum sized naturally occurring molecule showing minimum sized enzyme function"

It is a great honor and privilege for me to summarize and conclude this symposium. Prof. Umezawa asked me, or I should say kidnapped me in the first-class section of a Pan Am plane to Honolulu, to do this work because I am a layman in this field and as such was expected to observe everything quite objectively, while, on the other hand, all the other participants at this symposium were very sincere and eager to elucidate the "Science of Bleomycin." Indeed, in this symposium the chemical, physical, biological, and biochemical aspects of bleomycin have all been covered.

[Dr. Umezawa's general review[1] in the first session covered the central problem as well as the conclusion of the symposium. Therefore, I do not think it necessary to follow his presentation exactly.]

What I would like to do now is mention the points which I personally think important as well as other important points which have been missed and which, I think, should be considered in the future.

Dr. Takita presented a newly revised structure of bleomycin. The β-lactam ring, which should be labile as it is in penicillin, has been puzzling us for more than 10 years. The demonstration that it is an open ring throws new light on the problem of bleomycin both from the chemical and biological viewpoint. Since the mystery of this part of the molecule is now solved, we can now expect great new developments. For instance, both the partial and total syntheses of bleomycin should now be much easier than they seemed before. Moreover, the relationship between structure and function can also be considered without the necessity of introducing a hypothesis based on molecular lability, for instance, renal toxicity or the binding of bleomycin to the cell membrane or to DNA.

---

[1]*See* "Advances in Bleomycin Studies," pp. 24–36, this text.

Apart from this, I should like to thank Dr. Takita and, of course, Dr. Umezawa for elucidating the complete structure of bleomycin. Their pioneering work has taken many years, but without their efforts none of the studies reported at this symposium would have been possible.

Dr. Hecht clearly explained his elegant syntheses of three fragments of bleomycin. His method, devised on the basis of classical routes, is better than any others so far reported both with respect to mild conditions and to high yield. He is aiming at the total synthesis of bleomycin, but even the partial syntheses, which he described, should be useful for other work.

The partial synthesis of the β-lactam moiety, described by Dr. Ohno, has played an unexpected but significant role. By coincidence, the proposal of Dr. Takita that part of β-lactam is an "open ring" has been confirmed by this synthesis. The work of Dr. Takita and Dr. Ohno has utilized fully and clearly those elaborate and sophisticated technical findings presented by Drs. Naganawa and Muraoka. The standard patterns of bleomycin and its numerous derivatives on NMR and liquid chromatography have been described.

Dr. Terasima's presentation has made us reconsider the mechanism of action of bleomycin. The survival with time of various dividing cells, including squamous carcinoma cells, showed a biphasic upward-concave curve.

Dr. Cohen's work was presented by Dr. Grollman. Natural polyamines, such as putrescines, agmatines, spermidines, and others can antagonize the action of bleomycin in bacteria and HeLa cells. It seems probable to me that these actions take place, not on the DNA to be attached, but on the surface of the cell membrane. The same polyamines have been shown by Dr. Raisfeld to be toxic to rat lungs, and her technique should be useful in future studies. She found that a single intratracheal injection of bleomycin or other agents into rats resulted in strong pulmonary toxicity within 28 days. Conventional experiments to induce pulmonary toxicity required more frequent doses of agents and more than 50 days.

In painstaking work, Dr. Fujii has clarified the biogenesis of bleomycin, mainly by isolating many biosynthetic intermediates from the culture medium and testing their abilities to form bleomycin. Surprisingly, he found that natural amino acids are linked first and then modified after peptide bond formation. To confirm these results, he isolated a multi-enzyme system that formed the intermediates from these individual components. It seems to me that the finding of this kind of modification of peptides and the isolation of a multi-enzyme system for this conversion represent the first observations of this type and constitute brilliant advances in the field of natural products.

Many presentations (Drs. Strong, Grollman, Lown, Hori, Haidle, Horwitz, Hutchinson, Nunn) have emphasized the problem of DNA breakage upon formation of a metal complex of bleomycin. I should like to quote the summary of these presentations from the introductory review of Dr. Umezawa: "Bleomycin first binds to DNA. There are base sequences to which bleomycin can bind selectively. After the binding, the reaction which releases thymine occurs. If the concentration of bleomycin and the concentration of a sulfhydryl compound are high, other bases are also released. Strand scission occurs at the same time as the release of the base." This adequately summarizes all of the findings presented, and there is no need to comment further. However, in addition to these short but pertinent comments, I would like to mention a few important points. By coincidence, the DNA employed in almost all of these studies was the same covalently closed circular DNA. This was a lucky coincidence for all concerned. Dr. Grollman employed Gilbert's method in analyzing base specificity. With this method the base adjacent to a point of strand scission can be determined. As far as I know Dr. Grollman's results are the first definite results in this area although many workers in many places must have been studying this problem. But I dare say it will take a little more time before any definite conclusions can be presented.

Results on the mechanism of DNA breakage are very confusing. My personal feeling is this: many delicate chemical reactions, including enzyme reactions, do involve free radicals. However, work on free radicals is usually very tricky and sometimes not reproducible. In this respect the finding of Dr. Horwitz of the role of ferrous ion should be highly appreciated. However, the free radical of superoxide is not the only way to explain an available free radical. Therefore, the conflicting data on the presence of superoxide or of its dismutase should be carefully reexamined. One thing I should like to point out is that in these experiments light must be controlled. Please remember that most free radicals, except for that of superoxide, are obtained by light irradiation. Furthermore, precise studies on photochemical treatments should be very interesting.

The specificity of bleomycin for DNA, as opposed to RNA, has not been well explained. It seems to me that the significance of the presence or absence of the 2'-OH group has been neglected. Therefore, the degradation products derived from the sugar moiety should be more carefully examined.

Drs. Oppenheimer, Dabrowiak, Takita, and Sugiura have tried to elucidate the conformation of the metal complex in relation to its coordination sites. The most important result, it seems to me, is Dr. Sugiura's finding. He showed that like heme or vitamin B_{12}, bleomycin seems to have a catalytic function at the center of the metal complex. Dr. Takita demonstrated the rigid form of the metal complex with a CPK model. From his work we can now conveniently consider the structure as an ionophore in a broad sense.

Throughout this symposium I have continuously been thinking about the significance of research on bleomycin. In this connection, I should first like to point out the importance of bleomycin in the field of natural products. As far as I know, bleomycin is the only group of low-molecular-weight glycopeptides known to have an interesting function. Natural glycopeptides on the surface of the cell membrane are very important as binding sites of so-called biological receptors. The glycosidic linkage of receptors usually involves aspartic acid, lysine, or serine, but bleomycin is joined to its disaccharide through a β-OH-histidine moiety. The conformation of this junction should be carefully examined in the context of antimetabolites of biological receptors.

Another interesting point is the conformation of the metal complex, which can apparently absorb molecular oxygen. In this symposium bleomycin has been shown to be the third naturally occurring product with the ability to absorb oxygen that enters biological metabolism: the first of these compounds is heme, and the second is vitamin B_{12}. However, heme and vitamin B_{12} have no peptide and so no binding site for a biological substrate, whereas bleomycin has a glycopeptide moiety and probably a binding site for substrates. Using the term ionophore in a broad sense, bleomycin is a specific ionophore of copper, carrying copper into the cell. Conversely, we could say that bleomycin is carried into the cell by copper.

The mechanism of action of bleomycin is very interesting in that bleomycin *binds* to DNA, probably specifically, and breaks DNA to give a new *product* of DNA. This action itself is very much like that of an enzyme. Of course bleomycin is not an enzyme. But Prof. Umezawa suggested that bleomycin might still be intact ever after its breakage of DNA. So at least we can consider bleomycin as a model of an enzyme, the function of which is to nick DNA in a broad sense, including hydrolysis, oxidation, or reduction. The structure of bleomycin can be considered as equivalent to that of the active center of those conventional nicking enzymes. If this is correct, we can analyze the function of bleomycin on DNA in the same way as that of restriction enzymes. What is its base specificity? What is its binding constant to DNA? Which part of the molecule is the binding site and which the catalytic site? These questions have been answered to some extent in this symposium. But many questions still remain unanswered. In this model bleomycin is an incomplete enzyme. But if you fill in the missing parts, you can obtain a complete model. For instance, ferrous ion, as suggested by Dr. Horwitz, or some radical reactions, as often observed in the catalytic site of conventional enzymes, are candidates for these missing parts.

Fortunately, the molecular size of bleomycin is exactly comparable with that of the cleft of the active center in the enzymes so far analyzed by X-ray crystallography. Therefore, we can predict that the copper complex may represent the catalytic site and the bithiazole may represent the binding site. This point was indicated by Dr. Takita earlier in this session.

Thus, I am sure that future work on bleomycin, and especially on its structure–function relationship, will be fruitful for elucidating the mechanism of DNA-strand scission and should even throw light on the active center of enzymes in general. Again I should like to point out that while heme and vitamin B_{12} have no catalytic activity themselves, bleomycin does have catalytic activity just like an enzyme. This may be due to the presence of the glycopeptide moiety of bleomycin which is not present in heme and vitamin B_{12}. In this sense, we could say that *bleomycin is a minimum sized naturally occurring molecule showing minimum sized enzyme function*.

Apart from the mechanism of action of bleomycin, I should like to mention the importance of bleomycin in cancer chemotherapy. Bleomycin is preferentially toxic to squamous cell carcinomas. This selective toxicity has been explained by the absence of bleomycin hydrolase in these cells. However, this explanation is not adequate, since labeled bleomycin is accumulated in cells in the first stage of mitosis. That is to say, bleomycin can be expected to bind specifically to the surface of rapidly dividing cells. Such binding suggests the existence of a specific receptor of bleomycin on the surface of rapidly dividing cells. So far, there have been no reports of this kind of receptor, although the colicin receptor has been proposed to be equal to the bleomycin receptor in *Escherichia coli*.

Dr. Terasima showed a biphasic curve for the lethal effect of bleomycin on rapidly dividing cells with time. This phenomenon is very difficult to explain. But the specific binding of bleomycin on the surface of cells may provide some information in this connection. Dr. Cohen's work and Dr. Raisfeld's presentation, which demonstrated the toxicity of bleomycin or sometimes the anti-metabolic activity against bleomycin, can probably be interpreted in terms of this binding mechanism.

The reason for the preferential binding of bleomycin to rapidly dividing cells should be clarified as soon as possible since, based on these findings, other chemotherapeutic agents could then be developed that accumulated preferentially in other types of tumor cells, such as sarcoma cells.

Finally, since bleomycin is after all an antitumor agent, I should like to point out another advantage of bleomycin in cancer chemotherapy. Bleomycin does not harm resting cells; consequently, it does not suppress the immunological activity of lymphocytes. This is a well-known fact, and yet no attempts have been made to take advantage of this observation. In combination therapy with some immunopotentiator bleomycin might have a synergistic effect in cancer immunotherapy. That is to say, when some immunopotentiator such as BCG is given the toxicity and growth of tumor cells as antigen might be weakened by bleomycin without harming the immune-mediated cells, thus enhancing tumor immunology. Side effects of bleomycin, such as pulmonary toxicity, could be avoided by decreasing the dosage of bleomycin. This type of experiment should have already been done in animals as well as in clinical trials. I have heard that one such trial is now in progress at the Institute of Microbial Chemistry with bestatin as an immunopotentiator. I hope these trials will result in establishment of a better method of therapy in the near future.

Finally, I should like to express my hearty thanks to Dr. Umezawa, Dr. Hecht, and Dr. Takita for giving me the chance to present these concluding remarks. Without Dr. Umezawa we could not have held this symposium, since Umezawa's bleomycin is the main subject. Not only did he discover bleomycin but since then he has been supervising all fields of studies on bleomycin, indeed, the whole science of bleomycin. For this we owe him the highest respect and extend our gratitude to him. We are also very grateful to Dr. Hecht and Dr. Takita, the organizers of this very well-prepared, exciting, and stimulating symposium. On behalf of all the participants in this symposium I should like to express my cordial thanks to them.

A Practical Method for the Separation of Bleomycin Components

Akio Fujii

Ten grams of a mixture of bleomycins, isolated from a cultured broth by the conventional method[1, 2] (1492 units/mg, manufactured by Nippon Kayaku Co., Japan), was adsorbed on the top of a column packed with CM-Sephadex C-25 (Pharmacia Fine Chemicals, Sweden; bed volume of 6 liters, 8 × 119 cm) pretreated with 0.05 M aqueous ammonium formate. Each component was eluted by increasing the concentration of the salt linearly from 0.05 to 1 M (total volume 60 liters); 120 fractions of 500 ml each were collected. The main fractions (No. 43–48) containing BLM A_2 were adsorbed on a column of activated charcoal (Charcoal for Chromatography purchased from Wako Pure Industries Co., Japan; bed volume, 200 ml) packed with water after degassing, following by washing with distilled water to remove the salt and elution with acidic aqueous acetone (1:1 0.02 M HCl–acetone, v/v). The blue-colored fractions were collected, neutralized with Dowex 44 (OH⁻ form), and lyophilized to give 3.07 g of a blue, amorphous powder of BLM A_2 –copper complex. Other fractions containing bleomycin were combined as shown in Table 1, and subjected to the desalting process using activated charcoal to give amorphous powders of the copper complexes of each component[3].

Some modifications were found to be successful as follows: (i) The ammonium formate used for the column chromatography on CM-Sephadex C-25 could be replaced with other salts such as sodium chloride, ammonium chloride, or

Table 1: Isolation of Bleomycin Components from Mixture

Components	Fractions	Yields (g)
A_1 and demethyl A_2	19–24	0.530
B_1'	25–28	0.135
A_2	43–48	3.07
A_2'-a and b	51–54	0.480
B_2	56–59	1.35
A_5	81–83	0.330
B_4	86–88	0.150
A_6	107–109	0.040
B_6	112–119	0.055

phosphate etc. (ii) The activated charcoal for desalting could be replaced with Amberlite XAD-2 (Rohm and Haas). As an eluant in this case, acidic aqueous methanol (4:1 CH_3OH–0.01 M HCl, v/v) is recommended, permitting the chromatography to be monitored directly by UV. (iii) When a copper-free bleomycin mixture was subjected to the above process, addition of excess cupric ion prior to charging was required for better resolution.

References

1. H. Umezawa, K. Maeda, T. Takeuchi, and Y. Okami, *J. Antibiot. (Tokyo)*, **19**, 200 (1966).
2. H. Umezawa, Y. Suhara, T. Takita, and K. Maeda, *J. Antibiot. (Tokyo)*, **19**, 210 (1966).
3. A. Fujii, Ph.D. Dissertation, University of Tokyo (1971).

Preparation of Bleomycinic Acid from Bleomycin A$_2$

AKIO FUJII

Bleomycin Demethyl-A$_2$ from Bleomycin A$_2$ [1]

Twenty-two grams of bleomycin A$_2$ (copper-containing hydrochloride powder) was heated at 100°C under diminished pressure (15 mm Hg) for 16 hr and then dissolved in 200 ml of 0.05 M ammonium chloride. The resulting solution was applied to an 800-ml CM-Sephadex C-25 column pretreated with the same solvent. Then 690 ml of 0.1 M ammonium chloride was passed through the column, and the blue eluate was collected. The eluate was passed through a 500-ml column of activated charcoal, and the column was washed with water to remove inorganic salts. A 1:1 0.02 M HCl–acetone mixture was then passed through the column, and the blue eluate containing bleomycin demethyl-A$_2$ (copper-containing hydrochloride) was collected. A basic anion-exchange resin, Dowex 44 (OH$^-$ form) was added to the collected fraction and the pH was adjusted to 6.0. After removing the solvent by distillation and drying, 8.91 g of a blue powder of bleomycin demethyl-A$_2$ (copper-containing hydrochloride), mp 201–212°C (decomposion upon melting), was obtained.

The CM-Sephadex C-25 still contained undecomposed bleomycin A$_2$. Therefore, the column was again eluted with 0.1 M ammonium chloride solution to collect the blue fraction containing bleomycin A$_2$ (copper-containing hydrochloride). The collected fraction was treated in the same manner as in the case of bleomycin demethyl-A$_2$ (copper-containing hydrochloride) to recover 10.9 g of bleomycin A$_2$ (copper-containing hydrochloride).

Bleomycinic Acid from Bleomycin Demethyl-A$_2$ [2]

The hydrochloride of bleomycin demethyl-A$_2$ (Cu complex) (1.0 g) was dissolved in 20 ml of 1% aqueous trifluoroacetic acid, and 2.0 g of cyanogen bromide was added with stirring. The reaction mixture, pH 0.7, was sealed and the contents was stirred at 27°C for 18 hr. Excess cyanogen bromide was removed under diminished pressure. An aqueous 2 M sodium hydroxide solution was added to the reaction mixture with vigorous stirring and the solution was adjusted to pH 5.1.

The solution was applied to a 150-ml Amberlite CG-50 column (H$^+$ type) equilibrated with distilled water. The column was washed successively with 400 ml of 0.3% aqueous acetic acid and 200 ml of distilled water. After washing, the blue-colored fraction containing the bleomycin derivative was washed from the column with 1:1 0.02 M HCl–methanol. Methanol was removed from the product under diminished pressure, and the aqueous solution was adjusted to pH 4.5 by addition of pyridine. The treated solution was applied to a 100-ml Sephadex C-25 column which had been preequilibrated with 0.05 M pyridine-acetic acid, pH 4.5. The column was washed with 100 ml of 0.2 M pyridine-acetic acid buffer, pH 4.5, and then with 100 ml of 0.3 M pyridine-acetic acid buffer, pH 4.5. All unreacted bleomycin was eluted from the column by this treatment. The first fraction contained 254 mg of the starting material. After another 80 ml of eluate was collected, the column was washed with 115 ml of a 0.5 M pyridine-acetic acid buffer solution which effected elution of the 3-aminopropyl ester of bleomycinic acid. Two hundred milliliters of a blue-colored fraction was collected in this manner, concentrated under diminished pressure, and dried.

The product was reprecipitated from methanol-ether to yield 528 mg of the acetic acid salt of the 3-aminopropyl ester of bleomycinic acid (Cu complex) as a blue-colored amorphous powder. The ester was applied to a 20-ml Amberlite IR-45 (Cl⁻ form) column. The eluate was concentrated and dried to yield 497 mg of the hydrochloride of the 3-aminopropyl ester of bleomycinic acid (Cu complex) as a blue-colored amorphous powder.

The hydrochloride (1.0 g) was dissolved in 10 ml of distilled water and the solution was adjusted to pH 4.0 by addition of 0.1 M hydrochloric acid. The mixture was heated at 105°C for 6 hr in a sealed tube. The reaction mixture was adjusted to pH 6.5 by the addition of 0.1 M sodium hydroxide and was passed through a column packed with 20 ml of CM-Sephadex C-25 (Na⁺ form). After removing the eluate from the column, 50 ml of distilled water was passed through the column to elute all of the resulting bleomycinic acid in one fraction. The eluate was applied to a 100-ml Amberlite XAD-2 column equilibrated with distilled water. After washing the column with 200 ml of 1:1 CH_3OH–2.5 mM HCl. Bleomycinic acid was eluted in 70 ml of the last fraction. The eluate was adjusted to a pH of 6.5 with a Dowex 44 resin (OH⁻ form) and the solvent was concentrated. The resulting product was dried and precipitated from aqueous acetone to yield 780 mg of a blue-colored amorphous powder of bleomycinic acid (Cu complex). When the CM-Sephadex C-25 column was treated with 0.5 M NaCl, 185 mg of untreated material was recovered.

References

1. H. Umezawa, T. Takita, A. Fujii, and H. Ito, U.S. Patent No. 3, 960, 834 (1976).
2. H. Umezawa, T. Takita, A. Fujii, and T. Fukuoka, U.S. Patent No. 3, 886, 133 (1975).

A Practical Method for Preparing Copper-Free Bleomycin A_2

Akio Fujii

A solution of 3.0 g of copper-containing bleomycin A_2 hydrochloride (930 mcg potency/mg; 4.28% copper content) in 100 ml of distilled water was passed through a column packed with 1 liter of Amberlite XAD-2 in distilled water to effect adsorption of bleomycin A_2. Then 3 liters of a 5% aqueous solution of Na-EDTA was applied to the column to remove Cu-EDTA. The column was washed with 1 liter of a 5% sodium chloride solution and then with 2 liters of distilled water. Finally, 1.5 liters of a mixture (1:1) of 2.5 mM hydrochloric acid and acetone was allowed to flow through the column to elute bleomycin A_2. The latter half of the eluate, 700 ml in volume, was collected and concentrated under diminished pressure. The residue was subjected to reprecipitation from water–acetone, and dried to obtain 2.8 g of copper-free bleomycin A_2 hydrochloride (945 mcg potency/mg; 0.0083% copper content) in the form of a pale yellow powder; yield 93.3%, absorbance (aqueous solution) at 292 nm $E_{1 \ cm}^{1\%} = 103$.

The foregoing is a description of the basic procedure which may, of course, be modified to a certain extent if required. For instance, prior to adsorption, an aqueous solution of the antibiotics can be admixed with 1 to 3% of a chelating agent, e.g., Na-EDTA, adjusted to a pH of 2 or less with acid (e.g., hydrochloric acid), allowed to stand for several minutes, and then readjusted to pH 7. With this pretreatment, improved copper removal can be effected.

Reference

1. T. Takita, A. Fujii, Y. Muraoka, S. Mizuguchi, and H. Umezawa, U.S. Patent No. 3, 929, 993 (1975).

A Simple Assay for Bleomycin Activity: Degradation of Radioactive DNA to Acid-Soluble Products [1,2]

Susan B. Horwitz, Edward A. Sausville, and Masaru Takeshita

Principle

Bleomycin and Fe(II) together act to cause the efficient and rapid degradation of DNA to acid-soluble products when the drug and metal ion are present in a one- to twofold excess of DNA. The products include all four bases (pyrimidines are released preferentially in relation to purines), oligonucleotides, and an aldehyde product whose derivative with 2-thiobarbituric acid resembles that produced by malondialdehyde.

Reagents

Sodium phosphate buffer, $0.25\,M$, pH 7.0.
[3H] Thymidine-labeled DNA (see Radioactive DNA below).
$Fe(NH_4)_2(SO_4)_2 \cdot 6H_2O$, 1 m$M$, prepared immediately prior to use.
Bleomycin.
EDTA (ethylenedinitrilo)tetraacetic acid, $0.1\,M$ containing 5 mg/ml bovine serum albumin.
Perchloric acid, $5\,M$.

Procedure

Clean glassware is essential for reproducibility. Assays are done in 1×7.5-cm glass test tubes that have been cleaned in chromic acid–sulfuric acid followed by thorough rinsing with distilled H_2O. They are then washed with Haemo-Sol (Haemo-Sol, Inc., Baltimore) detergent followed by vigorous rinsing with distilled H_2O. This procedure is required to obtain good reproducibility ($\pm10\%$) in these assays.

Reaction mixtures (50 μl) contain 5 μM adenovirus [3H]DNA (3500 cpm), 50 mM phosphate buffer, 0.1 mM Fe(II), bleomycin (0–2.5 μg), and H_2O. Reaction mixtures are prepared in an ice bath with all components except

Fe(II). To each tube is added 10 μl of phosphate buffer and appropriate volumes of DNA, bleomycin, and H_2O. After equilibration at 37°C for 1 min, 5 μl Fe(II) is added and incubation at 37°C is continued for 15 min. Tubes are chilled in an ice bath and 50 μl EDTA and bovine serum albumin is added (final concentrations of 0.05 M and 1 mg/ml, respectively), then 10 μl perchloric acid (final concentration of 0.5 M). After 15 min in the ice bath, the precipitate is sedimented by centrifugation (6000 g for 25 min) and the radioactivity in the supernatant is determined. Controls containing no bleomycin are included in each set of assays. In the absence of bleomycin, less than 2% of the radioactivity should be acid soluble.

Radioactive DNA

Assays can be done with any purified radioactive DNA. It is most convenient to isolate DNA from viruses, bacteria, or mammalian cells that have been grown in the presence of [^{3}H] thymidine. In the assay described above adenovirus type 2 [^{3}H] DNA was isolated and purified from the virion[3]. The purified DNA was dialyzed against 2 mM Tris–HCl, pH 7.5, and the concentration of DNA was estimated by assuming an extinction coefficient of 6.6×10^3 M^{-1} at 260 nm. Radioactive DNA can also be prepared from bacteria[4] or mammalian cells[5].

References

1. M. Takeshita, A. P. Grollman, and S. B. Horwitz, *Virology*, **69**, 453 (1976).
2. E. A. Sausville, R. W. Stein, J. Peisach, and S. B. Horwitz, *Biochemistry*, **17**, 2746 (1978).
3. M. S. Horwitz, *J. Virol.*, **8**, 675 (1971).
4. C. W. Haidle, *Mol. Pharmacol.*, **7**, 645 (1971).
5. C. L. Schildkraut and J. J. Maio, *J. Mol. Biol.*, **46**, 305 (1969).

Index

A5033, 8, 9
Absorption of oxygen, 339
Absorption, spectra, 142
Action on bacterial cells, 263, 270-275
  DNA synthesis, 274
  toxicity, 270, 275
  toxicity inhibition, 272, 273, 275, 283
Action on DNA
  alkali labile bonds, 257, 258
  assay, fluorescence, 244, 246-249
  assay, electrophoresis, 244, 246, 247, 249
  assay, reproducibility, 255
  base release, 256
  double strand break, 244, 258, 259, 290
  effect of Fe^{2+}, 251
  inhibition, 251
  intercalative binding, 261
  rate, 248, 261
  single strand break, 244, 257, 260, 290
Action on mammalian cells
  assay, 288-290
  DNA, effect on, 280, 290
  effect of metal, 290-295
  RNA synthesis, 282
  toxicity, 277-279
  toxicity inhibition, 280, 283
6-Azauridine, 282
Alkali-labile damage of DNA, 224, 257, 258
Alkylation of BLM-demethyl A$_2$, 314-316
Alterations of DNA by BLM, 239
Amberlite XAD-2 chromatography, 77, 342, 344, 345
2'-(2-Aminoethyl)-2,4'-bithiazole-4-carboxylic acid, 48-51
  condensation with L-threonine, 52, 53
4-Amino-3-hydroxy-2-methylvaleric acid, 55-58

1-(p-Aminophenyl)-EDTA, preparation of, 314
β-Aminopropionamides
  ^{13}C-NMR, 69, 70
  pKa, 69, 70
  synthesis, 69
Arginine, 274, 276, 277
Bifunctional chelating agents, 312, 321
Binding to cell surface, 340
Biogenesis, 337
Biosynthesis of amino acid VI, 85
Bleomycin
  an enzyme model, 339
  biosynthetic intermediate P-3A, 1, 2
  "epimerization" of carbohydrates, 7
  ^{1}H-NMR coupling constants, 3
  pKa values of amino groups, 3, 94
  inactivation of, 200
  revised structure, 2
  stabilization via hydrogen bonding, 4, 5
  structure originally proposed, 1
  synthesis of bleomycin bithiazole moiety, 6, 7
  synthesis of bleomycin L-gulose, 7
  synthesis of model compounds, 5, 6
  treatment with 2,4-dinitrofluorobenzene, 4
BLM-^{57}Co, 163, 165
Bleomycin hydrolase, 31-33, 334
Bleomycin-induced lesions, 325, 326
Bleomycin metal complexes, 13-15
  proton release, 14
1-(p-Bromoacetamidophenyl-EDTA, 315
Cancer immunotherapy, importance of bleomycin in, 340
3-Carbamoyl-D-mannose, 40, 41
CD spectrum of BLM-Cu(II), 161
Cell binding—saturation model, 306-308
Cell sensitivity, 299
Cell survival, 297
Cell uptake of bleomycin, 305, 306

Chelates, as tumor visualizing agents, 309, 310
Chelate decomposition rates, effect of, 313
Chloramphenicol, 275
CM-Sephadex chromatogrpahy of natural BLM, 91, 341, 343-344
CO_2 in blood, 325, 326, 328
Co-Bleomycin chelate, 311, 312
Complex equilibria, 317, 318
Confluent fibrosis, 325-327
Conformation of metal complex, 338
Coordination sites of BLM-Cu(II), 158
Copper complex of BLM, 156, 292, 294, 314, 315
Copper-free BLM A_2, 345
[14]C-studies, 303-305
[67]Cu-Bleomycin chelate, 311
DC polarograms, 149, 151, 152
3-deazacytidine, 282
α,β-Diaminopropionic acid, 71
Distribution assay of Bleomycin-EDTA conjugate, 319
DNA,
  bleomycin binding, 181, 187, 188, 195-198
  Binding of tripeptide S to, 181
  Strand scission of, 29-31, 64, 170, 172, 195, 200-203, 226, 244, 258, 259, 290
    aldehyde release during, 179-182
    chromophore production during, 179-182
    effect of concentration, 196
    effect of free radical scavengers, 186
    effect of metal ions, 138-140, 153, 171, 172, 175, 184, 188, 189
    effect of oxygen, 173, 175, 177
    effect of reducing agents, 170, 171, 184, 185
    free base release during, 177, 186
    sensitivity to ionic strength, 173
    sensitivity to pH, 173, 174, 185, 191, 195, 196
DNA cleavage and repair, 11, 12, 305, 306
DNA employed in experiments, 338
Drug saturation and cell response, 300, 301
Electrochemistry, bleomycin and metal derivatives, 147

epi-bleomycin, 9, 10, 95, 97, 159, 202
Epithelial metaplasia, 325-328
ESR, Cu(II)-bleomycin complex, 140
  Fe(II)-bleomycin A_2 complex, 148
ESR spectrum of BLM-Co(II), 165ff
Ferrous ions and cell response, 308
Fluorescence quenching, 195, 203
Fragmentation of DNA by BLM, 209
Glycopeptides, function of, 338
L-Gulose, 40, 41, 58-61
Hae III, 208
Hind III, 232
Hinf I, 208
Hirudonine, 12, 13, 283-285
Hodgkin's lymphoma, 24
Hpa II, 233
^{111}In-BLM chelate, 311
Instillation of bleomycin by bronchoscopy, 334
Interaction of BLM with DNA, 215
Intermediate peptides P-5 and P-5m, 82
Intermediate peptides P-6m and P-6mo, 77, 83-84
Intermediate peptide P-7mo, 77, 84
Interstitial fibrosis, 325-327
Iron, complex formation with bleomycin, 172, 173, 213, 217, 218
iso-bleomycin, 9, 10, 95, 159, 202
β-Lactam, synthesis, 64-67
  infrared spectra, 67, 68
  X-ray structure, 68
Lung lesions, 326, 327
Lung pathology, survival relationship, 331
M5 196, 8, 9
Mammalian cells, dose-survival, 297
Mechanism of bleomycin action
  DNA breakage, 29-31, 64, 170, 172, 184
  DNA strand scission, 162, 218
  inhibition of DNA synthesis, 184
  interference with mitosis, 184
  radical intermediates, 185, 190, 191
  release of nucleosomes, 184
Metal binding selectivity, 316, 317
Metaplastic cells, 328
Nickel peroxide, 49-51, 52, 54
Neocarzinostatin, 192
^{13}C-NMR, Bleomycin and analogs, 43, 44, 108, 115, 122
  carbohydrate moiety, 114

chemical shift changes, 121
pulse Fourier transform parameters, 107
resonance assignments, 117-120
^{1}H-NMR, Bleomycin and analogs, 108, 109, 111, 112, 125-127, 130, 133, 135
carbohydrate moiety, 107, 109, 110
chemical shift changes, 129
Deglyco-bleomycinic acid, 110
resonance assignments, 113
Normal phase chromatography, 97
O_2 in blood, 325, 326, 328
Oligonucleotides, in limited digest of ^{32}P DNA, 178
end group homogeneity, 179
sensitivity to nucleases, 179
P-3A, 43-45, 156, 162
PEP (Pepleomycin), 8, 9, 34, 81
1-Phenyl-EDTA, 312, 313
Phleomycin, 24, 25, 173, 200
PM-2 DNA, 223
Potentiometric titrations, 140
Preparation of bleomycin analogs, 8
condensation by carbodiimide, 8
Fusarium acylagmatine hydrolase, 8
Preparation of bleomycin demethyl-A_2 from BLM A_2, 343
Preparative separation of BLM A_2'-a and b, 100, 105
Proposed biosynthesis of the pseudodipeptide moiety, 82, 84
Pseudotetrapeptide A, 38-40, 42
Pseudotetrapeptide B, 38, 39
Pseudotetrapeptide R, 39, 40
Pst I, 234
Pulmonary toxicity, 9, 13, 24, 34
and bleomycin structure, 331, 335
and dose levels, 333, 334, 335
and experimental limitations, 334
and mice studies, 324, 325, 328, 330
and intratracheal administration to mice, 325
and intratracheal vs. systemic administration, 334

in variety of animals, 333
model for bleomycin fibrosis, 333, 334
Pyrimidine moiety, synthesis of, 65, 70, 71
Pyrimidine release from DNA by BLM, 216, 217
Pyrimidoblamic acid, 28-30, 32
Rabbit tumors, 319, 320
Retention times of BLM's on Bondapak columns, 99, 101-103
Reverse-phase chromatography, 98-100
Scintillation scanning, 319, 320
Separation of BLM A_2, epi-BLM A_2 and iso-BLM A_2, 95, 96
Separation of BLM A_2'-c and depyruvamide BLM A_2'-c, 97
Separation of BLM B_2 and epi-BLM B_2, 98
Separation of BLM copper complexes, 92, 97, 98, 341
Separation of metal-free BLM's, 94, 95, 98, 342
Separation of natural Cu-BLM's, 104
SP-Sephadex chromatography, 96
Specificity of DNA cleavage, 216, 217 232
Spermidine, 272, 273, 280
Squamous cell carcinoma, 24, 31, 32, 34
Steptomyces verticillus, 170, 184
Structure of BLM-Cu(II), 162
Summary of Bleomycin Symposium, 336, 337
Superoxide anion, 184-194
Synthetic rates of L-cells, 302, 303
Tallysomycin, 10, 81
^{99}Tc-Bleomycin chelate, 309
Tetrapeptide S, 48, 52, 58
Time-inactivation, 300-301, 303
Toxicity of Bleomycin A_5, 12
Tripeptide A, 26
Tripeptide S, 25, 38-40, 42
Tumor survival, mice, 298
UV spectrum of BLM-Cu(II), 157

ERRATA

Bleomycin: Chemical, Biochemical, and Biological Aspects

(i)   **Page 6:** *The top part of the formula should read*

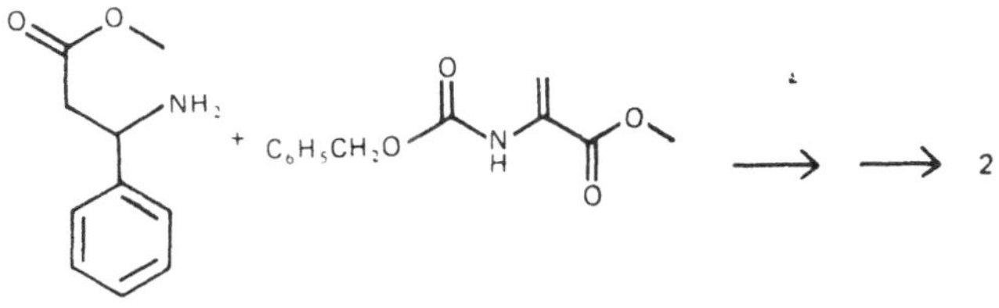

$$
\text{5} \quad + \quad C_6H_5CH_2O \text{--} \quad \text{6} \quad \longrightarrow \quad \longrightarrow \quad 2
$$

(ii)   **Page 9:** *The formula at the top of the page should read*

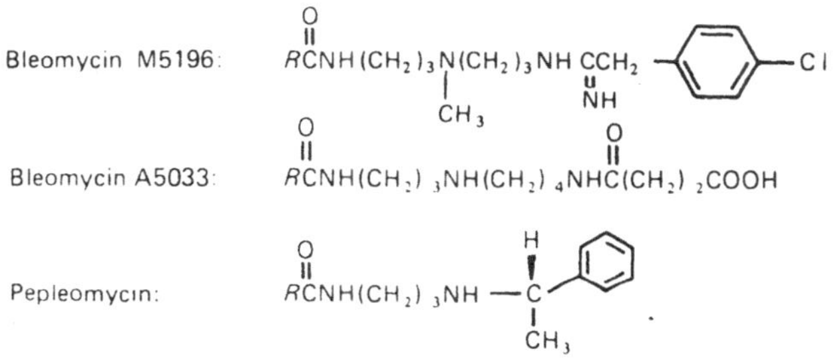

Bleomycin M5196:
$$\overset{O}{\overset{\|}{R}}CNH(CH_2)_3\underset{CH_3}{N}(CH_2)_3NH\,\overset{O}{\overset{\|}{C}}CH_2\underset{\overset{\|}{\underset{O}{NH}}}{\text{---}}\!\!\!-\!\!\!\text{Cl}$$

Bleomycin A5033:
$$\overset{O}{\overset{\|}{R}}CNH(CH_2)_3NH(CH_2)_4NH\overset{O}{\overset{\|}{C}}(CH_2)_2COOH$$

Pepleomycin:
$$\overset{O}{\overset{\|}{R}}CNH(CH_2)_3NH\text{---}\underset{CH_3}{\overset{H}{C}}\text{---}C_6H_5$$

$RCOOH = $ bleomycinic acid

(iii) **Page 10:** *The formula should read*

$R = H, CH_3$

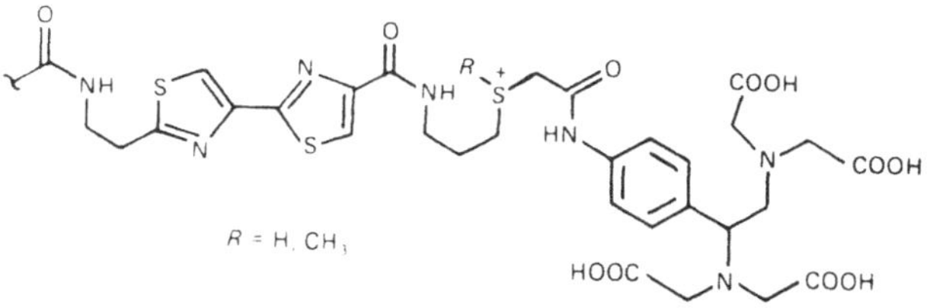

(iv)   **Page 137:** *Lines 9-11 should read*

of causing single and double strand breaks in DNA[4,5]. The *in vitro* degrading process, which leads to the production of free bases[6-9], is strongly metal ion dependent. Addition of Cu(II), Zn(II), and Co(II) salts[10,11] to the reaction

(v)   **Page 218:** *Equation (1) should read*

$$\text{Fe(II)--bleomycin} + O_2 \longrightarrow \text{Fe(III)--bleomycin} + {}^-O\text{--}O\cdot \tag{1}$$